# Neighbourhood Effects or Neighbourhood Based Problems?

David Manley • Maarten van Ham
Nick Bailey • Ludi Simpson
Duncan Maclennan
Editors

# Neighbourhood Effects or Neighbourhood Based Problems?

A Policy Context

*Editors*
David Manley
School of Geographical Sciences
University of Bristol
Clifton, Bristol, UK

Nick Bailey
School of Social and Political Sciences
Urban Studies
University of Glasgow
Glasgow, UK

Duncan Maclennan
Centre for Housing Research
School of Geography and Geosciences
University of St. Andrews
St. Andrews, Fife, UK

Maarten van Ham
OTB Research Institute
  for the Built Environment
Delft University of Technology
Delft, The Netherlands

Ludi Simpson
The Cathie Marsh Centre for Census
  and Survey Research
School of Social Sciences
University of Manchester
Manchester, UK

ISBN 978-94-007-9890-8         ISBN 978-94-007-6695-2 (eBook)
DOI 10.1007/978-94-007-6695-2
Springer Dordrecht Heidelberg New York London

© Springer Science and Business Dordrecht 2013
Softcover re-print of the Hardcover 1st edition 2013
This work is subject to copyright. All rights are reserved by the Publisher, whether the whole or part of the material is concerned, specifically the rights of translation, reprinting, reuse of illustrations, recitation, broadcasting, reproduction on microfilms or in any other physical way, and transmission or information storage and retrieval, electronic adaptation, computer software, or by similar or dissimilar methodology now known or hereafter developed. Exempted from this legal reservation are brief excerpts in connection with reviews or scholarly analysis or material supplied specifically for the purpose of being entered and executed on a computer system, for exclusive use by the purchaser of the work. Duplication of this publication or parts thereof is permitted only under the provisions of the Copyright Law of the Publisher's location, in its current version, and permission for use must always be obtained from Springer. Permissions for use may be obtained through RightsLink at the Copyright Clearance Center. Violations are liable to prosecution under the respective Copyright Law.
The use of general descriptive names, registered names, trademarks, service marks, etc. in this publication does not imply, even in the absence of a specific statement, that such names are exempt from the relevant protective laws and regulations and therefore free for general use.
While the advice and information in this book are believed to be true and accurate at the date of publication, neither the authors nor the editors nor the publisher can accept any legal responsibility for any errors or omissions that may be made. The publisher makes no warranty, express or implied, with respect to the material contained herein.

Springer is part of Springer Science+Business Media (www.springer.com)

# Preface

Governments have long been involved in interventions at the neighbourhood level, yet the perceived neighbourhood problems of crime, inequality, deprivation, poor health, and low educational outcomes continue to occupy national debates across the globe. Despite the commonality of problems, governments in different national contexts have pursued a wide variety of strategies in order to ameliorate or overcome the problems. In some contexts there is a long history of interventions including the demolition and regeneration of neighbourhoods (United Kingdom), the redistribution of households into different neighbourhood contexts (United States of America), or a combination of investments, redevelopment, and redistribution policies (The Netherlands). These interventions contrast with the relative lack of co-ordination in the Canadian context where neighbourhood level interventions are a relatively new phenomenon.

This book brings together a collection of chapters that discuss two main issues. The first chapters report on the links between the neighbourhood effects evidence base, neighbourhood problems, and individual outcomes. A focus on these problems is important because despite the vast number of publications related to neighbourhood based policies, the 'why' behind these policies is all too often forgotten. To provide a critical counterpoint, the approach of the neighbourhood effects literature is also questioned with a chapter which rotates the standard assumption that where you live affects your life chances and suggests instead that your life chances affect where you live. The second set of chapters provides details of the policy interventions that governments in different international settings have pursued in order to address the problems and effects that they perceive as issues. Here, examples are drawn from the United Kingdom, the United States of America, The Netherlands, Australia, and Canada.

In combining these two usually separate aspects of the literature within one volume we hope to stimulate a new debate into neighbourhood based policies and encourage policy makers to critically examine the assumptions that they make when developing area level interventions. This book will be of interest to academics and policy makers alike who want to know both why governments intervene in neighbourhoods in different national settings, and how these interventions differ across the international policy landscape. These insights are important for our understanding of cities and neighbourhoods as well as in the formulation of urban, housing, and social policy.

Many of the contributions in this book were presented at the seminar *Neighbourhood Effects or Neighbourhood Based Problems? A Policy Context* on 7 and 8 April 2011 at the University of Glasgow. The seminar was part of a wider ESRC Seminar Series, Challenges in neighbourhood effects research: does it really matter where you live and what are the implications for policy (RES-451-26-0704). The first book based on this seminar series, *Neighbourhood Effects Research: New Perspectives* appeared in early 2012 with Springer. The second book *Understanding Neighbourhood Dynamics: New Insights for Neighbourhood Effects Research* appeared early 2013 also with Springer. The seminar series, and the associated book series, is the result of a collaboration between researchers from the School of Geographical Sciences at the University of Bristol, OTB Research Institute for the Built Environment at Delft University of Technology, the Centre for Housing Research at the University of St Andrews, Urban Studies at the University of Glasgow, and the Cathie Marsh Centre for Census and Survey Research at the University of Manchester.

| | |
|---|---|
| Bristol, UK | David Manley |
| Delft, NL | Maarten van Ham |
| Glasgow, UK | Nick Bailey |
| Manchester, UK | Ludi Simpson |
| St Andrews, UK | Duncan Maclennan |
| February, 2013 | |

# Acknowledgements

The editors would like to acknowledge the financial support from the Economic and Social Research Council in the form of an ESRC Seminar Series grant (RES-451-26-0704). The editors also wish to acknowledge the generous financial and staff time support of OTB Research Institute for the Built Environment at Delft University of Technology and the Centre for Housing Research (CHR) at the University of St Andrews. The editors would like to express their thanks to all the authors for submitting their initial manuscripts on schedule, responding positively to comments and suggestions from the editors and delivering final versions of chapters with minimum delay. We are also grateful to Graeme Sandeman, the cartographer of the School of Geography and Geosciences at the University of St Andrews, who has redrawn the graphs and maps, and to Martine de Jong-Lansbergen from OTB Research Institute for the Built Environment at Delft University of Technology for all her work on the layout of the book.

# Contents

1 **Neighbourhood Effects or Neighbourhood Based Problems? A Policy Context**.................................................................................. 1
David Manley, Maarten van Ham, Nick Bailey, Ludi Simpson, and Duncan Maclennan

2 **Educational Area Based Initiatives: Issues of Redistribution and Recognition**.................................................................................. 25
Carlo Raffo

3 **Spatially-Concentrated Worklessness and Neighbourhood Policies: Experiences from New Labour in England**........................... 43
Stephen Syrett and David North

4 **The Role of Neighbourhoods in Shaping Crime and Perceptions of Crime**....................................................................... 67
Ian Brunton-Smith, Alex Sutherland, and Jonathan Jackson

5 **An Environmental Justice Framework for Understanding Neighbourhood Inequalities in Health and Well-Being**....................... 89
Jamie Pearce

6 **Capitalist Urbanization Affects Your Life Chances: Exorcising the Ghosts of 'Neighbourhood Effects'**.............................. 113
Tom Slater

7 **Social Mix: International Policy Approaches**...................................... 133
Keith Kintrea

8 **Neighbourhood Revitalization in Canada: Towards Place-Based Policy Solutions**................................................................. 157
Neil Bradford

9   **Neighbourhood Effects and Evidence in Neighbourhood Policy in the UK: Have They Been Connected and Should They Be?** ............................................................................. 177
    Rebecca Tunstall

10  **Neighbourhood Based Policies in the Netherlands: Counteracting Neighbourhood Effects?** ................................................ 195
    Gideon Bolt and Ronald van Kempen

11  **U.S. Assisted Housing Programs and Poverty Deconcentration: A Critical Geographic Review** ................................ 215
    George C. Galster

12  **Neighbourhood Effects and Social Cohesion: Exploring the Evidence in Australian Urban Renewal Policies** ......................... 251
    Kathy Arthurson

13  **Neighbourhoods: Evolving Ideas, Evidence and Changing Policies** ........................................................................... 269
    Duncan Maclennan

**Index** ......................................................................................................... 293

# Contributors

**Kathy Arthurson** Faculty of Health Sciences, Southgate Institute for Health, Society and Equity, Flinders University, Bedford Park, SA, Australia

**Nick Bailey** Urban Studies, School of Social and Political Sciences, University of Glasgow, Glasgow, Scotland, UK

**Gideon Bolt** Faculty of Geosciences, Utrecht University, Utrecht, The Netherlands

**Neil Bradford** Huron Universit College, Western University, London, ON, Canada

**Ian Bruton-Smith** Department of Sociology, University of Surrey, Guildford, Surrey, UK

**George C. Galster** Department of Urban Studies and Planning, Wayne State University, Detroit, MI, USA

**Jonathan Jackson** Department of Methodology, The London School of Economics and Political Science, London, UK

**Keith Kintrea** Urban Studies, School of Social and Political Sciences, University of Glasgow, Glasgow, Scotland, UK

**Duncan Maclennan** Centre for Housing Research, School of Geography and Geosciences, University of St Andrews, St Andrews, Fife, Scotland, UK

**David Manley** School of Geographical Sciences, University of Bristol, Bristol, UK

**David North** CEEDR, Middlesex University Business School, London, UK

**Jamie Pearce** Centre for Research on Environment, Society and Health (CRESH), Institute of Geography, School of GeoSciences, The University of Edinburgh, Edinburgh, UK

**Carlo Raffo** School of Education, Ellen Wilkinson Building, The University of Manchester, Manchester, UK

**Tom Slater** Institute of Geography, School of GeoSciences, The University of Edinburgh, Edinburgh, UK

**Ludi Simpson** The Cathie Marsh Centre for Census and Survey Research (CCSR), School of Social Sciences, University of Manchester, Manchester, UK

**Alex Sutherland** Institute of Criminology, University of Cambridge, Cambridge, UK

**Stephen Syrett** CEEDR, Middlesex University Business School, London, UK

**Rebecca Tunstall** Centre for Housing Policy, University of York, Heslington, York, UK

**Maarten van Ham** OTB Research Institute for the Built Environment, Delft University of Technology, Delft, The Netherlands

**Ronald van Kempen** Faculty of Geosciences, Utrecht University, Utrecht, The Netherlands

# Chapter 1
# Neighbourhood Effects or Neighbourhood Based Problems? A Policy Context

**David Manley, Maarten van Ham, Nick Bailey, Ludi Simpson, and Duncan Maclennan**

## Introduction

> "whenever there is widespread agreement or consensus that a certain policy, or set of related policies, should be pursued or enacted, it becomes necessary to step back and ask, why?" DeFilippis and Fraser (2010, p.135)

This book is about the ways in which governments try to intervene in neighbourhoods when they perceive things to have gone wrong: so-called area-based or neighbourhood-based policies. It is about the global and the local. It is about individual people and about the places in which they live and the ways in which they

---

D. Manley (✉)
School of Geographical Sciences, University of Bristol,
University Road, Clifton, Bristol BS8 1SS, UK
e-mail: d.manley@bristol.ac.uk

M. van Ham
OTB Research Institute for the Built Environment, Delft University of Technology,
Delft, P.O. Box 5030, 2600 GA, The Netherlands
e-mail: m.vanham@tudelft.nl

N. Bailey
Urban Studies, School of Social and Political Sciences, University of Glasgow,
25 Bute Gardens, Glasgow G12 8RS, Scotland, UK
e-mail: nick.bailey@glasgow.ac.uk

L. Simpson
The Cathie Marsh Centre for Census and Survey Research,
School of Social Sciences, University of Manchester, Humanities
Bridgeford Street, Manchester M13 9PL, UK
e-mail: Ludi.Simpson@manchester.ac.uk

D. Maclennan
Centre for Housing Research, School of Geography and Geosciences, University
of St Andrews, St Andrews, Fife KY16 9AL, Scotland, UK
e-mail: dm103@st-andrews.ac.uk

interact. It is about large scale structural socio-economic problems which, as the world becomes ever more globalised, are increasingly played out at the 'hyperlocal' level: our neighbourhoods. This book examines what policies are used to ameliorate problems perceived to originate at the neighbourhood level, and what outcomes are expected and for whom.

In this volume we make explicit links between the neighbourhood based polices and neighbourhood effects.[1] Whilst the diversity of neighbourhoods is not contested and inequalities are obvious for all to see, the importance of neighbourhood effects especially with regard to whether or not they have a casual impact on individual outcomes have never been more fervently debated than at present. This is despite the fact that, at the time of writing, the western world is experiencing some of the severest cuts in government spending in living memory and many of the neighbourhood interventions of the past decades have either come to a conclusion and not been renewed or have been cancelled mid flow. Against this back drop, western governments are increasingly looking to the private market as the stimulus for neighbourhood regeneration and change.

The debate that exists in the neighbourhood effects literature was reflected in the previous two books. The first volume was concerned with the theoretical foundations of the neighbourhood effects debate, and the examination of the state of the art in terms of empirical evidence relating to the identification of and search for such effects (van Ham et al. 2012a). Chapters were drawn from a range of national contexts, including the United Kingdom, the United States of America, The Netherlands, Australia, Sweden and Norway to provide evidence relating to neighbourhood effects on educational achievement, employment outcomes and teenage pregnancies as well as exploring the links between theory and practice in neighbourhood effects research, the problems with using evidence from the quasi-experimental settings in the United States and a discussion about how to look inside the "black-box" of mechanisms and processes that the phrase "neighbourhood effects" is usually used to cover.

Drawing on the findings of the first volume, the second investigated the processes of neighbourhood change and selective mobility into and out of neighbourhoods (van Ham et al. 2013). The primary focus was on one of the most significant challenges to the identification of real causal neighbourhood effects: selection bias resulting from non-random selection of people into neighbourhoods. Both of the previous volumes have been critical of the neighbourhood effects shibboleth (see also Manley et al. 2011) and have engaged in more cautionary discussions than is present in much of the literature. It is clear that neighbourhood effects research is at a crossroads and in order to move the debate forward there are many challenges that researchers must address head on (see van Ham and Manley 2012 for an overview

---

[1] A neighbourhood effect is defined as the idea that the neighbourhood in which an individual lives can negatively influence on their life outcomes across a vast range of domains including school dropout rates (Overman 2002); childhood achievement (Galster et al. 2007); transition rates from welfare to work (Van der Klaauw and Ours 2003; Simpson et al. 2006); deviant behaviour (Friedrichs and Blasius 2003); social exclusion (Buck 2001); and social mobility (Buck 2001).

of the challenges). Whilst looking to the future of neighbourhood effects research, it is important to also look at the policy prescriptions that have been made to attempt to improve neighbourhoods and the lives of individuals who live within them. That is the focus of this third and final edited collection.

The neighbourhood has long been a site of government intervention. This is because the neighbourhood represents a scale at which many government services and provisions are made (schooling, libraries and so on) and because political representatives are elected at this scale it represents a means to promote and enhance governance. The neighbourhood is a scale at which people can be persuaded to get involved and feel a sense of belonging (Pill 2012). In the long history of neighbourhood-based policies, there have been many incarnations of interventions. The most obvious developments have involved the construction (and reconstruction) of neighbourhoods and communities as a means to overcome the perceived social, economic or cultural problems experienced by individuals living in poor conditions, frequently in old industrial towns. In many Western countries there have been long traditions of constructing neighbourhoods as a means to developing better communities. In the UK, the Garden City movement of the early 1900s, and the overspill estates of the interwar period followed by the multiple waves of New Town developments in the post World War 2 period all placed the neighbourhood at the centre as a clearly defined space for individuals and households to live within. Since the 1980s, policies that specifically target neighbourhoods have commonly focused on the composition of the residents. These policies, frequently discussed under the rubric of social mix but which have more commonly been introduced using tenure mixing (co-locating social renters and owner occupiers in the same neighbourhood), have gone hand-in-hand with wider scale neighbourhood regeneration whereby dense social housing developments were knocked down and lower density low-rise properties were built in their place.

However, neighbourhood-based policies have not solely been focused on the development of physical housing infrastructure. Other aspects of neighbourhood and community life have also been targeted in the interventions specifically focusing on the main individual outcomes that concern researchers in the neighbourhood effects literature, and these form the topics in the first half of this volume (see below). These initiatives have targeted policy areas such as education, employment, crime, health and well-being. They have, in the UK, included Community Police Officers, to promote safety and crime reduction in specifically targeted neighbourhoods; investments in school buildings and other infrastructure including the rebuilding of poorly maintained and damaged buildings to provide newer, modern facilities; development of local employment projects through smaller scale local industrial units; and the investment in sport and leisure services.

In the previous two volumes research was presented which suggested that, even when casual mechanisms relating to neighbourhood disadvantage and individual outcomes were not present, because of selective migration or spatial exclusion, there is still be a case to be made for investments in neighbourhoods as a means to redistribute advantage and provide social facilities for communities. Thus, it appears logical that, in order to tackle neighbourhood inequalities, place- and person-based policies should go hand-in-hand. There are many multi-directional interactions between people, their neighbours and their neighbourhoods. One of the main problems in the neighbourhood effects literature stems from the difficulty of separating out all these different effects. Accepting that such links are present, it would also appear logical, at least to us, that policies designed to tackle the perceived problems that accrue because of concentrated poverty, spatial disadvantage and inequality should themselves include many multi-directional linkages. However, as many of the chapters in this book show, these links are rarely explored and frequently ignored.

Indeed, drawing on a recent comprehensive review of place- and people-based policies in the UK, a stark conclusion is drawn: "for the most part, person- and place-based policies have been developed separately and sometimes in isolation from each other. This reflects the responsibilities of government departments influenced by their different approaches and traditions" (Griggs et al. 2008, p.1). A further complication can arise with regard to the relative magnitude of place- and person-based effects. Time and again, the neighbourhood effects literature has shown that place-based effects are substantially smaller than person-based effects (especially for factors such as education, health, employment and household circumstances, see for instance van Ham and Manley 2010). Musterd and Andersson (2005) posed the question whether neighbourhood-based policies can ever be successful if neighbourhood effects are only ever found to be small in nature. If the neighbourhood only makes a small contribution to an individual's health, education or employment outcomes, then it would follow that interventions at that level can only make small changes. Maclennan (Chap. 13) counters this with a different view and suggests that the divide between place and person based intervention is, necessarily, a false one. The justification for place-based policy interventions arises not only from the potential effects of place but also through the advantage of having individuals gathered in a single neighbourhood or set of neighbourhoods and the resulting efficacy of being able to target specific resources at specific places (see Pill 2012). Nevertheless, in their admittedly partial review of policies between 1997 and 2008, Griggs and colleagues find very few policies initiatives that genuinely embrace the logical links between people and the places in which they live.

## From Effects to Policies

A persistent question that regularly surfaces in discussions about neighbourhood effects and neighbourhood-based policies is whether or not place-based policies remain relevant if neighbourhood effects do not exist? If irrefutable evidence was available that no causal links were presented between individuals and the contexts in which they live then would there still be merit in government pursuing place based initiatives? This is an important issue to address because, although it is true that there are large differences between neighbourhoods (variety in for example, wealth, health, and employment opportunities), it is less clear that living in poor neighbourhoods has a negative effect on residents. If the inequalities exist but do not cause significant differences for individuals living in the neighbourhood, or require a specific set of circumstances to cause a change, then the policy interventions are different compared to the policy interventions that occur if causal pathways between neighbourhood context and individual outcomes are persistent and repeated. The logic of this position is as follows: if the neighbourhood context can make a difference to an individual's life course (above and beyond that individual's personal characteristics), then an intervention at that same neighbourhood level should be able to either ameliorate the initial problem (for instance a concentration of unemployment leading to higher levels of unemployment amongst neighbours through negative socialisation) or remove it with a net gain in welfare for society as a whole. However, if the causal pathway is not present at the neighbourhood level, then a neighbourhood intervention is merely redistributing resources or opportunities to residents there at the expense of groups outside the neighbourhood – a zero sum game.

Discussing the outcome of neighbourhood regeneration in Scotland, Matthews (2012) notes that one of the reasons why large scale neighbourhood regeneration projects have had minimal success is because they are "inward-looking and failed to tackle the wider social forces that created and reinforced the neighbourhood's deprivation" (p.9, see also Hall 1997). Using the outcome of Australia social mixing policies, Arthurson (Chap. 12) notes that one of the reasons why Australian place-based policies have not had the impact that the policy makers expected was because problems in Australia were thought to be of the same magnitude and type identified in the American literature. However, as scholars elsewhere in the social sciences have highlighted, the adoption of "situated knowledges" is crucial to developing a better understanding of the local processes the produce local outcomes. Writing about economic geographies in general, Larner (2011, p.89) points out that "[w]e need to be clear with ourselves [....] that it is not good enough to simply study 'here' using the analytical tools of 'there'". For neighbourhood based interventions, it is logical that if the problems are comparatively smaller, then any gains from implementing area based policies are likely to be similarly smaller in magnitude.

It would be naïve to assume that neighbourhood effects are the only motivator behind the use of neighbourhood-based policies, although for many they provide the justification and rationale behind many area-based initiatives (Tunstall and Lupton 2010). In fact, there are many reasons why governments may wish to intervene at the neighbourhood level. Not least of these is the very fact that concentrations of poverty (and other so called social problems) bring together specific groups in specific places and the area level can be very useful for allowing the efficient targeting of resources. This can include provision of new services in neighbourhoods that previously had poor service provision (including health care, schooling or shopping facilities) or the provision of employment and skills training in neighbourhoods where large employers have closed, or it can include interventions such as policing where communities perceive issues with crime and safety.

Reading the neighbourhood effects literature, it would appear that in the minds of many academics the presence (or otherwise) of effects is a crucial element of the drive behind the development of placed-based policies (see for example: Platt 2011; Musterd and Andersson 2005; Lupton 2003; Tunstall and Lupton 2010). However, this view point is not reflected in the policy machinery and, as we will see below, in a number of place-based policy initiatives, the presence or otherwise of neighbourhood effects is irrelevant! However, for other policy initiatives it is crucial. One of the most commonly referenced neighbourhood-level interventions (although by no means the most prevalent or most substantial) has been poverty deconcentration through the creation of socially mixed neighbourhoods (See Chap. 7 by Keith Kintrea). Social mix has been created both indirectly, through policies in the UK and wider afield such as the Right-to-Buy (where former public housing tenants can become home owners by purchasing their property) and explicitly through the infilling or redevelopment of former social housing sites to include a proportion of private properties (either privately rented or owned through shared ownership, affordable housing schemes or mortgage and outright purchases). In theory, the links between social mix and neighbourhood effects should be very clear. For instance, the neighbourhood effects thesis suggests that individuals living in areas of economic disadvantage can become isolated, lacking links to groups outside their neighbourhood who can provide access to job markets and opportunities whilst the links that they do have within the neighbourhood increases exposure to 'negative' peer groups. Socially mixed neighbourhoods are thought to overcome this by enabling exposure to 'positive' peer groups who can provide access to previously closed social and informational networks. However, empirically, very few of these theoretical pathways for promoting advantage have been shown to operate; the empirical evidence remains much more sketchy than the theoretical literature would suggest (see for instance, Sarkissian 1976; Arthurson 2002; Galster 2007; Graham et al. 2009). Furthermore, it is immediately clear that this framework assumes that the flow of information and advantages gained are distinctly one sided. Individuals in concentrations of disadvantage require specific interventions to enable them to alter their life course (See Chap. 2 by Carlo Raffo for a discussion around why the pathologisation of individuals and groups based on a presumed collective experience may not be an appropriate model, using an educational example).

Conversely, new entrants from higher social groups (frequently owner occupiers) appear to have little to gain from the process, and certainly from the social renters. Other authors provide evidence that the presence or otherwise of neighbourhood effects was largely coincidental for the development of such policies. Using case studies from the United States, Joseph and colleagues (2007) concluded that in many cases the development of social mix was as much about local and national government's accessing the 'rent' that had accrued on desirable urban land where social housing was located as it was about the redevelopment of physical stock and the expected improvement to individual life outcomes.

Of course, it is important to remain critical of policy developments that intervene in neighbourhoods and individual life courses, to ensure that they do offer new opportunities and that they are genuinely targeted at real problems. The need to intervene in concentrations of poverty and the depiction of residents living in these areas have become unchallenged principles, that do not deserve such status (for discussion see Manley et al. 2011). Recognising this, Chap. 6 by Tom Slater provides an important and timely call to rethink the neighbourhood effects arena.

## Neighbourhoods Have Open Borders!

The whole of the second volume (van Ham et al. 2013) was devoted to the issue of selective mobility into and out of neighbourhoods. The rationale for this also lies behind one of the major challenges that neighbourhood policy makers face: neighbourhoods are not closed systems. Instead, they are open systems where individuals and households are (largely) free to flow in and out. Whilst this is a relatively simple statement to make, the processes underlying residential mobility and individual neighbourhood histories are incredibly complex (van Ham et al. 2012b). The neighbourhood mobility literature is diverse and reports on the econometric modelling of mobility at one end (through neighbourhood entry as a means of consumer modelling: see for instance Schelling 1969) through to understanding the cultural representations of space at the other (Clark 2009).

Because of the mobility processes, there is a widespread concern that even the most targeted area based policies may lose effectiveness because of 'leakage' with the argument running as follows: successful policies aimed at targeting inequalities may (for example) help individuals obtain better employment, raising their income, enabling them to move out of the neighbourhood, taking the (policy) resources that they have consumed with them. This leaves the neighbourhood with a vacancy and, as the residential sorting literature has shown, incomers into neighbourhoods tend to have very similar characteristics to those individuals who reside their already (see Bailey and Livingston 2008; Hedman et al. 2011) so that the neighbourhood remains the same. The empirical evidence for this simple model is relatively thin, however, and that which does exist suggests the picture may be a good deal more complicated; in particular, the leakage through selective migration may be much less than is generally assumed (Bailey and Livingston 2008; Bailey 2012).

Nevertheless, the concern with possible 'leakage' through residential mobility poses an important consideration for policy makers: whether they wish to help people or places. If the answer is people, then households moving out of neighbourhoods and taking their gains with them is not a problem. The vacancy they create in the neighbourhood is a positive outcome, representing a space into which another individual can move and potentially benefit from the policy interventions. However, if policy is designed to improve the neighbourhood, then any selective outflow would be of concern because it represents a loss to the area. In policy documents, this issue is rarely explicitly articulated.

The second aspect of population change in a neighbourhood that occurs as a result of the physical regeneration is displacement. Major regeneration projects frequently require the demolition and removal of the original dwellings so that new dwellings can be constructed. Over the 1990s and 2000s, and especially in combination with the promotion of socially mixed neighbourhoods, new dwellings were constructed at a lower density and with less public housing than was present in the neighbourhoods originally. During this process those households that were resident in the neighbourhoods were forced to move out to surrounding neighbourhoods either temporarily or, more commonly, permanently. In the second volume, Posthumus and colleagues (2012) investigated this process using data from The Netherlands and demonstrated that individuals were frequently moved to neighbourhoods that had at least as much and sometimes more deprivation. There are a number of important issues that arise from this idea which both the academic and policy literature must take account of: firstly, it must be recognised that that demolition of communities in this way and the displacement of households does not serve to improve the individual outcomes of the people in the areas targeted. More commonly, the perceived problems are pushed to other neighbourhoods – the so called waterbed phenomenon. Secondly, there is an issue of social and spatial justice, whereby former residents are excluded from the areas in which they use to live (Harvey 1973; Mitchell 2003; Soja 2010). Of course, place-based policies (of which there are many different types) and physical regeneration need not automatically lead to displacement and the loss of households from the community. In The Netherlands, regeneration policy during the 1980s and 1990s adopted an approach of regeneration for the people of the neighbourhood (an initiative explored in more detail Chap. 10 by Gideon Bolt and Ronald van Kempen; see also Bailey and Robertson 1997 for details of a comparable UK example).

## Looking Forward

One realisation that has become apparent through engaging with the papers presented at the three seminars, reading and editing the chapters that follow is that, the neighbourhood effects literature has been characterised by a lack of definition regarding what it is that actually concerns us. As Slater (2013, p.3) suggests, "[w]hilst it would be naïve to paint an impression that daily life in public housing is somehow a positive experience across the board, the tendency for outsiders to focus only on extreme and serious episodes occurring in public housing […] has played a significant part in the sorry trajectory of affordable housing provision in America and beyond". In short, the idea that urban areas have neighbourhoods with different characteristics, different levels of wealth, and differing degrees of infrastructure is not necessarily problematic. Indeed, neighbourhoods in urban areas need to be differentiated and heterogeneous partly to provide residential environments desirable and suitable for the diverse range of people that wish to live within the city and partly to provide accommodation for the individuals and households with different financial means. Cities need low cost neighbourhoods that provide entry points into the city as well as spaces for individuals and households who have become more established. When neighbourhood inequalities become starker, however, a range of negative consequences may ensue. For example, private finance may withdraw from the neighbourhood (See Chap. 6), denying the residents important services such as access to supermarkets or health and welfare services, and transport links can be broken. When this happens, vulnerable populations can become excluded from neighbourhoods in the wider urban environment making policy interventions necessary. It should go without saying that we all deserve to live in safe, healthy neighbourhoods and dwellings and that a major task of government should be to provide this, or at the very least facilitate the provision of these environments through regulation and policy. Unlike the tone of the debates at that are being played out at the time of writing in the UK, where Think Tanks such as the Policy Exchange are proposing that social housing in areas with high house prices should be sold off to facilitate the construction of new dwellings in cheaper areas (a sort of social 2 for 1 offer), government interventions in neighbourhoods should facilitate the opening up of neighbourhoods to populations disadvantaged and vulnerable populations, not the restrict their urban space.

To conclude, we would like to lay down a challenge to policy makers and governments involved in neighbourhood based policies: who are your policies designed to help and who will your policies disadvantage either intentionally or unintentionally? To address this question, we suggest policy makers should open themselves up and allow social researchers access to the policy structures. Crucially, they need to engage in a dialogue that allows the asking, not just the questions that conform to current government ideology, but also the more uncomfortable questions including those that challenge current beliefs and standpoints. Government policy makers and social scientists need to become open to the idea of experimental design and randomised trials with built in policy evaluation (Haynes et al. 2012). In a financial era where data collection is perceived as an additional an unnecessary governmental expense, built in and critical policy evaluation with full social science research backing is crucial. It is often said that experimenting on people's lives is unethical and immoral. Experiments carry risks and these need to be balanced against any possible benefits and acknowledgement concerning the inequitable distribution of who is exposed to the risks and benefits raises the spectre of more complex and difficult ethical issues. However, untested and ungrounded neighbourhood based policies borne out of beliefs, and which impact on individual lives, are equally as immoral and irresponsible.

## Book Structure and Contents

The remainder of the book is divided into two sections. The chapters in the first half of this book each tackle problems that are perceived to be the result of negative neighbourhood effects accrued from living in poverty concentrations. In turn the problems of poor educational attainment, worklessness, crime, and poor health outcomes are investigated and the potential links between neighbourhoods and policy interventions are explored. In the second part of the book attention is given more generally to the policy solutions that have been developed with regard to these problems in five national contexts: the United Kingdom, the United States of America, Australia, The Netherlands and as a contrast the largely aspatial policy context of Canada. These nation states were chosen because of their very different policy focuses with regard to addressing these issues through urban regeneration, social mixing, employment growth schemes and other Area Based Initiatives. Between Part I and Part II is a chapter that focuses on a critical view, that this introduction has also given prominence to and the case is made that who you are affects where you live rather than the oft cited where you live affects who you are of the neighbourhood effects literature. This counter point is very important in the wider neighbourhood effects debate and frequently one to which, frequently, insufficient space is devoted.

In Chap. 2, Carlo Raffo investigates the role that neighbourhood context plays in educational outcomes. There is a vast literature that links poor educational outcomes to disadvantage in the neighbourhood environment. In general, there has been a consistent policy drive to ensure that educational standards have risen across all areas in the UK. In places where this consistent upwards drive of standards has been less successful, Raffo shows that Area Based Initiatives (ABIs) have been used to provide extra resources and address more persistent inequalities. The chapter moves on to demonstrate that the vast majority of interventions have only been partially successful in raising attainment for some people. In the context of 40 years of ABI and a vast amount of money invested in improving outcomes such inconclusive results need to be interrogated. Raffo uses the framework of social justice to explain the lack of positive results and highlights that redistribution is about more than just financial resources. The vast majority of education ABIs did little to alter the causes of the inequalities including cultural injustices rooted in patterns of representation, interpretation and communication need to be addressed so that injustices where individuals from disadvantaged communities are rendered as deviant or dysfunctional and inappropriate to successful education are amended. Thus, many of the educational injustices that are linked to concentrations of poverty are actually based on the lack of politics of recognition. For example, the curriculum sets out and identifies standard cultural codes and assessment modes that dominate many mainstream class rooms. However, this standard set of codes can 'other' the experiences and cultures of pupils from a wide range of background and exclude them from the schooling process. To illustrate the point, a case study from Peterborough (UK) is presented. Here the curriculum is co-developed with external community partners so that a learning experience that values the pupil's backgrounds provides bridges between their external experiences and the learning environments.

In conclusion, Raffo reiterates that the perceived problems of educational achievement in disadvantaged areas are not solely about a lack of economic resources, but also about a lack of cultural recognition for the individuals living there. Thus, ABIs charged solely with tackling the economic injustice of educational inequality will never fully address the problems, and inclusion Raffo highlights the upcoming problems for educational inequality in the light of the post 2008 financial crisis and public spending reviews.

In Chap. 3, Stephen Syrett and David North document the links between concentrated poverty and what has become known as worklessness in the policy literature. They explore the policy initiatives that were instigated by the New Labour Government in the United Kingdom between the late 1990s and early 2010. Syrett and North make the link between the wider processes of labour market restructuring, and the negative cycles in the neighbourhood and persistent worklessness. The chapter considers the role that neighbourhood effects play in relation to the causes of worklessness and how the neighbourhood can mediate disadvantage. The major themes that are drawn out as mechanisms operating in neighbourhoods that could lead concentrations of worklessness forming including social capital and networks, the problems associated with neighbourhood stigmatisation and discrimination and the problem of physical isolation and poor public transport links preventing individuals from accessing opportunities for work when they do exist. Drawing on the work of Lupton and colleagues (2011) five different types of neighbourhoods where worklessness tended to be concentrated were identified.

The New Labour Government attempted to tackle the problems of worklessness using a wide range of Area Based Initiatives including the Action Team for Jobs Initiative, the Working Neighbourhoods Pilot Initiative and the National Strategy for Neighbourhood Renewal. Evaluation of these policies at the Local Authority Level (typically areas containing 250,000 people) demonstrated that employment levels had broadly increased when these initiatives had been in place. The problem for this analysis is that areas of this size cannot be regarded as a neighbourhood, and analysis at a more local level demonstrated a less positive picture. Neighbourhood level interventions are poorly positioned to address changes in labour market supply and demand which are likely to be the main causes of worklessness. But, they can provide a mechanism for the delivery of services. Finally, Syrett and North conclude that these ABIs existed in a disconnected policy arena and with many disparate agencies all competing to perform the same role in different places the lack of significant co-ordination between the policies and the wider economic environment meant that the effectiveness of the policies was muted.

In Chap. 4 Ian Brunton-Smith, Alex Sutherland and Jonathan Jackson tackle the issue of crime and perceptions of crime. They make direct links between the academic work on the causes of crime and many international policy initiatives including community policing and zero tolerance strategies. The historical development of neighbourhood context and crime is discussed with reference to neighbourhood deprivation in early work based in Chicago. However, they highlight that, in general, the experience of individuals in neighbourhoods is largely absent in the ways in which academic work has informed crime policy: early work relied on inappropriate analytical strategies and only in more recent work has the use of multilevel modelling techniques begun to overcome some of the more technical problems.

The authors give in-depth accounts of the mechanisms that are thought to be behind crime and the perceptions of crime starting with the idea of Social Disorganisation. Based on work from Chicago, it was hypothesised that higher levels of residential mobility and neighbourhood heterogeneity disrupt the formation of neighbourhood networks and prevent the development of community controls based on information. There are clear links between social disorder and the second mechanism listed, Neighbourhood Control. This emphasises the three domains of control private, parochial and public. All three levels need to function effectively for neighbourhoods to exercise the necessary controls on individuals and to influence public decision making sufficiently to ensure that neighbourhoods do not become disenfranchised. This also requires the neighbourhood to be able to mobilise the necessary resources from external agencies – such as the police – to establish the control of law and order. The third mechanism, Collective Efficacy, is based on positive control mechanisms. These include the process through which interpersonal trust can enable collective controls on individuals and also how efficacy can act as a mediator between the structural determinants of disorder and the fear of criminal behaviour. Low-level Disorder is identified as the fourth mechanism and this relates to relatively minor issues such as graffiti or vandalism which can act as signifiers that disorder is tolerated in a community and in turn lead to the fear of greater problems as well as the incidence of social disorder. The fifth mechanism is Subcultural Diversity which proposes a direct link between ethnic heterogeneity and variations in crime and concerns about crime. This theory focuses on conflict theory and suggests how inter-group tensions can lead to mistrust and external group fear. The sixth and final mechanism is Defensible Space and draws on ideas of territoriality and the physical design of the urban space. Critical to this mechanism is the way in which space is delineated and the boundaries through which a sense of ownership and therefore responsibility can be communicated.

The second part of the chapter deals with neighbourhood level policies for policing. In the UK context these have included neighbourhood policing programs, community support officers as part of larger regeneration initiatives, and the development of crime and disorder reduction partnerships. Using the police framework as a way to link into questions of neighbourhood effects and crime, the authors present a multilevel analysis looking at the components of the mechanisms listed above and data from the British Crime Survey. The model shows the importance of spatial autocorrelation in relation to the fear of crime and shows that neighbourhood characteristics represent an important driver in the development of an individual's fear of crime. In conclusion, the authors suggest that neighbourhood studies need to better reflect the ways in which individuals live in space and act out their daily lives in order to better understand the influences that they experience in developing their perceptions of crime.

One area where research into neighbourhood effects has been particularly prevalent is health research. Indeed as Jamie Pearce points out in Chap. 5 there is over 200 years worth of documentation on the subject. In this chapter, the evidence linking health and place is reviewed and three major problems with the previous work are identified: firstly few studies have developed a coherent picture of the processes operating in neighbourhoods, the historical development of these processes and the implications that they have for individual health and well-being outcomes. Secondly, little work has shed light on the ways in which the neighbourhood can mediate the associations between place and individual health outcomes. Thirdly, much of the previous work on health and well-being outcomes has adopted a 'deficit model' approach, whereby the problems of poor health and well-being are explained through the assumptions that those experiencing the problems are to blame and if only these people were given more knowledge they would adapt their behaviour to solve their problems.

One of the explicit problems that is explored in this chapter is the idea that many of the circumstances that lead to health and well-being problems are at the macro level and result from the decisions taken by multinational corporations (for instance in the form of not opening stores in less economically well off areas and reducing the supply of fresh food to the local residents) or in the form of macro level government policies in the provision of health care (whereby individuals living in less economically well off areas have to travel further to access doctor surgeries). As such, the neighbourhood is a wholly unsuitable level at which to analyse the problems that result from these interventions. Pearce contends that neighbourhood effects are an unsatisfactory conceptualisation of geographic health inequalities proposing instead an alternative framework known as "environmental justice" that extends the notion of social justice into the environmental arena. This framing enables three crucial aspects to be considered together: the social, health and environmental inequalities. This is in direct contrast to the current literature which isolates these interactions as single entities, or at best combines the social and health in one outcome to the determinant of the environment. The environmental perspective encourages a macro level evaluation of the processes that lead to ill health – not just the local ones traditionally associated with the neighbourhood effects literature but also the issues such as unequal investment in infrastructure, migration and mobility patterns which result in the concentration of lower income groups in areas that are less advantageous with regard to health and well-being outcomes.

Using this framework, Pearce provides details of the Multiple Environmental Deprivation Index (MEDIx), a small area measure of environmental characteristics thought to be related to health and well-being outcomes. Such an index is useful because it allows the environmental circumstances in which people live with to be related with their socio-economic circumstances, and it becomes apparent very quickly that places with social and economic disadvantage also experience environmental disadvantage highlighting the concentrations of disadvantage experienced by vulnerable individuals who frequently already have poorer health. In conclusion, Pearce calls for the neighbourhood effects literature to move beyond the deficit model, and to recognise the multiple influences that place can have on individual outcomes rather than isolating the social and economic from the environmental in order that we can move to a better understanding of how an individual's health can be influenced.

There is substantial debate in the neighbourhood effects literature about whether or not causal mechanisms can be identified through which individual life courses can be altered. Much of this debate is technical in nature and relies on increasingly complex econometric modelling. It is, however, rare that the foundations of the neighbourhood effects thesis are critically examined and the appropriateness of the framework as a mode of analysis called into question. In Chap. 6, at the pivot point in the book between Part 1 and Part 2, Tom Slater does just that and turns around the argument that where you live can affect your outcomes and presents the reverse case: your outcomes affect where you live. Highlighting what he calls the seductive simplicity of the neighbourhood effects thesis, he draws on the Marxian tradition of research to give precedence to the 'why' people live where they do aspect of neighbourhood effects research. Using Engels original work in Manchester, Slater demonstrates that the inequalities that were writ large in Manchester were the direct result of the system of private property rights. Engels provides a means to understand inner-city decline and the process of ghetto formation, neighbourhood decline and turnover as the consequence of successive reductions in capitalist investment in the infrastructure (property, parks, work places and services). This reduces the cost of entry into a neighbourhood which makes it available as a place to live for working class households.

In many ways, this chapter is uncovering one of the processes behind what the quantitative analysts have termed "selection effects", the idea that the distribution of individuals into their residential locations is a far from random process and that this structure matters. However, adding the Marxian perspective to this debate allows us to move beyond merely suggesting that the econometric models are incorrectly specified and, instead, allows us to reflect on whether the way in which we are approaching the investigation of neighbourhood effects is actually reinforcing the perceived problems of poverty that we wish to tackle. Thus, Slater shows that the very notion of a neighbourhood effect is an instrument of accusation, and that the neighbourhood effects literature has failed to engage with the wider socio-economic processes that occur outside the neighbourhood. Using educational dropout rates for teenagers in low socio-economic status neighbourhoods as an example the argument is made that, rather than blame the concentrations of low status individuals as the driving force behind the high incidence drop outs, the wider economic picture must be considered. Factors including the necessity of working to provide financial support to the wider household or to provide assistance to relatives in poor health (to cite two possibilities) should be integrated into the debate. By ignoring important structural aspects blame is laid at the door of the individual living in poverty preventing a fuller picture emerging, and the policy interventions that are prescribed are those that require individuals to be moved away from apparently negative neighbourhood environments as the solution to their problem, dealing with a symptom rather than a cause. Ultimately, Slater calls for the demolition of the neighbourhood effects thesis as a supportive prop for 'decision-based evidence-making' and the assumption that concentrations of poorer individuals automatically lead to reduced levels of place attachment, worse social networks and social capital and worse outcomes across a wide range of well-being and related outcomes

Part 2 of the book turns attention towards the specific policies that have been pursued to tackle the perceived problems highlighted in the first section and brings together a set of chapters that deal with different national contexts. Whilst the problems between countries may be strikingly similar the policy arrangements made to alter the perceived negative effects of concentrated poverty have been very different. However, one solution that has been pursued in multiple national contexts is that of mixed communities.

In Chap. 7 Keith Kintrea deals with the idea of mixed communities through the rubric of social mix. In essence social mix is a policy that seeks to incentivise the colocation of better off households in neighbourhoods previously dominated by poorer households. More often than not, social mix policies have been pursued as part of wider housing and regeneration programs. Whilst specific national contexts are explored further in the chapters that follow, this chapter provides a general overview of the policy. The chapter makes a direct link between social mix and neighbourhood effects as justifications for these policies include the idea individuals can become socially isolated when they live in deprived neighbourhoods and develop the 'wrong' sort of social capital. Mixing is a policy device through which outward looking social networks are thought to be enabled through the presence of wealthier residents. However, sceptics of the policy have pointed out that spatial proximity may not lead to physical mixing between the different social groups and is insufficient to create new links. Similarly, socially mixed communities have been described as communities without community with many frequently conflicting identities competing with each other. Lastly, social mix has been described as gentrification by stealth and the state-led destruction of communities in order to attract private investors into areas previously demarcated as state owned. It is rarely the communities of wealthier residents that are redeveloped for social mixing!

In his concluding comments, Kintrea asks what social mixing policies have achieved. He suggests that social mixing has (partially) been guided by ideological positioning and that the outcomes have been based more on hope than real expectations of change. In fact, there has been relatively little systematic evaluation of the majority of schemes and their impacts. To end, Kintrea notes that the social mix policies do little address the causes of inequality instead focusing on the symptoms. Nevertheless, improving the physical environment for households can provide benefits for the residents of the neighbourhoods.

In Chap. 8, by Neil Bradford, we begin the focus on national experiences of neighbourhood policy in Canada. Unlike the other countries in this volume, Canada does not have a history of national neighbourhood or even housing policy. This lack of spatial framework is compounded by the tensions between federal and provincial government policy claims which mean that there is intense competition over limited financial resources and there is little incentive to integrate or co-operate over these resources, or develop co-responsibilities or control. In his introduction, Bradford proposes that national-level policies that are enacted by national governments can be the source of neighbourhood effects. For instance, access to services and resources provided by the government are set by national policy, but the inequalities and challenges that individuals face as a result of these policies are played out in the local arena.

Within this context, three aspects of place-based policy have been developed: *Incrementalism* developing policies on a step-by-step basis; *Interscalar links* policy alone is not a panacea for urban poverty, and; *Learning from the local* the use of fine grained local knowledge. Using this framework, two cases studies are presented that show how urban revitalisation policies have been implemented over the last decade. The first, the Vancouver Agreement (VA), between the federal and provincial governments in Vancouver, was conceived as a city-wide policy for the targeting of resources in the Downtown Eastside of the city. It brought together multiple agencies to target social, economic and health priorities. The second case study is drawn from the Action for Neighbourhood Change (ANC) which operated in five cities across the country and was set out as a ground level engagement with communities to work with the people to tackle the problems that they face. This 2 year project was designed to test a resident lead regeneration project. The project brought together residents in the poorest neighbourhoods of Halifax, Toronto, Thunder Bay, Regina and Vancouver. The motivation for the ANC policy was a desire by policy makers to learn how they could further their mandates via collaborative place based work. Primarily, the focus was on how community led organisations could be used to address gaps in service provision or barriers to accessing national policy initiatives.

In conclusion, although Canada was a latecomer to place-based policies, it has embraced them with enthusiasm recently. As such, there is a desire in Canada to implement policies with the right mix of interventions. Moreover, Canadian policy makers are increasingly realising that local engagement is vital for the successful development of initiatives that intervene where the market failure has been observed. Bradford also notes that an important policy conclusion from the Canadian experience is that initiatives need to proceed on a case by case basis rather than assuming what works in one local will automatically work in another.

In Chap. 9 Rebecca Tunstall turns the policy focus towards the UK context. In this chapter Tunstall argues that evidence-based policy is attractive to policy makers as well as to researchers especially within the framework of impact-based research assessments. However, as yet, neighbourhood based policy interventions have not been linked explicitly to the neighbourhood effects literature. Tunstall argues that this disconnection is largely a function of the lack of UK specific studies on neighbourhood effects. Tunstall uses the UK government's "Treasury Green Book" – guidelines for policy appraisal and evaluation – as an illustration of how neighbourhood effects literature may influence government policy in the future.

The substantive part of this chapter consists of three examples of empirical work that the author has been involved with: an analysis that links individual personal and neighbourhood circumstances to a range of outcomes using the longitudinal British Cohort Studies and the Millennium Cohort Study. These studies enable the longer term outcomes of neighbourhood effects to be traced by looking at both childhood and early adult situations. In both cases evidence of (weak) neighbourhood effects were identified. The third example sought to understand neighbourhood reputation and stigmatisation as a barrier to employment using matched job applications for apparently identical (fictional) candidates where address was the only difference. Again, evidence of a neighbourhood effect was identified, whereby those individuals with addresses in stigmatised neighbourhoods were less likely to get offered job interviews compared with identical candidates from non-stigmatised neighbourhoods.

At the end of the chapter, Tunstall uses data from the participants of the ESRC seminar at which the original version of this chapter was presented as a means of conducting a participatory experiment to uncover what researchers think about the policy implications of neighbourhood effects research. Surprisingly, all participants believed that there was sufficient evidence that neighbourhood effects did exist, at least to a limited extent, and that the disagreement in the literature was not sufficient to render neighbourhood level policies ineffective.

Chapter 10 by Gideon Bolt and Ronald van Kempen focuses on the Dutch case drawing on policies aimed at deconcentrating poverty through desegregation. The Netherlands has a long tradition of neighbourhood level interventions and until the 1990s, the purpose of many neighbourhood level interventions was to improve the physical infrastructure for the residents. The 1980s brought with it a realisation that these policies did little to assist individuals and policy makers became convinced that concentrations of low income groups in specific places were the cause of societal ills. Consequently, the Dutch government refocused on the economic mix of residents in neighbourhoods. However, these policies changed focus post 2001 when ethnic mix became increasingly important and the discourse shifted towards ideas of assimilation and the explicit avoidance of ethnic minority segregation.

Government policy was directed at 'problematic neighbourhoods' and across the Netherlands, 40 neighbourhoods were target as areas that had an over-representation of low income, ethnic minority residents with excessive outflows of middle-class families and with few chances for labour market participation. More recently, the change in governmental priorities has resulted in a reduction in the urban and neighbourhood aspect of the integration and desegregation polices. A key policy introduced, initially in the city of Rotterdam, was the Special Measure for Urban Issues (and nicknamed locally the Rotterdam Law) which allowed municipalities to exclude residents from specific neighbourhoods when they could not meet strict criteria including the ability to financially support themselves independently or had not previously lived in the municipality for at least 6 years. Despite the vigorous adoption of the desegregation law, subsequent analysis has identified that the absolute difference in ethnic composition when comparing pre and post neighbourhood composition was nine households.

Bolt and van Kempen assess these policies against the empirical basis that exists in the academic literature. Citing literature using Dutch data the authors start by examining the applicability of Wilson's social isolation theory (Wilson 1987). The overall conclusion of the literature is that there is no evidence of social isolation of ethnic minority groups in the Netherlands and the living in concentrations of ethnic minorities does not hamper ties to the labour market. However, as Bolt and van Kempen note, research that only measures the number of ties cannot provide information about the quality of the social networks. Indeed, it is suggested that many of the ties the developed between households occur for reasons other than the fact that they live in relatively close proximity. In a modern society, social relations occur in a wide range of spaces and at a diverse set of scales, not necessarily just at the neighbourhood level.

In conclusion, Bolt and van Kempen suggest that the desegregation policies in the Netherlands that have sought to reduce the concentration of perceived social ills in neighbourhoods have not been very effective. Indeed, they suggest that the sectoral nature of the policies means that with highly restrictive access policies to the social sector combined with tight regulation and planning laws for new building the opposite effect may have occurred. That is, segregation may in fact be increasing. In sum, the authors point to the contradiction between policies that seek to desegregate communities being highly ineffective, while others enacted by the same government have exactly the opposite outcome.

There is a long history of neighbourhood level policy intervention in the United States, with the most (in)famous being the Gautreaux poverty deconcentration programs. However, although they have received the majority of attention, explicit poverty deconcentration policies only form a small part of a much wider raft of US policy initiatives, as examined in Chap. 11 by George Galster. Four major housing programs are discussed: (1) scattered-site public housing; (2) tenant-based Housing Choice Vouchers (HCV); (3) private developments subsidized through the Low-Income Housing Tax Credit (LIHTC); and (4) mixed-income redevelopment of distressed public housing estates (HOPE VI). Of these the third, LIHTC, is the largest and is outside the control of the Department of Housing and Urban Development (HUD). Galster observes 4 facts about these programs: residents of public housing in the United States live in more disadvantaged neighbourhoods compared with other people; that in site based programs (LIHTC) residents live in less disadvantaged areas than residents using HCVs; that HCV holders fare better than non-HCV holders even if in the same neighbourhoods; that HCV holders do not improve their neighbourhood circumstances with subsequent moves once they have left their initial neighbourhood. Galster notes that the first fact is obvious, and a consequence of planning policies, whilst the last three are because of: individual behaviours and constraints (including search strategies for housing); structural constraints including property availability and landlord participation, particularly for the HCV holders, and; program rules determining who could participate and where the administration system was governed. However, untangling which of these explanations underlies the outcomes observed as a result of the programs is very difficult and frequently the research into the outcomes has failed to provide answers.

Galster attempts to unpick how the various neighbourhood level programs have fared by reviewing the research outcomes. For instance, social capital of residents has been shown to be an invaluable source of support for residents in deprived neighbourhoods but also acts as a strong pull reducing the geographical extent of many residential searches. Properties advertised to HCV holders are often located in disadvantaged neighbourhoods, with landlords using HCVs to boost lower demand in harder to let areas. In combination with aggressive marketing tactics, knowledge about available properties as well as property availability serves to constrain the geographic extent of HCV holder moves. Finally, program administration details include the willingness of landlords to accept program members in areas that are easy to let compared with areas that are harder to let coupled with the fact that the HCV only cover a limited amount of rent and additional rental payments have to be met by the householders means that participants are frequently excluded from more desirable neighbourhoods. With regard to the LIHTC participants, properties were only available in areas that were deemed "Qualified Census

Tracts" which were defined as areas that were part of comprehensive redevelopment initiatives. In turn this excluded many non-disadvantaged areas simply because neighbourhoods had to have a low income status before they became recognised as areas that were suitable for LIHTC.

In conclusion, Galster provides a alternatives to the current US neighbourhood policies. The ideas include incentives to landlords in more desirable areas to accept HCVs, providing counselling to households in disadvantaged areas to assist them with the moving process and increasing the range of information (particularly about schooling) available to residents. However, as a caution, Galster notes that the American context is specific, and many of the issues faced in the US do not translate well to other contexts. As a result, detailed policy recommendations should not be based directly on the American experience. This is because US poverty is largely driven by the markets, where as in Western Europe, poverty is largely state driven. The fragmented federal structure of the US means that there is a lack of national co-ordination of programs in the US, and the politics of poverty and racism are local to the US. Finally, Galster challenges policy makers to consider on what basis should neighbourhood composition be judged? How much concentration is too much? Over what scale should the measures be judged and how quickly do the policies need to be progressed?

One of the national contexts in which US policies have been applied is that of Australia, which is the national context investigated in Chap. 12 by Kathy Arthurson. Set against the policy backdrop of post war social housing developments that are viewed as being increasingly problematic in terms of concentrations of unemployment, poverty and behavioural issues the Australian government has pursued policies of neighbourhood demolition and redevelopment. Starting in the 1980s, these first redevelopments renewed the physical infrastructure, frequently increasing the density of building. Recognising that the physical changes did little to address many of the perceived problems in the estates the programs were altered and increasing amounts of attention was paid to providing a social mix through selective redevelopment with policy makers arguing that, through social mix employment opportunities, educational achievements and service provision will all increase.

Arthurson draws on research conducted in Australia during the 2000s, investigating the level of social cohesion in 3 regenerated communities in Adelaide. Three dimensions are considered: the spatial scale at which the mix is implemented, the length of time that individuals are resident within the neighbourhood, and the stigma held by owner occupier residents towards their social renting counter parts. Social mix was identified as being less relevant to modern life, as individuals spent a lot of their time away from the neighbourhood. For some residents the age of neighbours was considered more important than their social status, and the broadening range of ages was a major barrier to the forming of friendships. However, the biggest tensions were reserved for the perceived differences in neighbours' standards and values surrounding behaviour. An important realisation here is the heterogeneous nature of the social renting group who are perceived as relatively homogenous groups in policy terms. One area in which individuals in different tenures did agree was around schooling, and the importance of having 'all walks of life' in the community school. However, this positivity needs to be tempered as home owners made specific judgements about the local community school and decided to send their children to schools elsewhere as a result. Consequently, those owners reporting positive feelings about the integration of children from different social backgrounds are a group who have specifically chosen that schooling route.

In conclusion, Arthurson suggests that the chapter highlights the processes, complexities and challenges that policy makers face. Importantly, the chapter shows that homogenous social housing communities do not have the exclusive rights to neighbourhood based problems. Neighbourhoods in which there is a large degree of social mix can face substantial challenges and problems. Whilst some residents recognise the diversity and plurality of residents' backgrounds in social groups, others stigmatise and point to the problems. Overall, Arthurson asks whether or not social mixing has become an outmoded concept: Wider networks beyond the residential neighbourhood have made the local environment less relevant for many residents. In conjunction with the clear contradictions between policies of social mix and providing housing for individuals with limited means has the consequence that social housing increasingly becomes a tenure for those in the greatest need alone, effectively increasing the isolation of low income groups and reducing mix in the very same tenure that the policy makers are attempting to reintroduce it to.

In the final chapter of this volume, Chap. 13, Duncan Maclennan contemplates how the policy environment has engaged the idea of neighbourhood effects. A difficulty for those interested in developing policy from research is that the vast majority of the academic contributions to the neighbourhood effects debates have come from work conducted in the United States of America, evolving from the Chicago school, where-as policy development requires more locally sourced examples as well.

In exploring why neighbourhood effects research has failed to have the expected impact on urban policy the first section of the chapter suggests a set of issues that need to be addressed in order for research to link directly with policy outcomes. Firstly, much of the neighbourhood effects research has essentially left the mechanisms of transfer as a black box. The broad area of work that is defined as neighbourhood effects consists of multiple disciplines researching from their own, often competing experiences and perspectives. This disagreement often makes it easy for policy makers to ignore research simply because the messages are inconsistent or inconclusive and lack guidance for developing policies. Secondly, researchers need to have a convincing story to tell policy makers. Despite the recent advances in neighbourhood effects theory and empirical research it is suggested that the ideas underneath the research are still sufficiently loose or fuzzy and that they do not relate back to the theoretical frameworks which they purport to investigate. Thirdly, the research needs to integrate the multiple aspects of individual life courses and the range of residential contexts through which people move. In this chapter, Maclennan suggests that, to date, the research undertaken in the name of understanding the urban residential environment has tended to be patchy and lacking in depth sufficient for policy makers to untangle the overall message that can be translated into direct policy interventions and initiatives.

Duncan then explores the suggestion that the lack evidence is sufficient to make neighbourhood based policies ineffective. Maclennan suggests that this assertion is incorrect in at least three ways: firstly, neighbourhood effects are not and never have been sole reason for area based interventions. Secondly, such a consideration places a false dichotomy between people and places. This means that successful policy interventions require a range of scales over which different aspects should be targeted. These scales include the local neighbourhood, but also include the sectoral and macro levels as well. In this conception the need, or otherwise, for strong neighbourhood effects to exist is not generally relevant. Third and finally, the link between academic evidence and policy is not as straightforward as a one-to-one relationship between evidence and policy development.

Ultimately, Duncan calls for a better understanding of the processes behind the phenomenon that are observed in the neighbourhood effects literature. This includes understanding better what can constitute a neighbourhood and neighbourhood space and whether they need to be spatially and temporally contiguous. Similarly, we need to know much more about how individuals choose their living environments, how they search for housing, what trade-offs they make and what cost structures they use when making their decisions. There are symmetries in the need to understand the effects of partial and missing information on these processes. Finally, we need to better understand the processes that are missing in the black-boxes that are used mediate neighbourhood effects. What mechanisms are important, for whom, when are they important and where. Only when we can thread all of these competing facets together will the academic discipline be in a better shape to deliver a more coherent story to policy makers and move beyond the policy mistakes of the past.

## References

Arthurson, K. (2002). Creating inclusive communities through balancing social mix: a critical relationship or tenuous link? *Urban Policy and Research, 20*(3), 245–261.

Bailey, N. (2012). How spatial segregation changes over time: sorting out the sorting processes. *Environment and Planning A, 44*(3), 705–722.

Bailey, N., & Livingston, M. (2008). Selective migration and area deprivation: evidence from 2001 Census migration data for England and Scotland. *Urban Studies, 45*(4), 943–961.

Bailey, N., & Robertson, D. (1997). Housing renewal, urban policy and gentrification. *Urban Studies, 34*(4), 561–578.

Buck, N. (2001). Identifying neighbourhood effects on social exclusion. *Urban Studies, 38*, 2251–2275.

Clark, A. (2009). Moving through deprived neighbourhoods. *Population, Space and Place, 15*(6), 523–533.

DeFilippis, J., & Fraser, J. (2010). Why do we want mixed income housing and neighbourhoods? In J. S. Davies & D. L. Imbroscio (Eds.), *Critical urban studies* (pp. 135–149). New York: Suny Press.

Friedrichs, J., & Blasius, J. (2003). Social norms in distressed neighbourhoods: testing the Wilson hypothesis. *Housing Studies, 18*, 807–826.

Galster, G. (2007). Neighbourhood social mix as a goal of housing policy: a theoretical analysis. *International Journal of Housing Policy, 7*(1), 19–43.

Galster, G., Cutsinger, J., & Lim, U. (2007). Are neighbourhoods self-stabilising? Exploring endogenous dynamics. *Urban Studies, 44*, 167–185.

Graham, E., Manley, D., Hiscock, R., Boyle, P., & Doherty, J. (2009). Mixing housing tenures: Is it good for social well-being? *Urban Studies, 46*(1), 139–165.

Griggs, J., Whitworth, A., Walker, R., McLennan, D., & Noble, M. (2008). *Person or place-based policies to tackle disadvantage. Not knowing what works*. York: Joseph Rowntree Foundation.

Hall, P. (1997). Regeneration policies for peripheral housing estates: inward- and outward-looking approaches. *Urban Studies, 34*(5), 873–890.

Harvey, D. (1973). *Social justice and the city*. London: Edward Arnold.

Haynes, L., Service, O., Goldacre, B., & Torgerson, D. (2012). *Test, learn, adapt: Developing public policy with randomised controlled trials*. London: Cabinet Office: Behavioural Insights Team.

Hedman, L., van Ham, M., & Manley, D. (2011). Neighbourhood choice and neighbourhood reproduction. *Environment and Planning A, 43*, 1381–1399.

Joseph, M. L., Chaskin, R. J., & Webber, H. S. (2007). The theoretical basis for addressing poverty through mixed-income development. *Urban Affairs Review, 42*(3), 369–409.

Larner, W. (2011). Economic geographies as situated knowledges. In J. Pollard, C. Mcewan, & A. Hughes (Eds.), *Postcolonial economies, chapter 4* (pp. 81–106). London: Zed Books.

Lupton, R. (2003). *'Neighbourhood effects': Can we measure them and does it matter?* (LSE CASE Paper 73). London: London School of Economics.

Lupton, R., Tunstall, R., Fenton, A., & Harris, R. (2011). *Using and developing place typologies for policy purposes*. London: Department for Communities and Local Government.

Manley, D., van Ham, M., & Doherty, J. (2011). Social mixing as a cure for negative neighbourhood effects: Evidence based policy or urban myth? In G. Bridge, T. Butler, & L. Lees (Eds.), *Mixed communities: Gentrification by stealth? chapter 11* (pp. 151–168). Bristol: The Policy Press.

Matthews, P. (2012). Problem definition and re-evaluating a policy: the real successes of a regeneration scheme. *Critical Policy Studies, 6*(3), 243–260.

Mitchell, D. (2003). *The right to the city: Social justice and the fight for public space*. New York: Guildford Press.

Musterd, S., & Andersson, R. (2005). Housing mix, social mix and social opportunities. *Urban Affairs Review, 40*, 761–790.

Overman, H. G. (2002). Neighbourhood effects in large and small neighbourhoods. *Urban Studies, 39*, 117–130.

Pill, M. (2012). Neighbourhood initiatives in Wales and England: Shifting purposes and changing scales. *People, Place and Policy Online, 6*(2), 76–89.

Platt, L. (2011). *Understanding inequalities: Stratification and difference*. Cambridge: Polity Press.

Posthumus, H., Bolt, G., & van Kempen, R. (2012). Urban restructuring, displaced households and neighbourhood change: results from three Dutch cities. In M. van Ham, D. Manley, N. Bailey, L. Simpson, & D. Maclennan (Eds.), *Understanding neighbourhood dynamics: new insights for neighbourhood effects research, chapter 5*. Dordrecht: Springer.

Sarkissian, W. (1976). The idea of social mix in town planning: an historical review. *Urban Studies, 13*(3), 231–246.

Schelling, T. C. (1969). Models of segregation. *The American Economic Review, 59*, 488–493.

Simpson, L., Purdam, K., Tajar, A., Fieldhouse, E., Gavalas, V., Tranmer, M., Pritchard, J., & Dorling, D. (2006) *Ethnic minority populations and the labour market: An analysis of the 1991 and 2001 census* (Rep. No. 33). London: DWP.

Slater, T. (2013). Expulsions from public housing: the hidden context of concentrated affluence. *Cities, 30*.

Soja, E. (2010). *Seeking spatial justice*. Minneapolis: University of Minnesota.

Tunstall, R., & Lupton, R. (2010). *Mixed communities: Evidence review*. London: Department of Communities and Local Government, HMSO.

van der Klaauw, B., & Ours, J. (2003). From welfare to work: does the neighborhood matter? *Journal of Public Economics, 87*, 957–985.

van Ham, M., & Manley, D. (2010). The effect of neighbourhood housing tenure mix on labour market outcomes: A longitudinal investigation of neighbourhood effects. *Journal of Economic Geography, 10*, 257–282.

van Ham, M., & Manley, D. (2012). Neighbourhood effects research at a crossroads: Ten challenges for future research. *Environment and Planning A, 44*(12), 2787–2793.

van Ham, M., Manley, D., Bailey, N., Simpson, L., & Maclennan, D. (Eds.). (2012a). *Neighbourhood effects research: New perspectives*. Dordrecht: Springer.

van Ham M., Hedman L., Manley D., Coulter R., & Östh J. (2012b). *Intergenerational transmission of neighbourhood poverty in Sweden. An innovative analysis of individual neighbourhood histories* (Discussion Paper No. 6572). Bonn: IZA.

van Ham, M., Manley, D., Bailey, N., Simpson, L., & Maclennan, D. (Eds.). (2013). *Understanding neighbourhood dynamics: New insights for neighbourhood effects research*. Dordrecht: Springer.

Wilson, W. J. (1987). *The truly disadvantaged: The inner city, the underclass, and public policy*. Chicago: University of Chicago Press.

# Chapter 2
# Educational Area Based Initiatives: Issues of Redistribution and Recognition

**Carlo Raffo**

## Introduction

In this chapter I want to examine the relationship between educational equity and educational area based initiatives (ABIs) in England. The reason for this particular focus is that, over time, the overwhelming evidence continues to point to educational inequalities being consistently concentrated in poor urban areas. In response to such evidence educational ABIs over the years have, to a lesser or greater extent, provided enhanced funding to schools in such areas. Given this additional funding why do these inequities still persist?

In answering this question I review and critique some of the arguments for why policy developments to improve educational inequalities continue to struggle with their stated aims. Based on the ideas of social justice developed by Fraser (1996, 2008), my argument is that current area based approaches to improving educational equity for those most disadvantaged have predominately focused on 'closing the attainment gap' through an affirmative redistribution of resources. This has been undertaken in the main without exploring wider economic disparities. Nor has there been a focus on status order issues that are associated with the way educational policy and educational institutions have, at times, culturally misrecognised specific groups of young people made poor. My argument is that effective educational area based initiatives are not possible without understanding how the interpenetration of economic distribution and cultural recognition act as a base for the way education is experienced by young people in such areas. My argument is that effective educational area based initiative are unlikely to succeed unless they, in effect, undertake two task simultaneously. Firstly ABIs need to provide appropriate and additional material and technical resources to schools and families to help disadvantage

C. Raffo (✉)
School of Education, Ellen Wilkinson Building, The University of Manchester,
Oxford Road, Manchester M13 9PL, UK
e-mail: carlo.raffo@manchester.ac.uk

students access more easily, and be supported more fully, in schooling. Secondly ABIs need to ensure that educational provision in schools is sympathetic to the cultural traditions, histories and values of students and families from the communities with which they engage. This recognition may take the form of area-based curriculum and pedagogy and also stronger democratic opportunities that enable representative and authentic engagement with students and families.

## The Development of Educational ABIs

Over the last 40 years educational policy has generally responded to educational inequality in ways that reflects two main perspectives. The first and dominant perspective suggests that, given appropriate and effective school leadership and teaching and learning, all schools, no matter what the intake or where they are located should be able to achieve broadly similar results. By pursuing strategies located in school improvement and effectiveness literature and ensuring that performance is driven by a cocktail of high stakes accountability measures, school choice and competition and detailed inspection regimes, the school system can overcome inequalities, or so the argument goes. Although dominant in policy discourse, such an agenda seems to run counter to much educational data that suggest that although improvements can be made through the school system, these tend to be sporadic, inconsistent and often difficult to sustain resulting at an aggregate level in disadvantaged students doing less well than their more affluent counterparts. This has resulted in a second general strand of policy development that has recognised that schools recruiting disadvantaged students often struggle to achieve set educational benchmarks. This second strand of policy response has generally been about providing additional compensatory resources to such schools to help them narrow the attainment gap. Given that the vast majority of these schools are located in disadvantaged urban contexts, policy initiatives have, over time, allocated these additional resources to: (a) those disadvantaged urban contexts, (b) the schools located in those contexts and (c) particularly underachieving groups of young people who live and attend schools in those contexts. Together these educational policies and interventions have become known as educational area-based initiatives (ABIs). Although these educational ABI policies in England have been of many kinds, frequently emerging and disappearing within the space of a few years, they can perhaps be best categorized under four main headings:

(i) ABIs targeted at schools in disadvantaged areas that predominately enrol a high proportion of disadvantaged or poor students;
(ii) ABIs co-ordinating policies in disadvantaged areas across education, health and social welfare;
(iii) ABIs in particular cities where poverty is heavily concentrated, and;
(iv) ABIs that focus on area regeneration initiatives that include an education component.

This classification of policies is perhaps far from perfect, and one might argue that there is some overlap between these different foci. Nonetheless, I would argue that it is a useful way of making sense of what otherwise might seem to be a complex and perhaps chaotic policy scene. So what do these ABIs look like?

If we were to examine the history of ABIs targeted at schools in disadvantaged areas evidence suggest that such initiatives have been around since the late 60s and early 70s. The first of these policy developments was the Educational Priority Areas (EPA) initiative (Smith 1987). The initiative was based on The Plowden Report (1967) that suggested that EPAs needed to provide additional resources for schools in designated priority areas. These resources were to be used for generating smaller classes, for more experienced and successful teachers, with salary incentives to attract them to work in EPAs; for priority in new or replacement school building, and in the expansion of nursery education; for teacher aides, teachers' centres and more school-based social workers. Building on such notions, later articulations of ABIs have included the Excellence in Cities (EiC) programme in England that was developed from the earlier Education Action Zones (EAZs). EAZs were guided by the principle of positive discrimination where compensatory and additional resources were provided to support schools working in the most challenging circumstances. EAZs were run by a small number of 'partners' including local authority, business, voluntary sector and community representatives. Such partnerships drew in local and national agencies and charities involved in, for example, health care, social care and crime prevention that also linked up to Health and Employment Zones and projects funded by the Single Regeneration Budget (SRB). A typical EAZ consisted of around 20 schools (usually two or three secondary schools plus their feeder primary schools). EAZs received government funding of up to £750,000 per annum for 3–5 years to support them in this task, which were supplement by £250,000 per annum sponsorship in cash or 'kind' from the private and/or voluntary sector. Given some of the operating problems of EAZs highlighted by programme evaluations (Halpin et al. 2004), the programme was merged with the Excellence in Cities programme in 1999. Paralleling the EAZ initiative, the aim of the EiC programme was to raise standards and promote inclusion in disadvantaged inner cities and other urban areas. Additional funds were provided to schools to improve leadership, behaviour, and teaching and learning. Initially just based in secondary schools, the programme quickly expanded to include primary schools. The programme attempted to tackle underachievement in schools through specific strands targeted at underachieving or disadvantaged groups. So: Learning Mentors worked with underachieving students in schools; Learning Support Units were established to provide for students at risk of exclusion from school for disciplinary reasons; a Gifted and Talented pupils programme was developed; and City Learning Centres were established to enhance adult learning opportunities (particularly through information technology) for local people. The programme lasted for much of the previous Labour government and is now to be replaced by the current coalition government pupil premium proposal. The general aim of the pupil premium is similar in many respects to its forerunners in that it proposes to target extra resources at schools with a high proportion of disadvantaged pupils. Although the above initiatives were developed by governments over different historical periods, the main focus for each of these particular ABIs was to provide additional resources to schools in the poorest urban areas in order to help engage young people and their families with education thereby improving their educational attainments.

Running parallel to schemes that focused primarily on schools, other ABIs' primary focus was the co-ordination of policies in disadvantaged areas across education, health and social welfare. These policy initiatives are perhaps best exemplified by New Labour's Sure Start initiative, Children's Centres and Full Service Extended Schools (FSES). Sure Start and Children Centres were set up in England to enhance the functioning of children and families living in disadvantaged areas by providing additional services in local programme areas. In many respects these centres reflected the research of early child development studies (Shonkoff and Phillips 2000) and of programmes such as Head Start in the US. Such programmes were aimed at providing additional resources to disadvantaged preschool children with the purpose of delivering programs and services that would prepare preschool children for elementary school. Typically services included parenting support, access to health provision and child care and educational facilities for young parents. Sure Start/Children Centres were strategically situated in areas identified as having high levels of deprivation and were designed to enhance the life prospects of young children in disadvantaged families and communities. FSES in England also constituted focal points at which strategies for raising educational standards overall were supported by additional resources. These resources were targeted at schools serving disadvantaged population and were utilised for developing strategies for tackling neighbourhood and family problems. Hence FSES were expected to intervene in the multiple problems which beset children, families and communities living in disadvantage. However, at the heart of these interventions was a commitment to education as the pathway to achievement and hence to employment and social inclusion – and to raised expectations as a necessary precondition of raised achievement. The FSES initiative therefore focused on both the educational development needs of young people and the requirements for enhanced family and community engagement. These latter requirements were supported in FSES via the provision of parenting classes, crèches and skills development programmes that recognised the need for a more integrated multi-agency approach to delivering core public in one accessible location. Once again, however this strand of ABI provision was about diverting additional funds to schools and other agencies to help improve the integrated support for both disadvantaged young people and families' engagement with, and attainments in, education.

Whereas many of the initiatives highlighted above provided additional resources to targeted areas across England, the London Challenge was an example in England that recognised the distinctive difficulties facing schools in the capital. These difficulties included, high levels of disadvantage, low levels of educational achievement, the challenges of a multi-ethnic population, and the balkanisation of governance of London education. The Challenge deployed a range of strategies to address these issues, including programmes aimed at increasing teacher recruitment and retention, a gifted and talented programme, targeted intervention with low-performing schools, developments in vocational education and support to local authorities in managing their education systems.[1] Similar programmes have also been set up in other major cities in England. The major focus of all the city challenges has been the diversion of additional funds to schools and educational systems in those cities in order to raise educational attainment for young people most disadvantaged in those cities.

As intimated above in the discussion about EAZs, although one can readily trace the development of area-focused educational interventions in England, this policy approach has historically been repeated across many aspects of government social policy. For example schools and early years centres in disadvantaged areas participated in interventions that were developed by government departments and agencies other that the Department for Education (DfE). For instance, the early experiments with extended schools arose out of the cross-departmental National Strategy for Neighbourhood Renewal. Likewise, the Single Regeneration Budget and New Deal for Communities managed by the Department for Communities and Local Government (DCLG), had dedicated education strands that were often a source of additional funding for schools and early years centres.

## What Has Been the Success of Educational ABIs?

Evidence seems to suggest that ABIs of the sort described above have perhaps been only partially effective in raising attainment. And yet they have been successful in focusing educational attention on particular groups, areas and institutions. In doing so they have stimulated educational endeavour by acting as catalysts for considerable activity on the ground that have generated some improved outcomes at which they were targeted. For example, the rate of increase in GCSE (national examination) performance for EiC areas has been around twice that of non EiC schools for a number of years in succession (see Kendall et al. 2005). This means there has been a narrowing of the achievement gap between EiC and non EiC areas from 12.4 % in 2001 to 6.9 % in 2005. Moreover, there have been improvements for targeted groups of young people. However Melhuish and colleagues (2005) in their review of Sure Start suggest that such benefits appear to accrue predominately to those moderately disadvantaged rather than for those more severely disadvantaged - echoing other evaluations of similar interventions (Love et al. 2002). Such groups appear to be better placed to make use of any resources available and often do so through their engagement with better placed support networks. The evidence for FSES produced by Cummings et al. (2005) evaluations suggest a partial break in the cycle of disadvantage with positive outcomes in relation to increased pupil engagement with learning, raised attainment, and a growing trust and support between home and school. There was also improved multi-agency working that brought some benefits to children and their families.

---

[1] See Department For Education and Skills website: http://www.dfes.gov.uk/londonchallenge/

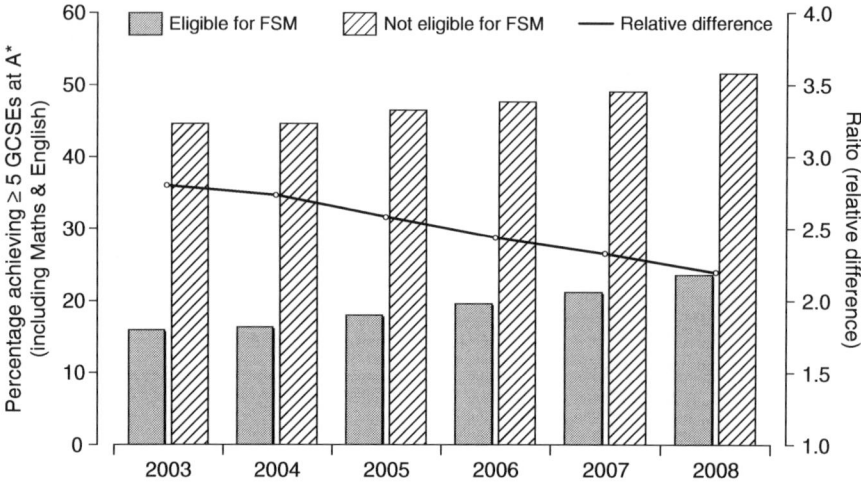

**Fig. 2.1** Percentage of Children achieving 5 or more GCSE's at A*-C (including English and Maths) by Free School Meal eligibility, 2003–2008 (Originally published as Figure 1 in Goodman et al. (2011) and reproduced here under Creative Commons Attribution License)

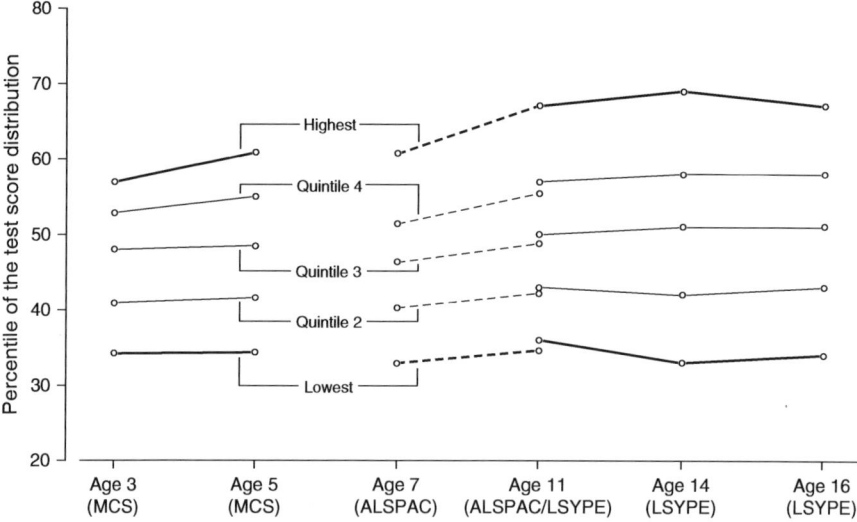

**Fig. 2.2** Cognitive achievement outcomes by socio-economic position quintile, across surveys and ages (Originally published as Figure 2 in Goodman et al. (2011) and reproduced here under Creative Commons Attribution License)

The graph below perhaps provides an indication of the relative success of such interventions. When taken together, there is every reason to suggest that ABIs, with other school improvement and effectiveness interventions, have partially narrowed the gap in educational attainment between disadvantaged pupils (as measured by their eligibility for free school meals (FSM)) and rest of the student population (Figs. 2.1 and 2.2).

Although this evidence is encouraging there is a need to perhaps examine the linked claims in the light of a more sophisticated set of data that examine current educational performances of young people at different ages and in relation to the socio-economic quintiles to which they belong (see table above). Here we see a constant gap in educational performance between the socio-economic quintiles and over the age classifications. Based on this potentially more comprehensive data there is a suggestion that although there have been numerous investments in ABIs over a number of decades, the basic premise still holds in England that the more disadvantaged an individual's background the less well he/she will achieve in education (Chitty 2002) and the more that social hierarchies are reproduced through generations (Blanden and Machin 2003). Given the evidence highlight above, why have ABIs only had a limited impact on educational equity?

## Educational ABIs and Issues of Redistribution and Recognition

In order to examine some of the reasons for the very partial success of ABIs the chapter builds on some of the social justice theorising of Nancy Fraser. Fraser (1996) outlines two types of social justice that focus on either notions of redistribution or recognition. Social justice claims that focus on redistribution emphasises a more just distribution of resources and goods. In the case of education, redistributive justice is about ensuring a fairer allocation of educational attainments across classes, ethnicities, gender and places. It is about narrowing the attainment gap between those most and least privileged in society. Educational interventions for equity that focus on redistribution attempt to ensure that resources are appropriately apportioned to enable this narrowing of the gap to occur. According to Fraser the range of redistribution spans from the affirmative to the transformative. Affirmative redistribution attempts to resolve issues of maldistribution by altering aspects of the allocation. However this is done without disturbing the underlying mechanisms that generate the inequalities of distribution in the first place. Given the tentative nature of affirmative redistribution Fraser warns of its dangers:

> ....because they leave intact the deep political-economic structures that generate injustice, affirmative redistribution reforms must make surface reallocations again and again. The result is often to mark the beneficiaries as "different" and lesser, hence to underline group divisions (Fraser 1996, p.46).

In contrast to affirmative redistribution, transformative redistribution seeks to re-dress distributional injustices by altering the underlying structures and frameworks that generate them.

From the details provided above about educational ABIs, one can see how the vast majority of them can be categorised as examples of affirmative redistribution. Firstly, educational ABIs have provided enhanced funding for schools in disadvantaged and challenging urban contexts. In addition ABIs such as FSES and Sure Start/Children Centres have redistributed additional resources for out of school support for families and parents through, for example, parenting classes or improved crèche facilities. Finally, ABIs have been clearly linked into the wider regeneration of areas where educational developments have been planned alongside wider redistributive activities such as enhanced economic opportunities and support for young people and their families and communities. In addition the underpinning core rationale for ABIs is economic. Firstly ABIs are economic in the sense that economic resources are being redistributed in particular ways to support schools, young people and families in disadvantaged communities achieve higher and more equitable levels of educational attainment. Secondly they are also economic in that the aim of improved educational attainment is about generating a fairer distribution of human capital that can translate into enhanced levels of economic activity and social mobility for those most disadvantaged. One might argue, therefore, that educational ABI policy focuses on improved educational outcomes as a way of reducing economic marginalisation through enhanced labour market participation.

However as the evidence documented in this chapter suggests, these interventions have only been partially successful in achieving their aims. Part of the reason for the comparative and historic failure of ABI policies, I would argue, is due to their continued focus on affirmative redistribution at mainly the school level with little concern for macro structural inequalities that might require transformative levels of redistribution. As Anyon notes, while educational policy interventions need to address the response of the education system to the academic challenges posed by learners living in poverty, they also need to tackle what she calls the "macroeconomy" (Anyon 2005).

Perhaps an equally compelling argument is that a focus on redistributive equity per se is insufficient. Building on Fraser's arguments I will suggest the need for educational ABIs to focus as much on the politics of recognition as they do on the politics of redistribution.

A politics of recognition in the field of education manifests itself in progressive educational principles that focus on schools being at the centre of communities that serve all children from those communities. A politics of recognition is also about school leadership forging inclusive democratic school communities dedicated to personal growth through active involvement with others. As Fraser states, a politics of recognition targets cultural injustices which are rooted in social patterns of representation, interpretation and communication that include:

> .... cultural domination (being subjected to patterns of interpretation and communication that associated with another culture and are alien and/or hostile to own's own); nonrecognition (being rendered invisible via authoritative representational, communicative, and interpretative practices of one's culture; and disrespect (being routinely maligned or disparaged in stereotypic public cultural representations and/or in everyday life interactions) (Fraser 1996, p.7).

Fraser argues that resolving these types of injustice requires a revaluing of disrespected identities and the cultural products of marginalised groups. Rather than classifying inequalities in class based ways that reflect a politics of redistribution, a politics of recognition perceives the sufferers of injustice through focusing on the relations of recognition they experience and specifically the lesser honour, esteem and prestige that they benefit relative to other groups in society.

A politics of recognition is about campaigning about the injustices that those in positions in power can inflict on particular groups in society in the way that they orchestrate discourses of deficit, unworthiness and lesser capability in their understandings and relations with these groups. In terms of education, the politics of recognition elucidate the many cultural injustices that different and particular groups of young people and families experience through the education system. Educational injustices based on a lack of politics of recognition manifest themselves through the way identities, funds of knowledge, and educational desires of these groups of young people and their families are silenced by dominant educational discourses that operate at national policy and school practice level. At school practice level, it is about how teachers understand and relate to their students. Ideally this understanding would recognise that young people enter schooling from different structural positions that reflect social habitats that embody distinctive and different qualities of cultural disposition, or 'habitus'. It would also recognise that these dispositions can be affected by place-based social relations, structures and cultures that encompass the way "individuals, families and groups shift and manoeuvre within the diverse habitats … adapting to local conditions" (Barker 2010, p.34). Furthermore it would recognise that these dispositions in turn operate selectively in schools as 'cultural capital' of stronger or weaker species (Bourdieu 1986). However, far too often these understandings are not evident in schools. Instead it is teachers' 'standard' cultural codes of curriculum, pedagogy and assessment that dominates mainstream classrooms. This results in disadvantaged young people's educational dispositions acquiring lesser 'capital' value because of how far they stand apart in relation to teachers' and schools' dominant codes (Bourdieu and Passeron 1997). In many, if not all, poor urban communities the cultural habits brought to school by significant proportions of students are basically not utilised or scaffolded to traditional school learning methods and contents. Simply put, their 'virtual schoolbag' (Thomson 2002) is not unpacked. Rather, their lack of fit with the cultural selections that are valued by school become individualised and internalised as 'failure'. As Riddell states:

> Some young people ….*will* have a background of rich learning in community, but *transfer* may be weak if the school's expectations and environment cannot "recognise" the learning and its conventions, particularly if the dominant ones are of a different type from those in school (Riddell 2007, p.1033).

Compounding these school based educational injustices of lack of recognition are the broader recognition injustices that young people's families and communities experience. For example Lawson's study (Lawson 2003) in the US addresses teachers' and parents' perceptions of the meanings and functions of parental involvement in children's education. Lawson found that in many respects teachers' prevailing orientations toward disadvantaged parents were based on deficit perceptions that portrayed them as having poor parenting skills and poor supportive capacities with regards to education. This approach contributed to what Lawson argued was a systematic silencing of the strengths, struggles, and 'communitycentric' worldviews evident in the parents' perspectives. He argues that parental involvement is a limited and limiting concept in low-income ethnically concentrated communities because the notion is largely the domain of teachers that hold a narrow 'schoolcentric' view. In addition Bauder (2002) examines the ideological underpinnings of the discourse of deficit neighbourhood effects that permeates much dominant educational discourse about the challenges that schools in such contexts face. He suggests that the language of deficit neighbourhood implies that the demographic context of poor neighbourhoods instils 'dysfunctional' norms, values and behaviours into youths, thus triggering a cycle of social pathology that creates a educational disengagement and disaffection. The problem is that these essentialist conceptions (fixed traits that do not allow for variations among individuals or over time) then imbue the way that schools and teachers think, act and react to young people in such contexts that then further contribute to the neighbourhood effects phenomenon.

By reflecting on the politics of recognition, it becomes clear how both historic and current approaches to ABIs have given very little import to the cultural identities, agency and viewpoints of disadvantaged young people, families and communities. So what would ABIs look like if they were underpinned by a politics of recognition?

A politics of recognition supports progressive educational principles that give primacy to young people's educational aspirations, actions and choices in pursuing educational processes and outcomes that they value. In that sense they reflect Sen's shift of attention away from material deprivation per se towards the freedom and ability to convert resources into valued achievements (Sen 1999) – in other words a shift in focus from issues of material redistribution to ideas about recognising an individual's freedom to choose and engage with what he/she values. What therefore becomes important for educational policy and practice is to recognise and understand the way social relations that young people experience influence their educational choices and actions. This would then enable schools and teachers to build and develop appropriate bridging strategies that connect the informal learning experiences of young people to the codified learning of the curriculum. In a previous paper I have argued that these choices and actions stem from educational identities that are premised on the intersectionality of space/place, class, ethnicity and gender that together impact in the way disadvantaged young people's educational capabilities and outcomes are brought into existence. Educational ABIs therefore need to recognise and build on young people's educational identities. This requires that schools, colleges and other educational institutions understand in a detailed anthropological way the manner in which these educational identities are formed and developed. To facilitate this process schools and teachers need to engage in a whole host of curricular, pedagogical and relational strategies that give voice, choice and independence to young people (Hattam et al. 2009) and their identity narratives (Goodson 2008). For example this may mean schools and teachers, in partnership with young people and their communities, developing area or placed based curricula (Facer 2010) that reach out and enable the different "funds of knowledge" (Gonzalez et al. 2005) of young people and communities to be respected and utilised in schools. Currently the Royal Society of Arts (RSA) is involved in developing such an approach with schools in Peterborough.[2] The Peterborough curriculum builds on the experience and recommendations that emerged from the RSA sponsored Manchester pilot and seeks to create sustainable and dynamic links between schools and external partners, generating new networks of learning outside schools. In particular, the Peterborough curriculum attempts to develop a city wide approach that draws on the funds of knowledge and resources of parents and immediate local communities, not just the economically successful and historically significant ones. In addition it is working to address city wide logistical challenges facing school and partners as well as encouraging reflection and clarity around the concepts of 'area', 'place', and 'local' in this context. As part of the broader Citizen Power project in Peterborough, the area based curriculum aims to provide Peterborough with a model for a curriculum that encourages participation, attachment and innovation in its young people, in its schools, and in the wider community of the city. The objectives for the curriculum are for partnerships between schools and partners to be created and sustained; for the curriculum to be co-developed and have positive learning outcomes for young people and/or the wider community; and that curriculum frameworks based on place can have positive impacts on attachment, place creation, and civic engagement in young people and/or the wider community. The curriculum is also about ensuring that appropriate bridges are created from young people's narratives and informal learning experiences of Peterborough into the codified curriculum knowledge of schools and the wider experiences of life (Lingard 2005). The desired result is that young people are enabled to develop skills and capabilities that will help them to operate effectively and appropriately at both the global and within the local (Gruenewald and Smith 2008; Raffo 2006). In essence the Peterborough curriculum is about educational provision being aligned with the needs of local communities that at the same time reflect the identities and aspirations of young people from those communities.

The Peterborough curriculum is therefore about education policy that engages with issues of recognition so that schools move from being what Sanchez-Jankowski (2008) defines as enterprise orientated institutions delivering centralised bureaucratic educational targets that potentially misrecognise communities and young people to being neighbourhood orientated that appropriately respect those communities and give democratic voice and governance back to those communities. The transformative importance of strong democratic links between schools and communities enables teaching and learning to increase the self-esteem and self-efficacy of both parents and young people and increases the emphasis on personal growth and the on young people's own interests, creativity and expression.

Given current economic difficulties, one might argue that a re-working of educational policy around a politics of recognition as exemplified through the Peterborough curriculum could be viewed as a cost effective way of effectively engaging nationally with issues of educational inequality. By embracing the diversity and identities

---

[2] For a more detailed explanation of the Peterborough curriculum visit http://www.thersa.org/projects/citizen-power/the-peterborough-curriculum2.

of disadvantaged young people and their families, educational practice might loosen its ties with a policy agenda that is focused on a narrow redistribution of academic educational credentials that appear to often to have little meaning in the lives of many poorer young people and their families. Instead it would provide opportunities for the way schools "should work with their communities, aiming to facilitate mutual relationships that develop people and create conditions that build confidence and commitment. This is how schools and teachers can enrich, improve and even transform lives" (Barker 2010).

Although potentially attractive at one level what an educational politics of recognition perhaps underemphasises is how the educational identities of young people are also influenced by the economic conditions within which these young people live. Educational agency is restricted by poverty with, for example, aspects of educational access and horizons limited in what young people may value. As Sen (1999) notes, agentic freedoms to pursue valued educational outcomes are important but are not real freedoms if constrained by material poverty and disadvantage. Educational policy therefore needs to recognise how the distribution of economic resources impact on disadvantaged young people, families and communities. At the macro level, for example, educational policy should be aware of how increased levels of globalization supported by neo-liberal economic policies have had deleterious impacts on many urban communities, particularly in relation to high levels of family and environmental poverty. It also has to be aware that poverty has indirect influences on educational outcomes because of the way factors such as low levels of economic resources, inadequate housing and estate designs, noise pollution, environmental pollution and dirty living conditions, inadequate local resources and infrastructures mediate young people's identity and engagement with educational processes. The impact of these economic and environmental factors suggests the need for ABI educational policy to, not only embrace a politics of recognition, but at the macro- and meso-levels to align itself more fully with general redistributive public policy on poverty eradication. Brady (2009), in his international comparative analysis of welfare reform, suggests that poverty eradication is most successfully brought about through focusing on the widespread improvements in welfare benefits and the structural improvements in employment and neighbourhood renewal. This is about investing in places through improved health and transport infrastructures, the development of neighbourhood employment opportunities and the re-imagining of assets in the community. It is about re-energised attempts at reducing child poverty statistics through the appropriate combination of welfare and tax reforms. It is also about ensuring that places, families and young people are not stigmatised by poverty but instead are provided with the assets and resources to help engage in civic aspects of life, including education, with dignity and pride. These arguments are closely aligned to the American scholar Jean Anyon who argues for a reconceptualisation of what policy to tackle the urban poverty-education link might mean:

> Policies such as minimum wage statutes that yield poverty wages, affordable housing and transportation policies that segregate low-income workers of color in urban areas and industrial and other job development in far-flung suburbs where public transport does not reach, all maintain poverty in city neighbourhoods and therefore the schools. In order to solve the systemic problems of urban education, then, we need not only school reform but

the reform of these public policies. If, as I am suggesting, the macro-economy deeply affects the quality of urban education, then perhaps we should rethink what "counts" as educational policy. Rules and regulations regarding teaching, curriculum and assessment certainly count; but, perhaps, policies that maintain high levels of urban poverty and segregation should be part of the educational policy panoply as well ... (Anyon 2005, p.2–3)

In arguing for both issues of redistribution and recognition to be included in the way we make sense of, and deal with, educational disadvantage and equity my approach mirrors Fraser's "bivalent conception of justice" (Fraser 2008). In terms of finding a practical and normative way of perusing a bivalent conception of justice, Fraser argues for the ideal of participatory parity. Applying her ideals of participatory parity to education would require two conditions to be satisfied. Firstly objective preconditions that preclude young people from opportunities to interact fairly with peers in schools due to certain levels of material inequality and economic dependence would need to be eradicated. These might include social and economic arrangements that institutionalise deprivation, exploitation and gross disparities in wealth, income and leisure time. As I have argued above, the interconnections that generate environmental, economic and social poverty for many young people would need to be eliminated. Both macro social and economic policy and ABIs more generally would need to work on achieving this elimination. Secondly, an 'intersubjective' condition for educational participatory parity requires that ABIs express equal respect for all young people in order to ensure equal opportunity for achieving social esteem through their educational experiences. This condition precludes cultural patterns of school activity that systematically depreciates particular categories of disadvantaged young people and the qualities associated with them. Precluded, therefore, are school values systems and organisational and managerial mechanisms that deny some young people the status of full interacting school citizens. In particular this means focusing on the way schools classify, categorise and preclude young people either by excessively ascribing them as different or by failing to acknowledge their distinctiveness. For example this might refer to how streaming, banding and other forms of setting help to deprecate particular groups of young people. Precluded also are school market mechanism that exacerbate school segregation by class and ethnicity that also help to reinforce educational inequalities. Fraser recognises, however, that the ideal of participatory parity may never be achieved but by maintaining it as a primary analytical focus it can highlight the extent to which change may be required in order for a more equitable state of affairs to exist. To assist the process of improving participatory parity Fraser in her later writings (Fraser 2008) suggests the need for equity to include issues of representational justice. This recognises that in pursuing both distributional and relational justice, disadvantaged families and young people often lack a voice in the deliberations and the democratic decision-making about the direction and form of educational policy and practice they may end up experiencing. The field is therefore reminded that without appropriate representation disadvantaged groups can become marginalised from the educational mainstream that results in little understanding and hence investment in such practice and provision.

## Conclusion

Although educational ABIs might gain considerably by focusing on the progressive principles and politics of recognition, my argument, in line with that of Fraser's, is that favouring any one particular approach over another is problematic for educational equity. What I have argued is that most groups of educationally disadvantaged young people are so positioned because they are disadvantaged both economically and culturally. My argument is that given the effects of socioeconomic maldistribution and cultural misrecognition on poor young people, educational ABIs cannot be informed solely by a politics of redistribution or a politics of recognition but instead require both to operate simultaneously in order to bring about fundamental shifts in the way young people are both motivated and are enabled to engage with education.

But what are the probabilities of this happening? At one level of analysis Rees and colleagues provide a clear rationale for why area based policy is unlikely to engage in more radical transformative redistributive interventions:

> Once social and economic disadvantage is redefined as an aspect of the wider inequalities which are characteristic of British society, then these limitations become apparent. The state is not in a position to engage with issues of social inequality, structural shifts in the organisation of economic activity and their consequences, except at the margins. The kinds of redistribution which would be necessary to do so simply do not appear on the policy agenda. ABIs and the conceptualisations of disadvantage on which they are based reflect this. They provide a means of presenting the promise of 'active government', but within the highly restricted policy repertoire which in reality is available. (Rees et al. 2007, p.261).

Rees and colleagues analysis of the need for government to continually develop and re-implement ABIs reflects Fraser perceptions that affirmative redistribution results in the state needing to make surface reallocations again and again. And given that much educational policy, including past and current educational ABI repertoires, draws from a neo-liberal project – a project which protects the market and the accumulation of capital, that sees education as supporting the needs of capital and yet at the same time requires education to 'fix' the social and educational problems that the project then generates - underlying structural inequalities are unlikely to be accommodated in the policy theorising of these initiatives. Emerging educational policy developments such as pupil premiums may provide some affirmative redistribution of additional funding to schools attracting poor students, but it will not compensate for all the additional cuts and increased levels of unemployment that poorer families are likely to now experience during these difficult economic times. The limited nature of these redistributed measures are also highlighted by the fact that part of the pupil premium is likely to be funded by the cutting of previous affirmative educational redistribution policies such as the City Challenge and educational maintenance allowances.

Given this situation what is the likelihood of progressive educational principles and a politics of recognition informing changes in educational policies? Once again current government pronouncements do not seem to bode well. Very little in current educational policy suggests that the problems of educational inequalities lie with the lack of a credible curriculum and/or set of appropriate assessment strategies that recognise the values and orientations of disadvantaged young people. Nor do policy announcements suggest new ways in which schools might engage with communities and neighbourhoods in more democratic and inclusive ways. Instead the explanation, much in line with aspects of past New Labour policy discourse, focuses on the paucity of schooling in our most disadvantaged areas where much of the remedy is to be found in organisational solutions and leadership strategies based on the practices of Academies and Free schools.

So how would I sum up the prospects for the future? Clearly what I have provided in this chapter is not a particularly hopeful analysis. And yet at the same time my arguments are not just a policy scholarship critique without positive suggestions for change. I do highlight what could or perhaps should be done. However the questions posed by my analysis ask whether policy makers engaged in the development of educational ABIs really do want to:

(a) give power back to schools and communities to develop educational interventions that might engage with the needs and desires of those most disadvantaged
(b) side with those least well off economically and financially to enable them to access appropriately that provision and hence engage with wider societal opportunities.

The current spending review does not provide much hope that the new coalition government will in fact respond positively to either issue in the near future. In terms of what might be done, there will be schools and local authorities with their communities and neighbourhoods (and with the support of particular strands of government educational policy such as pupil premiums) that will continue to embrace aspects of both a politics of a redistribution and recognition that will enable some young people and families to benefit from education. These hard fought victories need to be supported and applauded and the initiatives and interventions that underpin them developed further. However, given much of what I said in this article I do not see that they will be scaled up to inform major strands of educational area based policy initiatives. My fear is that the link between education and disadvantage will continue into the next generation and perhaps beyond.

# References

Anyon, J. (2005). *Radical possibilities*. New York: Routledge.
Barker, B. (2010). *The pendulum swings: Transforming school reform*. Stoke on Trent: Trentham Books.
Bauder, H. (2002). Neighbourhood effects and cultural exclusion. *Urban Studies, 39*(1), 85–93.
Blanden, J., & Machin, S. (2003). *Educational inequality and the expansion of UK higher education*. London: London School of Economics, Centre for Economic Performance.
Bourdieu, P. (1986). *Distinction: A social critique of the judgement of taste*. London: Routledge & Kegan Paul.

Bourdieu, P., & Passeron, J.-C. (1997). *Reproduction in education, society and culture*. London: Sage.
Brady, D. (2009). *Rich democracies, poor people: How politics explain poverty*. Oxford: University Press US.
Chitty, C. (2002). Education and social class. *The Political Quarterly, 73*(2), 208–210.
Cummings, C., Dyson, A., Papps, I., Pearson, D., Raffo, C., & Todd, L. (2005). *Evaluation of the full service extended schools project: End of first year report* (DfES Research Report RR680). Nottingham: DfES Publications.
Facer, K. (2010). *Towards an area-based curriculum: insights and directions from the research*. London: RSA.
Fraser, N. (1996). *Social justice in the age of identity politics: Redistribution, recognition and participation, the tanner lectures on human values*. Stanford: Stanford University.
Fraser, N. (2008). *Scales of justice. Reimagining political space in a globalizing world*. Cambridge/Malden: Polity.
Gonzalez, N., Moll, L. C., & Amanti, C. (2005). *Funds of knowledge*. Mahwah: Lawrence Erlbaum.
Goodman, A., Gregg, P., & Washbrook, E. (2011). Children's educational attainment and the aspirations, attitudes and behaviours of parents and children through childhood. *Longitudinal and Life Course Studies, 2*(1), 1–18.
Goodson, I. (2008). Schooling, curriculum, narrative and social future. In C. Sugrue (Ed.), *The future of educational change: International perspectives*. London: Routledge.
Gruenewald, D., & Smith, G. (2008). Making room for the local. In D. A. Gruenewald & G. Smith (Eds.), *Placed-based education in the global age* (Local diversity). Mahwah: Lawrence Erlbaum.
Halpin, D., Dickson, M., Power, S., Whitty, G., & Gewirtz, S. (2004). Area-based approaches to educational regeneration. *Policy Studies, 25*(2), 75–85.
Hattam, R., Brennan, M., Zipin, L., & Comber, B. (2009). Researching for social justice: contextual, conceptual and methodological challenges. *Discourse: Studies in the Cultural Politics of Education, 30*(3), 303–316.
Kendall, L., O'Donnell, L., Golden, S., Ridley, K., Machin, S., Rutt, S., McNally, S., Schagen, I., Meghir, C., Stoney, S., Morris, S., West, A., & Noden, P. (2005). *Excellence in cities: The national evaluation of a policy to raise standards in urban schools 2000-2003* (*DfES* Research Report RR675A). Nottingham: DfES. Available at: http://www.dfes.gov.uk/research/data/uploadfiles/RR675A.pdf. Accessed 26 April 2006.
Lawson, M. A. (2003). School-family relations in context. Parent and teacher perceptions of parent involvement. *Urban Education, 38*(1), 77–133.
Lingard, B. (2005). Pedagogies of indifference. *International Journal of Inclusive Education, 11*(3), 245–266.
Love, J. M., Kisker, E. E., Ross, C. M., Schochet, P. Z., Brooks-Gunn, J., Boller, K., Paulsell, D., Fuligni, A. A., & Brady-Smith, C. (2002). Making a difference in the lives of infants and toddlers and their families: The impacts of Early Head Start. Washington, DC: U.S. Department of Health and Human Services. http://www.mathematicampr.com/publications/PDFs/ehsfinalvol2.pdf
Melhuish, E., Belsky, J., & Leyland A. (2005). *Early impacts of sure start local programmes on children and families: Report of the Cross-sectional Study of 9-and 36-Month old children and their families* (DfES Research Report NESS/2005/FR/013). London: DfES.
Raffo, C. (2006). Disadvantaged young people accessing the new urban economies of the post-industrial city. *Journal of Education Policy, 21*(1), 75–94.
Rees, G., Power, S., & Taylor, C. (2007). The governance of educational inequalities: The limits of area-based initiatives. *Journal of Comparative Policy Analysis, 9*(3), 261–274.
Riddell, R. (2007). Urban learning and the need for varied urban curricula and pedagogies. In W. Pink & G. Noblit (Eds.), *International handbook of urban education*. Dordrecht: Springer.

Sanchez-Jankowski, M. (2008). *Cracks in the pavement: Social change and resilience in poor neighborhoods*. California: University of California Press.

Sen, A. (1999). *Development as freedom*. Oxford: Oxford University Press.

Shonkoff, J. P., & Phillips, D. A. (Eds.). (2000). *For the committee on integrating the science of early childhood development from neurons to neighbourhoods. The science of early childhood development*. Washington, D.C.: National Academy Press.

Smith, G. (1987). Whatever happened to educational priority areas? *Oxford Review of Education, 13*(1), 23–38. Plowden Twenty Years On (1987).

The Plowden Report. (1967). *Children and their primary schools. A report of the Central Advisory Council for Education (England)* (London: Her Majesty's Stationery Office) *in the lives of infants and toddlers and their families: The impacts of early head start. Vol. 1: Final technical report*. Princeton: Mathematica Policy Research Inc.

Thomson, P. (2002). *Schooling the rustbelt kids – Making the difference in changing times*. Staffordshire: Trentham Books.

# Chapter 3
# Spatially-Concentrated Worklessness and Neighbourhood Policies: Experiences from New Labour in England

**Stephen Syrett and David North**

## Introduction

The issue of work and worklessness[1] has been at the centre of policy concerns relating to the need to tackle deprivation concentrated within particular neighbourhoods (Syrett and North 2008). Yet although worklessness concentrated in particular neighbourhoods is a widespread and enduring feature of the contemporary employment landscape, attempts to address this challenge have frequently had limited success. In part this reflects a recurrent failure of policy interventions to understand how the causes of concentrated worklessness are rooted within the interaction between person and household factors, the workings of labour and housing markets operating at wider spatial scales, and the characteristics of particular neighbourhoods. Critically, although the challenge of concentrated worklessness is manifested at the neighbourhood level, effective policy responses require integrated actions across a variety of spatial levels. This requires understanding that the role of neighbourhood effects in causing concentrated worklessness is limited but that recognising the distinctive characteristics of neighbourhoods is central to the design and implementation of effective policy interventions.

In this chapter we explore the causes of worklessness concentrated at the neighbourhood level and the effectiveness of policy attempts by successive New Labour governments to reduce the high levels of worklessness that characterise England's most deprived neighbourhoods. The chapter first analyses how localised negative

---

[1] The term 'worklessness' relates to those individuals involuntarily excluded from the labour market and claiming out-of-work benefits, including various forms of unemployment and incapacity/disability benefit.

S. Syrett (✉) • D. North
CEEDR, Middlesex University Business School,
The Burroughs Hendon, London NW4 4BT, UK
e-mail: s.syrett@mdx.ac.uk; D.North@mdx.ac.uk

cycles combine with broader processes of labour market change to reinforce patterns of concentrated worklessness. The particular role of neighbourhood effects within these processes is considered and the important differences in the nature of employment deprivation between deprived neighbourhoods recognised. The chapter then considers the development and effectiveness of work-related neighbourhood policies. Through an examination of the wide-ranging employment-related initiatives developed in relation to deprived neighbourhoods between1997 and 2010 in England, the chapter considers the aims, outcomes and effectiveness of these initiatives and their relationship to the developing evidence base. The chapter concludes by identifying the limitations of this neighbourhood policy agenda and why it was unable to transform the employment fortunes of England's most deprived neighbourhoods.

## Causes of Concentrated Worklessness in Deprived Neighbourhoods

The persistence of worklessness concentrated in particular neighbourhoods has led to much debate as to the precise role of neighbourhood effects in explaining this phenomenon (van Ham et al. 2012, 2013). Most studies of work and worklessness in relation to deprived neighbourhoods have concluded that neighbourhood effects have only a minimal additional impact over and above the characteristics and circumstances of individuals and households (Fieldhouse and Tranmer 2001; Buck and Gordon 2004; Sanderson 2006; Nunn et al. 2010). Yet these studies also recognise that certain neighbourhood effects can compound problems of worklessness through direct effects that flow from the characteristics of the resident population as well as, the characteristics of the place itself, notably in terms of relative levels of physical isolation

In fact understanding the existence and persistence of worklessness in particular areas requires analysis of the interaction between two key factors: first, how broader processes of labour market restructuring produce patterns of job loss and sectoral change which impact upon the numbers and types of jobs available within particular local economies within which neighbourhoods are located; second, how a series of place-based interactions operate to create and maintain concentrated worklessness in particular neighbourhoods.

### *Economic Restructuring and Labour Market Change*

The evolving geographies of labour markets associated with global economic shifts in production, consumption and exchange and a changed territorial basis for economic competitiveness, relate to the existence of concentrated worklessness in deprived neighbourhoods in complex ways. Many neighbourhoods characterised by high levels of employment deprivation are associated with older industrial areas left redundant and in need of reinvention by processes of economic change. But high levels of worklessness concentrated within particular neighbourhoods are also a feature of prosperous high growth city-regions. Central to understanding this relationship between neighbourhoods enduring high levels of worklessness and wider contexts of economic change, are the related processes of sectoral restructuring and the changing nature and type of employment.

The sectoral restructuring that has taken place through processes of deindustrialisation and service sector growth has created new geographies of job loss and growth (Turok and Edge 1999; Webber and Swinney 2010). Within the UK, neighbourhoods with high levels of worklessness are disproportionally located within areas of former manufacturing and coalmining activity, situated in inner city areas of large metropolitan cities, northern towns and cities, 'one-industry' towns and outer urban areas. The loss of traditional manufacturing jobs has negatively impacted not only on particular areas but also on specific groups (e.g. older men, single parents, ethnic minorities).

However the existence of concentrated worklessness relates not only to the geography of job loss but also the geography of new sources of employment and the type of jobs being created. During the period of sustained employment increase in the UK from the early 1990s until the 2008 economic downturn, a mismatch was evident between the former geography of employment and the newly emergent geographies of job growth. New and existing enterprises have frequently favoured locations towards the edge of towns or in smaller towns. Many new employment opportunities have developed on the edge of urban areas, rather than inner cities, and new firm formation, retail activity and emerging markets frequently have been attracted to accessible areas outside large urban areas. Migration and commuting flows have to some extent compensated for this mismatch, but these processes remain a weak means of adjusting residential patterns of the working age population to employment opportunities (Beatty et al. 1997). This is because, for those with few or no skills, competing for jobs elsewhere is difficult, whilst for many individuals their ties to families, friends and places and related issues of quality of life, makes out migration unattractive. For those on low wages with restricted transport mobility, the possibilities of longer range commuting are limited.

However the issue is not only one concerning the changing geographies of job loss and job creation but also the types of jobs being created. Recent job growth has been strongly focused within the service sector and characterised by an increasing polarisation between well-paid and high-skilled professional working in the 'knowledge-based' sectors and a large flexible workforce comprising low paid workers in insecure and low-grade service sector employment (Goos and Manning 2007). In labour market areas experiencing higher levels of worklessness, the new service sector jobs have been concentrated primarily within consumer, personal and public services, rather than in business and producer services. In consequence, for residents in poorer neighbourhoods who are less competitive in the labour market, it is low-skill, low wage insecure jobs that are most likely to provide a route into the labour market. Yet such jobs are often unattractive in terms of their pay, conditions and career development potential, and for those in receipt of benefits, the low level of wages may result in a minimal, or no, increase in overall household income (Beatty et al. 2009a).

## *Concentrated Worklessness, Cycles of Decline and Neighbourhood Effects*

The geography of concentrated worklessness within particular neighbourhoods only partially reflects the wider patterns of changes in employment and jobs discussed so far. Localities that endure concentrated worklessness are commonly characterised by multiple dimensions of deprivation (e.g. in relation to health, education, housing, environment and crime etc.) that act to mutually reinforce one another to create and reinforce spatially concentrated deprivation. These interrelated causes of deprivation are frequently conceptualised in terms of 'vicious cycles' that lock neighbourhoods into ongoing poverty and deprivation. In relation to work, a series of vicious circles can be identified that link local unemployment to local social outcomes and then further reduce the employment prospects of residents in the short and longer term (Gordon 2003). Issues such as access to job information, short-term jobs creating interrupted work histories, health deterioration, family fragmentation – notably the impact of lone parenthood – and educational underachievement are particularly important here. Taken together, these largely social forces tend to reproduce spatial concentrations of unemployment, even if the original reason for high unemployment was in fact something quite different.

In seeking to explain localised concentrations of worklessness, a number of interrelated processes need to be considered. First, between these areas and areas with higher levels of employment and job opportunities, there is a lack of effective equilibrating processes, via migration and commuting, which reinforces existing concentrations (Gordon 2003). Second, spatial externalities in the housing market lead to residential sorting processes (Cheshire et al. 2003). These act to concentrate workless populations where low-cost rented property and the location of social housing predominate within certain neighbourhoods (Maclennan 2000). Third, those with a history of a lack of employment and job loss suffer a 'scarring effect' making them more vulnerable to further unemployment which is reinforced when concentrated in particular neighbourhoods (Burgess et al. 2003). Finally, although problems of worklessness are rooted principally within the characteristics and circumstances of individuals and households, various neighbourhood effects can be identified which act to compound these problems (Buck and Gordon 2004; Sanderson 2006). Such neighbourhood effects relate principally to socialisation processes, the nature and extent of social capital, stigmatisation and discrimination, and the relative physical isolation of deprived neighbourhoods.

## Socialisation Processes: Cultures of Worklessness

The most contentious area of debate related to neighbourhood effects and worklessness relates to whether deprived neighbourhoods are characterised by localised 'cultures of worklessness'; that is a distinctive set of attitudes, norms and values relating to work that lie outside those of mainstream society. Such cultures are said to be characterised by an expectation of welfare benefits, a normalisation of illegal behaviour and the emergence of a set of values at variance to those of mainstream society (Murray 1996). Yet there is little evidence to support notions of the existence of 'cultures of worklessness' within deprived neighbourhoods within the UK (Lupton 2003; ODPM 2004). In fact Lupton (2003) concluded in her study that what was most remarkable was the extent to which excluded communities endorse, rather than reject, mainstream societal values.

However there is evidence of particular perceptions, attitudes and aspirations towards work that reflect the nature of socialisation that takes place within a particular milieu characterised by prolonged periods of intergenerational worklessness and limited employment opportunities for a high proportion of the resident population (Bauder 2001; CLG 2010a). Where cultures of worklessness are said to exist, they are characterised by lowered incentives to work – in a context where peers are also unemployed and the informal economy has a strong pull factor – and a view of joblessness as unproblematic given circumstances of lowered aspirations and short-term horizons (Ritchie et al. 2005). Yet critically, these attitudes and expectations do not pervade all residents of deprived neighbourhoods, nor are they confined only to these areas. In this respect they are different by *degree* rather than *kind*, reinforced by material circumstances and restricted social networks (Syrett and North 2008).

Such attitudes and aspirations may result from peer pressure, a lack of role models (i.e. of those in employment, or more importantly still, those in good jobs with career advancement possibilities), low self-esteem and expectations (of individuals themselves and externally from employers), and limited experience, direct or indirect, of the world of work (CLG 2010a). Differences in local perceptions and behaviours are often characterised by a narrow, insular and highly localised view of the labour market (Green and White 2007), often reinforced by local stigma, which produces narrow travel horizons and compounds exclusion though a loose sense of attachment to the mainstream labour market (Fieldhouse 1999).

## Social Capital: Contacts and Networks

The importance of social capital and social networks for the processes of finding, securing and maintaining employment, has been an influential theme within the policy discourse over recent years (Putnam 2000; SEU 2000; Taylor 2002). This discourse is predicated upon contested concerns that deprived neighbourhoods are characterised by declining levels of social capital (SEU 2000; Forrest and Kearns 2001). Although the operationalisation of notions of social capital has proved problematic, there has been work on area effects relating to social contacts and networks in the labour market. This research has drawn attention to the importance of networks of families, friends, and social contacts not only in obtaining information about jobs, but also in being successful in competing for them (e.g. Shuttleworth et al. 2003; Meadows 2001). The resources provided through social networks are particularly significant given the importance of informal recruitment processes. As Watt (2003) demonstrated in his study of the work histories of local authority tenants in Camden, 'reputation' needs to be transmitted by word of mouth to employers so that being enmeshed in the appropriate social networks proved crucial in providing the routes by which information about jobs and workers' reputations could be circulated. Watt concluded that having the right reputation and social contacts were probably as important as the possession of training certificates.

A further issue affecting access to jobs relates to the information that workless people have about the jobs that are available within commuting distance. Some studies (e.g. Lawless 1995; Atkinson and Kintrea 2001) have found that the unemployed tend to have poor knowledge of job opportunities within the local labour market. This may be partly the result of there being inadequate information available at the local neighbourhood scale, for example if there is poor access to employment services located in city centre locations (Speak 2000).

Within neighbourhoods where access is primarily to employment in low paid low skill jobs, a mismatch is evident between the informal recruitment methods of employers, particularly smaller employers who tend to rely on 'word of mouth' methods, and the job search routes of residents where the networks and contacts for obtaining information about job vacancies are poor (Hasluck 1999). The disadvantaged, therefore, are likely to be more dependent on family and friends as they have fewer ties to paid work and less access to job information. Yet if members of their family and friends are also out of work, this is going to separate them further from the kind of information that they need and make it more difficult to obtain employment. Dickens (1999) suggested that this kind of 'network failure' is an important factor underlying the problems in deprived neighbourhoods, reinforcing other processes creating inequalities in labour market outcomes and thereby 'tipping' deprived neighbourhoods further into a vicious cycle of decline.

## *Stigmatisation and Discrimination*

There are persistent suggestions from those who live and work in deprived neighbourhoods that some employers discriminate against job seekers from such areas. The existing evidence to support employer discrimination in recruitment practices on the basis of address or postcode has to date been fairly limited. In large part this reflects the considerable challenges in researching and isolating discriminatory practice. Past studies in England have found some evidence of implicit discrimination by employers against long-term unemployed residents of a deprived neighbourhood in Sheffield which had a poor reputation (Lawless 1995) and postcode discrimination against lone parents from stigmatised areas in Newcastle (Speak 2000). It would appear that those seeking work often perceive that they are disadvantaged by where they live. For example, in a study of the young unemployed in Newham, Roberts (1999) found that almost a third of the interviewed young people from the most deprived parts of the borough thought that employers were put off by the area in which they lived.

The most comprehensive study undertaken specifically on this issue (Nunn et al. 2010) concluded that there was evidence from both its qualitative and quantitative analysis that postcode selection of address-based discrimination plays a modest and secondary role to personal characteristics, within very specific conditions, as a contributor to worklessness in deprived areas. However isolating such area effects from other disadvantages that individuals and groups face in the labour market is clearly very difficult. The issue of separating out place-based discrimination from other forms of discrimination is one issue here. For example, as many deprived neighbourhoods are characterised by concentrations of particular ethnic groups it is necessary to consider the relationship between racial discrimination and place based discrimination. As racial and area discrimination can be closely associated with each other, red lining certain areas is, in the minds of some employers, tantamount to shutting out certain groups of people.

## *Physical Isolation and Poor Public Transport*

Many deprived neighbourhoods are characterised by a degree of relative physical isolation from centres of employment. The move of many employers away from areas close to town and city centres to retail and business parks which are invariably on the edge of urban areas, has in many cases exacerbated this situation. The increased separation between residential and employment areas has made it more difficult for those without private transport to access jobs. Residents of deprived neighbourhoods typically have much lower levels of vehicle ownership and are hence more reliant on often inadequate public transport (DETR 2000; SEU 2003). Bus routes often do not provide good links between residential and employment areas and many new jobs in the service sector involve working in the evenings and at weekends when bus services are more limited (Lucas et al. 2008; Thickett 2011). Furthermore the relatively high cost of public transport fares can provide a major disincentive to travel to take up low paid employment. As a result the relative location of deprived neighbourhoods in relation to employment and the lack of adequate and affordable public transport can be a significant institutional barrier to improving employment prospects for residents in these areas (Sanderson 2006).

To summarise therefore, the starting point for understanding concentrated worklessness are the characteristics of workless individuals and their households. Yet as the previous discussion has demonstrated, such characteristics are frequently bound up with neighbourhood effects that act to exacerbate and reinforce problems of labour market exclusion in certain neighbourhoods. Localised work cultures, restricted social networks and employer discrimination against stigmatised neighbourhoods produce direct effects rooted within the areas' population characteristics, whilst place based effects emanate from the relative physical isolation of these neighbourhoods, particularly in terms of limited mobility to access jobs reflecting reliance upon often poor public transport provision.

## Worklessness and Deprived Neighbourhoods: Identifying Difference

The interaction between wider labour market change and localised conditions is constituted in the experiences of high levels of worklessness within particular neighbourhoods. This results in neighbourhoods having quite distinctive characteristics and experiences of employment deprivation (CLG 2010b). The centrality of person and household factors to understanding worklessness means that the population characteristics of neighbourhood residents and their degree of mobility and related population churn, is of critical importance (Robson et al. 2008). Whilst ultimately each neighbourhood is unique, it is possible to identify key similarities and differences between neighbourhoods and the identification of certain common types (CLG 2009a).

In their study of employment-deprived neighbourhoods in England, Lupton et al. (2011) identified five groups of neighbourhoods on the basis of a number of characteristics related to claimant rates, housing, qualifications, types and sectors of employment, ethnicity, population change and per capita wealth.[2] The main groups they identified comprised:

- *Highly deprived social housing neighbourhoods*: neighbourhoods where social housing predominated along with extreme multiple deprivation
- *Older workers in declining areas*: consisting of more stable neighbourhoods, characterised by older workers and steady employment
- *High churn neighbourhoods with younger workers*: high turnover, socially mixed neighbourhoods in self-contained labour markets, with younger workers in vulnerable employment and high levels of private rented housing
- *Ethnically mixed neighbourhoods in stronger labour markets*: neighbourhoods characterised by their mixed social housing and location in buoyant cities with stronger labour markets and their young, socially and ethnically mixed populations

---

[2] This report identifies both a 5 group and more nuanced 10 group classification of employment-deprived neighbourhoods. These were derived using a number of selected characteristics comprising: Jobseeker's Allowance claim rate; Incapacity Benefit index; percentage social rented dwellings; percentage private rented dwellings; percentage with no qualifications; percentage employed in manufacturing; percentage employed in hotels; percentage in elementary occupations; percentage Black Caribbean ethnicity; neighbourhood population turnover; neighbourhood population change (2001–2007); Gross Value Added per capita (£k).

- *Inner London*: a type of neighbourhood found only in inner London, where values for key characteristics varied substantially from those found elsewhere.

Mapping these different types demonstrated significant regional differences in the distribution of employment-deprived neighbourhoods reflecting, amongst other factors, important differences in regional labour market supply, demand and mediating institutional factors. Such regional variations illustrate the different types of labour market contexts and processes in operation which have implications for the nature and extent of the varied neighbourhood effects related to concentrated worklessness discussed previously.

Analysis of this type points to the need for policy interventions that address concentrations of worklessness to be sensitive to the differences between deprived neighbourhoods. In addition it also points to the possibilities of learning between neighbourhoods which may not be geographically proximate but experience similar conditions and outcomes – for example those located in former mining areas, seaside towns or out of town social housing estates (Lupton et al. 2011). The differentiated spatial manifestations of the problem of concentrated neighbourhood worklessness indicate the need for different policy mixes and delivery mechanisms; ones tailored to particular neighbourhood needs but situated within an understanding of the changing nature of supply and demand in the wider local and regional labour market. The next part of the paper analyses the extent to which policy interventions developed in England under the New Labour governments (1997–2010) in relation to the problem of concentrated worklessness in deprived neighbourhoods were able to recognise and respond meaningfully to meet this policy need.

## Policy Responses to Tackling Worklessness in Deprived Neighbourhoods

A central objective of successive New Labour governments was to move workless people into employment. This was driven by the goal of achieving "full employment", with the aim of raising the employment rate to 80 % (from 72.5 % in 1997), and rooted in the dominant belief that employment was the best route out of poverty. Although the number of working age people registered as unemployed fell during the first two NL administrations (from 1.9 million in 1997 to 1.4 million in 2005), this was dwarfed by the number who were economically inactive and which grew over the same period (from 7.6 million to 7.9 million), supporting arguments of a growth of 'hidden unemployment' (Beatty and Fothergill 2002). Against this background, tackling the causes of worklessness became an urgent priority of government policy during the second and subsequent NL administrations.

New Labour's analysis of the causes of worklessness and the tendency for this to be concentrated in small geographical areas centred on the barriers to work associated with various personal and household characteristics. These were seen as adversely affecting not only the employability of individuals, but also their attitudes to looking for work in the first place. A study by the Social Exclusion Unit (SEU 2004) emphasised the multiple causes of worklessness including the role of various place and people based area effects, stressing the interplay between different factors (e.g. lack of educational qualifications, work experience, basic and social skills, social networks, high levels of lone parenthood) that reinforced an individual's disadvantage in the labour market.

In addition to these individual and household barriers, NL thinking was also influenced by research showing how institutional factors contributed to concentrations of worklessness. Of particular significance here was the interconnection between the housing and labour markets, as housing status was considered to be the principal factor influencing where workless people live (Cheshire et al. 2003). The operation of the benefits system was also increasingly seen as a disincentive to entering the labour market, with survey evidence indicating that many people on disability and housing benefits were concerned that if they took a job they might lose the financial security and income levels that benefits provided. Other recognised institutional barriers to work included the lack of affordable transport and childcare, the latter particularly affecting lone parents.

Informed by the assumption of a relatively tight national labour market, NL's approach to tackling worklessness comprised a number of supply-side programmes and measures aimed at overcoming the barriers that workless individuals faced in entering the labour market and changing their attitudes towards obtaining employment. Through a combination of training and job readiness initiatives it was assumed that individuals would be in a better position to compete for jobs, whilst tackling institutional barriers such as childcare and public transport provision would facilitate better access to employment. In addition it also introduced changes to the benefits system to ensure that those recipients capable of working did seek work. Ideologically, this was consistent with the shift towards a work-focused welfare state that had already been set in train by previous Conservative governments (Evans 2001). NL's welfare programme demonstrated an incremental shift towards a 'conditional' regime whereby claimants were increasingly required to undertake some work-related or training activity in exchange for benefits or face sanctions.

The centrepiece of NL's programme was the New Deal, introduced immediately upon coming to power in 1997, and comprising a suite of 'new deals' targeting specific groups (i.e. the young unemployed, the long-term unemployed aged 25 and over, those aged 50 and over, lone parents and disabled people). Whilst the evaluation of the New Deal showed it was relatively successful for those on the margins of entering the labour market, it was less successful for those facing multiple barriers (DWP 2008). A further notable characteristic of policy development was the introduction of a series of area-based initiatives (ABIs) to augment mainstream provision in an attempt to better reach more marginalised individuals and groups living within deprived areas. This marked ongoing recognition of the relative failure of mainstream policies to reach effectively the most disadvantaged living in poor neighbourhoods and the limitations of a highly centralised policy agenda focused primarily on national-level analysis of aggregate supply and demand. These developed through two key phases: first a focus explicitly upon the neighbourhood level, and latterly attempts to tackle concentrated worklessness within wider local and sub-regional strategies.

## Neighbourhood-Based Worklessness Policies

Awareness that other kinds of interventions were required in order to target those furthest from entering the labour market led to a succession of area-based programmes, invariably run as pilot schemes, aimed at localities experiencing persistently high levels of worklessness (Syrett and North 2008). These comprised ABIs which targeted neighbourhoods at a policy delivery level, such as the 'Action Team for Jobs' initiative (2000–2006),[3] and others premised upon the existence of negative neighbourhood effects. For example in the latter case, the 'Working Neighbourhoods Pilot' initiative (2004–2006), aimed to counter localised 'cultures of worklessness' through providing intensive support to help people find work and incentivise them to stay in work through retention payments.

A key dimension of this plethora of activity was the introduction of neighbourhood-oriented policies where issues of worklessness were addressed as part of more comprehensive strategies to improve living and working conditions across a range of dimensions (employment, health, crime, education and skills, housing and the physical environment) with a view to narrowing the gap between these areas and the rest of the country. However one consequence of these holistic neighbourhood approaches was a degree of confusion as to whether policy was aimed at the neighbourhood itself – based upon recognition of significant place based effects – or at the communities and citizens who lived within the neighbourhoods, or indeed a mixture of both (Atkinson 2007), Two major initiatives under New Labour were particularly significant in this respect: the National Strategy for Neighbourhood Renewal (NSNR) and the New Deal for Communities (NDC).

## National Strategy for Neighbourhood Renewal

The National Strategy for Neighbourhood Renewal was launched in 2001 with the objectives of lowering rates of worklessness and crime and improving health, skills, housing and the physical environment within England's poorest neighbourhoods, and to narrow the gap in relation to these elements between these neighbourhoods and the rest of the country. It was supported by dedicated funding (mainly the Neighbourhood Renewal Fund later reformed into the Working Neighbourhood Fund) but also aimed to improve the delivery of mainstream public services within the poorest neighbourhoods. This funding supported a range of local authority-led worklessness programmes targeted at hard-to-reach clients, predominantly in the forms of advice, guidance and support, but also supporting some transitional employment schemes and business and enterprise support.

---

[3] The Action Team for Jobs initiative aimed to increase employment rates among disadvantaged groups in deprived areas based on outreach work in local communities and the involvement of community and voluntary organisations as well as employers.

The impact of the NSNR on levels of worklessness appears to have been broadly positive, although much depends upon the spatial level at which the analysis was conducted. The Government's own evaluation of the NSNR considered change at the levels of the local authority district and the neighbourhood (CLG 2010b). At local authority district level, in the period after 2001, there was a consistent improvement against key worklessness indicators in those districts that qualified for the Neighbourhood Renewal Fund. This was not just in absolute terms, as would be expected in a period of national employment growth (the worklessnes rate in England fell from 9.8 % in 2001 to 8.9 % in 2007) but also relative to the national average (CLG 2010b). Employment rates also improved and the gap with the national average narrowed to 75 % of the 2001 figure, although there was considerable variation in performance across the nine English regions.

At the neighbourhood level, the worklessness rate of the most deprived 10 % Lower Super Output Areas (LSOAs)[4] improved in relation to the national average of all LSOAs, although it remained more than seven times that of the least deprived neighbourhoods. However, when rates within the most deprived 10 % LSOAs within each district (for both NSNR and non-NSNR areas) were compared with their respective district averages, the gap widened slightly, indicating the stubbornness of the worklessness problem in the most deprived neighbourhoods. Moreover, the analysis also showed that as economic growth slowed from 2006, it was these most deprived neighbourhoods that were the most vulnerable to rising levels of worklessness (CLG 2010b).

In analysing evidence of the contribution of the NSNR to tackling issues of worklessness locally, the national evaluation local research report concluded that in relation to employment there were 'variable rates of improvement and limited impact' (CLG 2010a, p.5). This report pointed out that NSNR interventions generally had a "more consistently positive impact upon the symptoms of neighbourhood deprivation (for example crime, environmental factors and aspects of public health) as opposed to its root causes (including worklessness and low educational attainment)" (CLG 2010a, p.8). In the majority of the case study NRF districts evaluated, there had been a narrowing of the gap in relation to employment. But evidence of progress was mixed (e.g. there was also evidence of a decline in the ethnic minority employment rate and/or increases in long-term claimants) and local perceptions of worklessness often remained pessimistic.

Overall, the NSNR appears to have had a marginal positive impact on employment outcomes when integrated with wider worklessness strategies (CLG 2010a, p.30). Where positive improvement in relation to worklessness was evident in case study NRF districts, this was due to favourable national economic conditions producing job opportunities and rising employment rates and supported by the availability of sometimes significant levels of focused long term investment (not just via the NRF but also from other area based programmes), which permitted local authority led employment services targeted at hard to reach client groups to be supplemented.

---

[4] A Lower Super Output Area (LSOA) is the smallest geographical area designed for the collection and publication of small area statistics within England and Wales.

Where worklessness remained a considerable problem, key barriers identified included neighbourhood effects related to embedded cultural factors which constrained the uptake and impact of employment programmes (comprising a lack of role models, inter-generational unemployment, low aspirations, resistance to travelling to work, poor work ethic), combined with a reduction in the availability of job opportunities and 'access to decent work' and the negative impact of population churn, and organisational barriers (e.g. the failure of key organisations such as Jobcentre Plus to develop targeted modes of delivery) (CLG 2010a).

## *New Deal for Communities (NDC)*

The need to tackle unemployment and economic inactivity in turning around the poorest neighbourhoods was a critical element of New Labour's flagship New Deal for Communities (NDC) programme, launched in 1998 as a 10-year £2 billion programme focused on 39 designated areas. The programme was designed to improve outcomes in the NDC areas in relation to 'place-based' issues (crime, the community, housing and physical environment) and 'people-based' issues (education, health and worklessness).

The severity of the worklessness problem found in the NDC areas was substantial. In 1999 there were an estimated 50,710 workless people (defined as being involuntarily excluded from the labour market and claiming out-of-work benefits) in NDC areas, representing 23 % of the total working-age population (CRESR 2005). By 2008, there were still 45,800 workless residents representing an average of 18.4 % substantially higher than 8.9 % national average. However this average figure masks considerable variation across NDC areas from a lowest rate of 10.8 % to a highest of 29.8 % (Beatty et al. 2009b).

Across the programme, 11 % of total NDC expenditure was allocated to tackling worklessness over the 2000–2006 period (CRESR 2005; Beatty et al. 2009b). This resource was used in the development of local strategies for tackling worklessness (analysing needs, objective setting, targeting of priority groups), working with other local partners, particularly public bodies, such as Jobcentre Plus, as well as voluntary bodies and to a much more limited extent, the private sector. Typically interventions focused upon supply-side interventions comprising combined job brokerage and information advice and guidance projects, recruitment and job matching services with local businesses, and skill development projects (often sectorally focused). On a much more limited scale were demand-side projects including Intermediate Labour Market (ILM) projects (that sought to create short term jobs to develop participants skills and experience), the creation of jobs for local people through Section 106 Agreements,[5] and business support projects promoting enterprise activity (Beatty et al. 2009b).

---

[5] A Section 106 Agreement permits a local planning authority to enter into a legally-binding agreement with a landowner/developer such that the granting of planning permission is dependent upon the provision of certain services and infrastructure, such as highways, recreational facilities, education, health and affordable housing.

Despite this considerable activity, the overall impact upon aggregate worklessness rates within NDC areas was limited. In absolute terms the worklessness rate (comprising both Job Seeker Allowance and Incapacity Benefit/Severe Disability Allowance claimants) across the NDC Programme did fall significantly by 4 % points from 22 to 18 % across the 1999–2008 period. However, in comparison to similarly-deprived comparator areas in the same local authorities, the decrease in worklessness in NDC areas was only marginally greater. As Beatty et al. (2009b, p.15–16) conclude: "There is no evidence as yet to indicate that NDC areas were seeing more in the way of improvement to worklessness than were similar neighbourhoods in the same local authority". Yet the NDC evaluation also provided evidence of very positive responses from local informants and beneficiaries that the development of flexible employment-related services tailored to address the needs of local people and area effects at the neighbourhood level had been highly beneficial (CLG 2008). And for individual participants there was evidence that participation in such projects did increase the likelihood of making the transition from not being in employment in 2002 to being in employment by 2004 (CLG 2009b).

The difference between positive neighbourhood experiences of worklessness interventions and limited programme-wide impacts within the NDC programme – a difference evident in many other ABIs too – illustrates a major measurement challenge (Beatty et al. 2009b). The programme-wide change data reflects the considerable changes affecting the NDC areas, as people move in and out of both employment and the NDC areas, thereby disguising some of the impacts of the programme (CRESR 2007). It was found that out-movers from NDC areas were more likely to be employed (71 % of those of working age) than in-movers (47 %) or stayers (55 %). In this context of wider flux, individual-level changes and gains that result from neighbourhood interventions seeking to move individuals closer to the labour market are lost or difficult to pick up within the wider data collection.

## *Workless Neighbourhoods in Their Local and Sub-regional Contexts*

As the neighbourhood policy agenda developed, in terms of addressing issues of worklessness within deprived neighbourhoods, a substantial policy and governance disconnect emerged (Syrett and North 2010). Much neighbourhood policy became focused upon issues of public service delivery and was poorly integrated with the market-led regional and urban policies focused upon strengthening competitiveness which, in turn, paid little or no attention to how these activities might benefit the most deprived neighbourhoods (North et al. 2009). In response, central government sought to refocus neighbourhood-level policy more directly upon jobs and enterprise (ODPM 2004; PSMU 2005), while Regional Development Agencies (RDAs) were issued guidance to refocus their role to take greater account of the needs of their most deprived areas (DTI 2005; PMSU 2005).

At the base of these changes was increased recognition that the roots of employment problems that beset deprived neighbourhoods lay within the wider local and regional labour markets within which they were embedded rather than as the result of neighbourhood effects, and hence effective policy action required integrated activity across spatial levels and greater freedom for locally and sub-regionally managed interventions (CLG 2006). This view was reflected in the major Treasury-led review of subnational economic development and regeneration policy (SNR) which saw a major change in the direction of policy development and governance arrangements (HM Treasury 2007). In relation to deprived neighbourhoods the SNR marked a shift in the relative importance of spatial levels, away from the neighbourhood and the region, towards an emphasis upon the local and the sub-regional alongside a greater economic focus within neighbourhood renewal policy (CLG 2009c). In term of tackling worklessness, and as part of the government's aspiration to achieve an 80 % employment rate, this line of policy thinking led to the introduction of the City Strategy initiative, as a first attempt to develop a more sub-regionally based and locally-managed approach to tackling high levels of economic inactivity within major cities.

The City Strategy (CS) initiative was intended to combat issues of worklessness and poverty in urban areas by empowering local stakeholders to develop policy interventions tailored to specific local circumstances (Green and Orton 2012). The key objectives comprised significantly improving employment rates, particularly among the most disadvantaged, and ensuring individuals were better able to find and remain in work as well as improve their skills so they could progress in work. Fifteen cities and city-regions[6] with employment rates below the national average were selected to be the pilot 'pathfinders' for the 2007–2009 period, and the initiative was extended until March 2011. The City Strategy initiative was not primarily about the provision of new money but rather focused upon getting better value from existing service provision. Strategies were seen as a way of pooling resources and funding streams, and of integrating a range of employment, training and health provision targeted at disadvantaged groups and neighbourhoods. Each area received seedcorn money to establish consortia made up of government agency providers, local government and Local Strategic Partnerships (LSPs), the private sector, the voluntary sector and the Trade Union Congress (TUC).

In practice the City Strategy Pathfinders (CSPs) provided support to mainstream provision by plugging gaps and offering supplementary services to specific individual or client groups. Many CSPs chose to target their resources either by area or by sub-group of benefit claimants, with activities focused upon client and employer engagement. The CSPs demonstrated considerable variation in relation to the extent and nature of spatial targeting towards those areas of concentrated worklessness and did not explicitly recognise neighbourhood effects.

---

[6]These comprised: Birmingham, Coventry and Black Country; Blackburn with Darwen; East London; Greater Manchester; Leicester; Merseyside, Nottingham; South Yorkshire; Tyne and Wear; West London; Dundee; Edinburgh; Glasgow; Heads of the Valleys; Rhyl.

However most did use client engagement strategies that involved community based outreach work which necessarily had some element of spatial organisation (Green et al. 2010).

Unfortunately the extent to which spatial targeting within the CS impacted upon worklessness within deprived neighbourhoods was difficult to ascertain, Firstly, the scale of the CSs varied significantly as did their relative focus upon strategy and delivery, making comparison between them difficult. Secondly, the CSs operated in a period of dramatic change both in terms of labour market conditions and policy. The onset of a deep and prolonged recession made moving long-term workless individuals into employment much more difficult and led to a shift in emphasis towards the newly unemployed. In addition this period saw large changes in policy particularly the introduction of a major welfare reform process as well as policy initiatives designed to address the consequences of the recession. Consequently measuring the effect of CS upon levels of worklessness through quantitative analysis has proved difficult with issues of attribution and value-added remaining largely unanswered (Green et al. 2010).

## The Impact of Policy Interventions

Although in the period prior to the impact of the economic recession (1997–2007), there was some success in getting more people into work and raising the employment rate for lone parents, those with a health condition or disability and those from ethnic minority groups (DWP 2008), there was no significant diminution in the gap in levels of worklessness between the most deprived neighbourhoods and the rest. Analysis of spatially-disaggregated evidence relating to changes in the number of people in receipt of out-of-work benefits (based on DWP longitudinal data) demonstrates that the gap between the most deprived areas (defined as the 10 % of Super Output Areas with the highest concentration of claimants) and the least deprived areas remained largely unchanged between 2000 and 2008, indicating the lack of success of NL's policies in this respect (The Poverty Site 2010). The impact of the recession exacerbated this situation. Analysis of Job Seeker's Allowance (JSA) rates demonstrated that the 10 % of areas with the highest JSA claimant rates experienced a greater absolute increase in claim rates from 7 to 9 % in the 2005–2009 period whilst areas with the lowest rates saw an increase from 1 to 2 % (Tunstall 2009).

In terms of neighbourhood-level approaches to tackling worklessness, the evidence from programmes such as the NDC and NSNR demonstrates that intensive neighbourhood-level interventions do not have a major impact upon the objective of reducing worklessness (Beatty et al. 2009b; CLG 2010a). Such interventions demonstrate that tackling the worklessness problems of deprived neighbourhoods requires understanding the linkages between these neighbourhoods and the wider labour and residential markets in which they are embedded. For example in relation to population mobility, there is some evidence that residents of deprived neighbourhoods who improve their labour market skills through training or mentoring initiatives and get a new or better job then leave the area, the so-called 'get on and get out' scenario. However whilst there is evidence that higher levels of mobility are marginally associated with poorer outcomes (CRESR 2007), what is more apparent is that processes of population mobility, employment and neighbourhood change are related in multiple and complex ways (CLG 2009a, 2010a; Lawless 2011).

The level of the neighbourhood is not the best level for seeking to understand and respond to these wider changes in local labour markets, and their relationships with housing markets and changing employer requirements. Given the only minor impacts of neighbourhood effects in causing concentrated worklessness, there is little reason to expect interventions targeted at such effects would result in any significant changes in levels of worklessness. In contrast bodies operating at a wider scale, such as local authorities and sub-regional bodies (e.g. city-regions), are better placed to devise strategies within which neighbourhood level interventions can then be developed (Beatty et al. 2009b; Lawless 2011). This is particularly important with respect to the key role of demand-side conditions for tackling problems of concentrated worklessness. Studies have consistently demonstrated the importance of labour demand (Syrett and North 2008; Beatty et al. 2009a; Green et al. 2010) but the policy agenda largely ignored this issue, particularly in the period of employment growth up until 2007. NL governments consistently maintained that employment growth in the national economy meant that there was no shortage of job opportunities in most places and the vast majority of policy activity targeted at worklessness had a narrow supply-side focus. Yet this ran counter to research evidence (e.g. Webster 2000; Beatty and Fothergill 2002; Coombes and Raybould 2004) showing that there continued to be insufficient jobs within commutable distances in areas with the highest levels of worklessness, especially in those areas that had borne the brunt of de-industrialisation.

Where there had been at least a modest growth in the numbers of jobs (such as in low value added services), the sustainability of these jobs was often questionable. Frequently these jobs were not sufficiently well-paid or attractive in terms of hours, security and future prospects to make movement off welfare benefits a rational choice (Beatty et al. 2009a). The ability of job seekers to compete for jobs was also affected by the competition that they faced from in-migrant workers, particularly from the A8[7] countries, as it was found that employers often preferred workers from elsewhere in the EU because they were perceived to have a stronger work ethic (Green 2007).

In practice there is little evidence of neighbourhood-level interventions oriented towards these demand side issues. This is not surprising, given that institutions operating at this level are poorly placed to develop such interventions that are likely to be highly costly, to require wider strategic overview and result in considerable leakage

---

[7] The A8 countries comprise eight of the ten countries that joined the European Union in 2004 from Eastern Europe (Czech Republic, Estonia, Hungary, Latvia, Lithuania, Poland, Slovakia and Slovenia).

from the neighbourhood scale. However, neighbourhood-oriented employment policies variously developed under an array of area-based interventions have demonstrated an ability to respond to the particular problems of those living within areas of concentrated worklessness. These initiatives were effective in filling the gaps in mainstream service provision through delivering or developing localised, flexible schemes to support workless individuals back into employment (Beatty et al. 2009b) and proved better able to provide the outreach and more intensive levels of personalised and holistic support required by those people facing the most severe and multiple barriers to employment. In this respect, such initiatives sought to deal simultaneously with a range of individual and household factors and any compounding neighbourhood effects. Other evidence, based on a review of various evaluations of government employment programmes, also concluded that 'place-based' policies aimed at getting people into work tended to be noticeably better in terms of outcomes achieved than mainstream 'person-targeted' policies (Griggs et al. 2008). However, it is important to recognise that the implementation of place-based approaches is time-consuming, resource intensive and relatively expensive.

One outcome of the period of neighbourhood-focused policy under New Labour was a better practical understanding of 'what works' in tackling worklessness (Sanderson 2006; Meadows 2008; Policy Research Institute 2007; Syrett and North 2008; Beatty et al. 2009b). More effective initiatives were characterised by an emphasis upon outreach activities that proactively engaged with the most disadvantaged groups furthest from entering the labour market. Clients were found to be more responsive to voluntary initiatives and services were more likely to be effective if located in familiar and accessible community-based facilities and delivered by trusted local voluntary or community-based organisations.

Also important given the diverse and multiple barriers to employment that individuals face, personalised and holistic approaches enabled the provision of specialist help (in relation to issues of health, drug or alcohol abuse, debt, housing and family breakdown) alongside employment-related support on issues such as skills, language difficulties, job search and making applications. In this respect the key role for trusted and motivated personal advisors or mentors was frequently identified. Such advisors can operate flexibly in relation to an individual's needs, providing continuity of support and guidance to appropriate sources of specialist help at the right times, and build up self-esteem and provide contact with positive role models.

Provision which gave support throughout a long-term process of labour market engagement – starting with pre-employment training and confidence building and continuing through to support for job search and interview preparation and ongoing training both in and out of employment – was found to be more successful. Critical here was an emphasis upon job retention and progression and not just getting workless people into work, so as to ensure a period of sustained employment, which entailed continuing support for people once they had obtained work.

The active involvement and good relations with employers was also crucial given their role in controlling access to job opportunities, so that initiatives were informed of available job vacancies and what employers were looking for in order to help workless people become job ready and make them able to compete for the jobs on offer.

Being able to influence employers' recruitment practices in favour of disadvantaged groups and to redress discriminatory practices required building trusted relationships over time (Nunn et al. 2010). Yet studies also show that meaningful employer engagement frequently remains restricted in practice (Green et al. 2010).

## Conclusions

Interventions conceived at the neighbourhood level are poorly positioned to address changes in labour market supply and demand operating at wider spatial scales which are the primary causes of concentrated worklessness. However, they can play a vital role in tailoring supply-side initiatives to meet local circumstances and significantly improve policy delivery to disadvantaged individuals and communities. In this respect they can address both the various neighbourhood effects that compound high levels of worklessness – for example in relation to improving information flows, developing employment networks and job linkages and addressing discriminatory practice and localised cultural attitudes – and locally-constituted institutional barriers, for example in relation to transport and childcare.

Indeed given the scale of the problem of worklessness within deprived areas and the major structural changes taking place within labour markets and their contribution to rising levels of inequality (National Equality Panel 2010), any expectation that modestly-resourced neighbourhood level initiatives would generate major changes in aggregate levels of worklessness appears somewhat misplaced. As Beatty et al. (2009b) point out, even in the example of the NDC Programme which was considered 'well funded', the total spend on the worklessness outcome amounted to about £380 per workless individual per year, a scale of spending unlikely to have a major impact upon localised worklessness. Significantly, the global financial crisis and ensuing economic downturn led to a rapid growth in unemployment nationally from 2008 and created an environment in which tackling concentrated worklessness is considerably more difficult than in comparison to the period of employment growth which provided the context for NL's policy initiatives.

For neighbourhood-level actions to have any significant impact on levels of concentrated worklessness, these need to be integrated with wider economic strategies that impact upon the availability of appropriate employment opportunities within the local and regional economy. Yet such integration requires strong partnership working in tackling worklessness and in practice the extent of this is variable. There are a number of well-established and long-standing barriers to more effective policy integration within and across spatial scales (North and Syrett 2008; Green and Orton 2012). The complex governance system that evolved in England over the NL years combined with a plethora of central government initiatives made co-ordination and integration of policy difficult to pursue in practice. The highly centralised nature of the vast majority of this policy activity and lack of integration between central state departments which dominated governance arrangements, meant sub-national institutions lacked the power and resources to develop and manage local and regional worklessness strategies. As labour markets operate predominantly at local and sub-regional scales and are best addressed at this level – for example in terms of identifying skills needs, providing appropriate education and training, developing employer engagement and new employment sites and linkages to deprived communities – the lack of strategic and delivery capacity at this level presented a major constraint.

Yet whatever the strength and nature of the subnational governance system and the wider global economic changes, in relation to labour market regulation and welfare provision the central state retains a key role regarding work and worklessness within deprived neighbourhoods. Given the primacy of people-based characteristics in understanding supply-side causes of worklessness, mainstream policies relating to skills and education are of critical importance, as are those that address key institutional barriers such as childcare, transport and housing. In relation to welfare payments and 'making work pay' – a key issue within deprived communities – the current reforms of benefits being pursued by the Coalition government are likely to have profound, and potentially highly damaging, impacts upon the nature and constitution of concentrated worklessness, far greater than any neighbourhood based initiatives. There remains considerable scope for effecting change in relation to pay and working conditions for low-income workers, through stronger regulation at the bottom end of the labour market. However given the commitment of successive governments to the promotion of labour market flexibility there has been little appetite to pursue further this type of regulatory activity.

# References

Atkinson, R. (2007). Under construction – The city-region and the neighbourhood: new actor in a system of multi-level governance? In I. Smith, E. Lepine, & M. Taylor (Eds.), *Disadvantaged by where you live? Neighbourhood governance in contemporary urban policy* (pp. 65–82). Bristol: The Policy Press.

Atkinson, R., & Kintrea, K. (2001). Disentangling area effects: Evidence from deprived and non-deprived neighbourhoods. *Urban Studies, 38*(12), 2277–2298.

Bauder, H. (2001). Work, young people and neighbourhood representations. *Social and Cultural Geography, 2*(4), 461–480.

Beatty, C., & Fothergill, S. (2002). Hidden unemployment amongst men: A case study. *Regional Studies, 36*(8), 617–630.

Beatty, C., Fothergill, S., & Lawless, P. (1997). Geographical variation in the labour-market adjustment process: The UK coalfields 1981–91. *Environment and Planning A, 29*(11), 2041–2060.

Beatty, C., Fothergill, S., Houston, D., Powell, R., & Sissons, P. (2009a). A gendered theory of employment, unemployment and sickness. *Environment and Planning C: Government and Policy, 27*(6), 958–974.

Beatty, C., Crisp, R., Foden, M., Lawless, P., & Wilson, I. (2009b). *Understanding and tackling worklessness: Lessons and policy implications. Evidence form the New Deal fort Communities Programme*. London: Department of Communities and Local Government.

Buck, N., & Gordon, I. (2004). Does spatial concentration of disadvantage contribute to social exclusion? In M. Boddy & M. Parkinson (Eds.), *City matters: Competitiveness, cohesion and urban governance* (pp. 237–254). Bristol: The Policy Press.

Burgess, S., Propper, C., Rees, H., & Shearer, A. (2003). The Class of '81: The effects of early-career unemployment on subsequent unemployment experiences. *Labour Economics, 10*(3), 291–311.

Centre for Regional Economic and Social Research (CRESR). (2005). *New Deal for Communities 2001-05: An interim evaluation*. Neighbourhood Renewal Unit (Research Rep. No. 17). London: ODPM.

Centre for Regional Economic and Social Research (CRESR). (2007). *The moving escalator? Patterns of residential mobility in New Deal for Communities areas'* (Research Rep. No. 32). London: Department for Communities and Local Government.

Cheshire, P., Monastiriotis, V., & Sheppard, S. (2003). Income inequality and residential segregation: Labour market sorting and the demand for positional goods. In R. Martin & P. Morrison (Eds.), *Geographies of labour market inequality*. London: Routledge.

CLG. (2006). *The dynamics of local economies and deprived neighbourhoods*. London: Department for Communities and Local Government.

CLG. (2008). *Tackling worklessness in NDC areas – A policy and practice update. Some lessons from the New Deal for Communities Programme*. London: Department for Communities and Local Government.

CLG. (2009a). *A typology of the functional roles of deprived neighbourhoods*. London: Department for Communities and Local Government.

CLG. (2009b). *Four years of change? Understanding the experiences of the 2002-2006 New Deal for Communities Panel*. London: Department for Communities and Local Government.

CLG. (2009c). *Transforming places, changing lives: Taking forward the regeneration framework*. London: Department for Communities and Local Government.

CLG. (2010a). *Evaluation of the national strategy for neighbourhood renewal: Local Research Project*. London: Department for Communities and Local Government.

CLG. (2010b). *Evaluation of the national strategy for neighbourhood renewal. Final report*. London: Department for Communities and Local Government.

Coombes, M., & Raybould, S. (2004). Finding work in 2001: Urban-rural contrasts across England in employment rates and local job availability. *Area, 36*(2), 202–222.

DETR. (2000). *Social exclusion and the provision of public transport: Main report*. London: Department of the Environment, Transport and the Regions.

Dickens, W. (1999). Rebuilding urban labour markets: What can community development accomplish? In R. Ferguson & W. Dickens (Eds.), *Urban problems and community development*. Washington, D.C.: Brookings Institute Press.

DTI. (2005). *Guidance to RDAs on regional strategies*. London: Department of Trade and Industry.

DWP. (2008). *Transforming Britain's labour market: Ten years of the New Deal*. London: Department for Work and Pensions.

Evans, M. (2001). Britain: moving towards a work and opportunity-focused welfare state? *International Journal of Social Welfare, 10*(4), 260–266.

Fieldhouse, E. (1999). Ethnic minority unemployment and spatial mismatch: The case of London. *Urban Studies, 36*(9), 1569–1596.

Fieldhouse, E., & Tranmer, M. (2001). Concentration effects, spatial mismatch or neighbourhood selection? Exploring labour market and neighbourhood variation in male unemployment risk using Census microdata from Great Britain. *Geographical Analysis, 33*(4), 353–369.

Forrest, R., & Kearns, A. (2001). Social cohesion, social capital and the neighbourhood. *Urban Studies, 38*(12), 2125–2145.

Goos, M., & Manning, A. (2007). Lousy and lovely jobs: The rising polarization of work in Britain. *Review of Economics and Statistics, 89*(1), 118–133.

Gordon, I. (2003). Unemployment and spatial labour markets: Strong adjustment and persistent concentration. In R. Martin & P. Morrison (Eds.), *Geographies of labour market inequality* (pp. 55–82). London: Routledge.

Green, A. (2007). Local action on labour market integration of new arrivals: Issues and dilemmas for policy. *Local Economy, 22*(4), 349–361.

Green, A. E., & Orton, M. (2012). Policy innovation in a fragmented and complex multilevel governance context: Worklessness and the City Strategy in Great Britain. *Regional Studies, 46*(2), 153–164.

Green, A., & White, R. (2007). *Attachment to place: Social networks, mobility and prospects of young people*. York: Joseph Rowntree Foundation.

Green, A., Adam, D., Hasluck, C. (2010). Evaluation of Phase 1 City Strategy (Research Rep. No 639). Department of Work and Pensions.

Griggs, J., Whitworth, A., Walker, R., McLennan, D., & Noble, M. (2008). *Person or place-based policies to tackle disadvantage? Not knowing what works*. York: Joseph Rowntree Foundation.

Hasluck, C. (1999). *Employers, young people and the unemployed: a review of research* (Research Rep. ESR 12). Sheffield: Employment Service.

HM Treasury. (2007). *Review of sub-national economic development and regeneration*. London: Department for Business, Enterprise and Regulatory Reform/HMT & Department of Communities and Local Government.

Lawless, P. (1995). Inner-city and suburban labour markets in a major English conurbation: processes and policy implications. *Urban Studies, 32*(7), 1097–1125.

Lawless, P. (2011). *Regeneration- what are the problems and what can we achieve in addressing them? Neighbourhood level perspectives from the NDC programme, Discussion paper for the regeneration futures roundtable*. London: Department of Communities and Local Government.

Lucas, K., Tyler, S., & Christodoulou, G. (2008). *Assessing the 'value' of new transport initiatives in deprived neighbourhoods in the UK*. York: Joseph Rowntree Foundation.

Lupton, R. (2003). *Poverty street: The dynamics of neighbourhood decline and renewal*. Bristol: The Policy Press.

Lupton, R., Tunstall, R., Fenton, A., & Harris, R. (2011). *Using and developing place typologies for policy purposes*. London: Department for Communities and Local Government.

Maclennan, D. (2000). *Changing places, engaging people*. York: Joseph Rowntree Foundation/York Publishing Services.

Meadows, P. (2001). *Young men on the margins of work: An overview report*. York: Joseph Rowntree Foundation.

Meadows, P. (2008). *Local initiatives to help workless people find and keep paid work*. York: Joseph Rowntree Foundation.

Murray, C. (1996). The emerging British underclass. In R. Lister (Ed.), *Charles Murray and the underclass: The developing debate* (pp. 23–52). London: Institute for Economic Affairs.

National Equality Panel. (2010). *An anatomy of economic inequality in the UK: Report of the National Equality Panel*. London: National Equality Panel.

North, D., & Syrett, S. (2008). Making the links: Economic deprivation, neighbourhood renewal and scales of governance. *Regional Studies, 42*(1), 133–148.

North, D., Syrett, S., & Etherington, D. (2009). Tackling concentrated worklessness: Integrating governance and policy across and within spatial scales. *Environment and Planning C: Government and Policy, 27*(6), 1022–1039.

Nunn, A., Bickerstaffe, T., Hogarth, T., Bosworth, D., Green, A.E., Owen, D. (2010). *Postcode selection? Employers' use of area- and address-based information shortcuts in recruitment decisions* (Research Rep. No. 664). Department for Work and Pensions.

ODPM. (2004). *Jobs and enterprise in deprived areas*. London: Office of the Deputy Prime Minister.

Policy Research Institute. (2007). *Towards skills for jobs: What works in tackling worklessness?* Coventry: Learning and Skills Council.

Prime Minister's Strategy Unit (PMSU). (2005). *Improving the prospects of people living in areas of multiple deprivation in England*. London: Cabinet Office.

Putnam, R. (2000). *Bowling alone: The collapse and revival of American community*. New York: Simon and Schuster.

Ritchie, H., Casebourne, J., Rick, J. (2005). *Understanding workless people and communities: a literature review* (Research Rep. DWPRR255). London: Department of Work and Pensions.

Roberts, F. (1999). *Young people out of work in South West Newham* (Commentary Series No. 83). London: Centre for Institutional Studies, University of East London.

Robson, B. T., Lymperopoulou, K., & Rae, A. (2008). People on the move: Exploring the functional roles of deprived neighbourhoods. *Environment and Planning A, 40*, 2693–2714.

Sanderson, I. (2006). *Worklessness in deprived neighbourhoods: A review of evidence, Report to the Neighbourhood Renewal Unit*. London: Department for Communities and Local Government.

Shuttleworth, I., Green, A., Lavery, S. (2003, May 13). *Young people, job search and local labour markets: the example of Belfast*. Seminar paper presented at Employability: Lessons for Labour Market Policy. Warwick: Seminar, University of Warwick.

Social Exclusion Unit (SEU). (2000). *National strategy for neighbourhood renewal: A framework for consultation*. London: Cabinet Office.

Social Exclusion Unit (SEU). (2003). *Making the connections: Final report on transport and social exclusion*. London: Social Exclusion Unit.

Social Exclusion Unit (SEU). (2004). *Jobs and enterprise in deprived areas*. London: Office of the Deputy Prime Minister.

Speak, S. (2000). Barriers to lone parents' employment: Looking beyond the obvious. *Local Economy, 15*(1), 32–44.

Syrett, S., & North, D. (2008). *Renewing neighbourhoods: Work, enterprise and governance*. Bristol: Policy Press.

Syrett, S., & North, D. (2010). Between economic competitiveness and social inclusion: New labour and the economic revitalisation of deprived neighbourhoods. *Local Economy, 25*(5/6), 476–493.

Taylor, M. (2002). Community and social exclusion. In V. Nash (Ed.), *Reclaiming community*. London: Institute for Public Policy Research.

The Poverty Site. (2010). *UK: concentrations of worklessness*. http://www.poverty.org.uk/43/index.shtml. Accessed 10 June 2010.

Thickett, A. (2011). *Putting passengers first? An examination of the role of urban bus transport in enhancing the everyday mobility and life chances of lone parents living in deprived areas*, Unpublished PhD thesis. London: Middlesex University.

Tunstall, R. (2009). *Communities in recession: The impact on deprived neighbourhoods*. York: Joseph Rowntree Foundation.

Turok, I., & Edge, N. (1999). *The jobs gap in Britain's cities: Employment loss and labour market consequences*. Bristol: The Policy Press.

van Ham, M., Manley, D., Bailey, N., Simpson, L., & Maclennan, D. (2012). Introduction. In M. van Ham, D. Manley, N. Bailey, L. Simpson, & D. Maclennan (Eds.), *Neighbourhood effects research: New perspectives* (pp. 1–22). Dordrecht: Springer.

van Ham, M., Manley, D., Bailey, N., Simpson, L., & Maclennan, D. (2013). Understanding neighbourhood dynamics: New insights for neighbourhood effects research. In M. van Ham, D. Manley, N. Bailey, L. Simpson, & D. Maclennan (Eds.), *Understanding neighbourhood dynamics: New insights for neighbourhood effects research* (pp. 1–22). Dordrecht: Springer.

Watt, P. (2003). Urban marginality and labour market restructuring: Local authority tenants and employment in an inner London Borough. *Urban Studies, 40*(9), 1769–1789.

Webber, C., & Swinney, P. (2010). *Private sector cities: A new geography of opportunity*. London: Centre for Cities.

Webster, D. (2000). The geographical concentration of labour-market disadvantage. *Oxford Review of Economic Policy, 16*(1), 114–128.

# Chapter 4
# The Role of Neighbourhoods in Shaping Crime and Perceptions of Crime

**Ian Brunton-Smith, Alex Sutherland, and Jonathan Jackson**

## Introduction

Crime and perceptions of crime are not randomly distributed across local areas – this much we know from criminological research (see inter alia, Shaw and McKay 1942; Baldwin and Bottoms 1976; Brantingham and Brantingham 1981; Skogan and Maxfield 1981; Wikström 1991; Sampson 2012). Starting with the early finding that higher levels of crime are evident in more socially disadvantaged neighbourhoods, a growing number of academic studies has emphasised the effect on crime and individual perceptions of the local environmental context (e.g. Sampson and Groves 1989; Sampson et al. 1997; Wikström and Sampson 2003). Drawing on ever more detailed sources of contextual data – and utilising advanced statistical approaches such as multilevel modelling – these studies are providing an increasingly convincing account that neighbourhood context has an important role to play in shaping levels of crime and individual perceptions.

This broad range of academic work has – in turn – informed a number of policy initiatives, including the US emphasis on community policing (Skogan 2003) and zero tolerance strategies (Dennis 1997). The UK has seen the rise of the neighbourhood policing initiative and a growing emphasis on community centred programmes (Singer 2004; Morris 2006). The use of Police Community Support Officers as a

---

I. Brunton-Smith (✉)
Department of Sociology, University of Surrey, Guildford, Surrey GU2 7X, UK
e-mail: i.r.brunton-smith@surrey.ac.uk

A. Sutherland
Institute of Criminology, University of Cambridge, Sidgwick Avenue, Cambridge CB3 9DA, UK
e-mail: as2140@cam.ac.uk

J. Jackson
Department of Methodology, The London School of Economics and Political Science,
Houghton Street, London WC2A 2AE, UK
e-mail: j.p.jackson@lse.ac.uk

more direct link between local communities and police has also followed the increasing recognition of the potential role of local neighbourhood interactions for crime reduction (Hughes and Rowe 2007).

Yet, despite the increasing number of academic studies in this area, a detailed conception of the complex ways in which individuals variously experience 'neighbourhoods' has been largely absent from the empirical assessments of neighbourhood effects that have informed recent policy. Multilevel models have provided an efficient methodology to incorporate neighbourhood level processes alongside individual data. But studies have still had to rely on administrative geographies that often bear little resemblance to the lived experiences of residents of particular neighbourhoods. Moreover, these studies have generally assumed that individuals are only influenced by their immediate surroundings, leaving open the possibility that other neighbourhood effects may be in operation that originate from surrounding neighbourhoods. And while there is a growing recognition in the methodological literature that when estimating neighbourhood effects, one needs to pay attention to selection bias and people's selective mobility (van Ham et al. 2012, 2013), there needs to be more criminological work in this area, with Robert Sampson's (2012) call for a broader conception of neighbourhood effects particularly important in this regard.

In the first part of this chapter we provide a brief overview of the range of different ecological theories that have been advanced to explain the link between socially disadvantaged neighbourhoods and levels of crime and perceptions of (or fear of) crime. In the second part of the chapter we consider the two main weaknesses of neighbourhood effect studies – definition of neighbourhood and selection bias – and we demonstrate, by way of empirical example, how understanding of the ways in which neighbourhoods affect local residents can be further extended by allowing for the additional influences of surrounding local areas, using data from the British Crime Survey. This draws upon methodological work incorporating spatial autocorrelation in neighbourhood models (see for example Morenoff et al. 2001). In the third and final part of the chapter, we discuss the extent to which this range of neighbourhood effects research has influenced policing and crime reduction policy initiatives in England and Wales. Here we discuss the growing importance that neighbourhoods and local communities have held in policy strategies to reduce crime and raise public confidence. We highlight a number of initiatives including neighbourhood watch, neighbourhood policing, new deals for communities, community wardens (latterly Police Community Support Officers) and Crime and Disorder Reduction Partnerships.

## Neighbourhood Effects on Crime and Perceptions of Crime

The spatial patterning of crime across local areas is a consistent finding in criminological research, with higher crime generally identified in more social disadvantaged areas. For example, recent Home Office data finds that approximately

19 % of households in the most deprived local areas were the victim of crime within the last year, compared with 14 % of those from the least deprived areas (Flatley et al. 2010). Similarly, perceptions of crime differ systematically across neighbourhood contexts (e.g. Brunton-Smith and Sturgis 2011), leading many to explore possible neighbourhood level mechanisms that exert an effect on levels of crime and perceptions within local areas.

Serious attempts at explaining this link with deprivation stemmed from the early urban sociology work of the Chicago school that had begun to emerge in the 1920s and 1930s, and in particular the work of Park and Burgess (1924) and Thrasher (1927). Examining the impact of urbanisation and social mobility on a range of outcomes, these studies emphasised the importance of the physical and social environment in shaping human behaviour and social outcomes, leading to discussions of the role of local context over and above individual motivations, as well as an increasing interest in possible neighbourhood mechanisms.

## *Social Disorganisation*

These ideas were most famously formalised in Shaw and McKay's (1942) social disorganisation theory. In a 20 year study of the spatial distribution of delinquency across urban areas in Chicago, Shaw and McKay linked local delinquency rates to measures of population change, substandard housing, and economic and racial segregation. They identified the highest rates of delinquency in areas of low socioeconomic status, which is unsurprising. But they also demonstrated considerable consistency in these neighbourhood problems across time, despite complete changes in the populations occupying those areas. Rather than viewing delinquency as a direct result of competition over a lack of economic resources, they suggested that this occurred in conjunction with the impact of residential change and high levels of ethnic heterogeneity, limiting the ability of informal social control mechanisms available to community's to control their residents. This obstruction of informal social controls was primarily reflected through restrictions on residents' abilities to develop strong friendship networks within their community, reduced participation in local organisations, and a limited set of social resources available to supervise teenage peer groups. Skogan (1986) argues that these locally based social networks were instrumental to a community's capacity for informal social control by making them better able to recognise strangers and more apt to engage in guardianship activities against potentially disruptive behaviour. A reduction in the availability of community ties was thus an important source of reduced informal control over residents.

Drawing on aggregate data from the British Crime Survey, Sampson and Groves (1989) provide a detailed empirical test of the central tenets of social disorganisation theory, demonstrating significant relationships between heterogeneity, mobility, neighbourhood economic status, and the levels of neighbourhood disorganisation. They also introduced direct measures of the neighbourhood level of organisational participation, friendship networks, and unsupervised teen groups, which they demonstrated mediated the relationship between the structural measures of disorganisation and the rate of criminal victimisation. Extending the original theory, Sampson and Groves (1989) also argued for the inclusion of measures of family disruption, with less active supervision resulting in more problems with low-level disorder from teen groups. In addition to viewing them as important sources of supervision for their own children, Sampson and Groves identified parents as important agents of informal social control of other young people within a neighbourhood, further limiting the level of deviant behaviour. Specifically, they suggested that if the parents of children know one another so called 'inter-generational closure' is achieved which increases the net amount of informal control for that neighbourhood. Further exploring the role of inter-generational closure Sampson and colleagues (1999) report that informal social control of children accounted for nearly half of the relationship between residential stability and levels of delinquency.

Bursik and Grasmick (1993) also expanded on the original theory, highlighting the overlapping and conflicting sources of organisation in a given community. Pointing to the existence of neighbourhoods that have extensive personal networks facilitating informal social controls, but which nevertheless have relatively high rates of crime, they emphasised the need to incorporate the wider context of formal controls. In particular they highlighted the influence of external market forces instigating community changes that can have an influence on levels of crime in addition to the effects of informal local controls. To account for this, Bursik and Grasmick (1993) incorporated the broader, public level of control, more explicitly recognising the wider context in which informal social controls operate. This public level of control is directly related to a local neighbourhood's ability to obtain public goods and services that are allocated by agencies external to the community (e.g. the levels of community policing and the resources provided to implement local crime control initiatives), which are instrumental in limiting levels of crime and reducing fear (Herbert 2005). A local community's ability to organise effectively against crime problems will thus be partially dependent on their ability to influence the public decision making agencies that are responsible for delivering these resources to the community (Carr 2005).

Sampson and colleagues (1997) introduced an evolution of social disorganisation theory, focusing on the part that 'collective efficacy' has to play in influencing levels of crime, and indirectly fear. Here they argue for the important roles of the level of mutual trust and cohesion amongst residents within a community, which interact with a neighbourhood's capacity for informal social control. This occurs by enhancing the capacity of a neighbourhood for mutual cooperation amongst residents (Sampson et al. 1999). Collective efficacy was introduced as a neighbourhood level consequence of social capital (Putnam 2000), highlighting the importance of levels of trust between residents for facilitating informal social control mechanisms. While a community may have strong social networks facilitating the informal control of disorderly behaviour, without strong feelings of trust and cohesion amongst residents, they may be unwilling to confront people that are disrupting public spaces. Similarly, lower levels of collective efficacy will limit residents' willingness to tackle low-level physical signs of disorder within the community. Sampson and colleagues (1997, p.919) argue that more socially cohesive neighbourhoods are "fertile contexts for the realisation of informal social control" and that considerable variation in the extent of cohesion across communities is instrumental in explaining the variations evident in levels of crime and fear. In their study of variations in levels of crime across 343 Chicago neighbourhoods, they find that collective efficacy does effectively mediate the relationship between social composition and levels of violence. This is true even when controlling for friendship and kinship ties, organisational participation and neighbourhood services.

Although it was originally introduced to explain variations in levels of crime across areas, researchers have also drawn on social disorganisation theory to explain neighbourhood differences in fear of crime. There are two dominant ways that social disorganisation has been linked to levels of fear in existing research. The first views fear as a direct response to the levels of crime in the neighbourhood, thus implying a similar relationship between the structural determinants of disorganisation and fear through reduced mechanisms of formal and informal social control (Bursik 1988). This relationship has since been extended by viewing fear as both a reaction to higher levels of crime in more disorganised neighbourhoods, and as another dimension of disorganisation that may lead people to withdraw from community life, further increasing the extent of crime as informal social controls are weakened (Carr 2005; Woldoff 2006). The second proposes that lower levels of community involvement in more heterogeneous and unstable neighbourhoods limits the number of familiar people known to each resident, leading to higher levels of anxiety and further withdrawal from the local community (Krannich et al. 1989). In contrast, more socially integrated neighbourhoods are expected to have stronger networks of local support, alleviating the levels of fear from residents (Hale 1996). This approach also highlights the important part that informal social control has on levels of fear, with residents in more disorganised neighbourhoods perceiving themselves to have less influence on the behaviour of others, leading to increased fear (Taylor and Covington 1993). It thus implies that social disorganisation can influence fear of crime largely independently from its impact on levels of crime.

## *Subcultural Diversity*

Researchers have also emphasised the direct link between neighbourhood ethnic heterogeneity and variations in concerns about crime, arguing that this reflects the impact of subcultural diversity (Merry 1981a). The subcultural diversity thesis can be viewed as a specific application of 'conflict' theory, which has primarily been used to explain lower levels of trust in more ethnically diverse neighbourhoods (Putnam 2007). This characterises diversity on the basis of 'insider' and 'outsider' groups within an area, with insiders those that share the dominant ethnicity of the community, and outsiders identified as those that are of a different ethnicity. This leads to inter-group tensions and fosters out-group suspicions that reduce levels of social solidarity. These inter-group tensions have also been linked with a reduced likelihood of effective socialisation within a neighbourhood, making residents less likely to intervene to solve local neighbourhood problems, promoting higher levels of crime and fear (Taylor and Covington 1993).

Proponents of subcultural diversity argue that fear of crime will be higher amongst those living in close proximity to people from different cultural backgrounds, with the manners and behaviours of other groups identified as fear inspiring (Covington and Taylor 1991). This is also closely linked to levels of community involvement, with subcultural diversity promoting feelings of social isolation amongst those living in neighbourhoods with high proportions of residents from cultural backgrounds different to their own. In this way, subcultural diversity can also be linked with collective efficacy as an important restriction on community cohesion (Adams and Serpe 2000).

In contrast to this negative impact of ethnic diversity, 'contact' theory suggests that ethnic diversity may actually have a positive effect on community relations. Proponents of contact theory argue that the presence of 'outsider' groups might actually result in higher levels of social solidarity, by fostering increased tolerance of those identified as 'outsiders' (Putnam 2007). From this perspective, increased contact with those that are different actually serves to enhance the community bonds within the neighbourhood, strengthening the available informal social controls within the community to deal with problems of crime.

## *Low Level Disorder*

Another explanation for spatial variations in levels of crime and perceptions of crime can be found in the role of low level signs of disorderly behaviour and physical deterioration in the local area (Skogan 1990). Most famously discussed in Wilson and Kelling's (1982) seminal 'broken windows' thesis, this emphasises the importance of local environmental 'cues' including signs of vandalism, abandoned buildings, graffiti, and unchecked litter that signal to potential offenders that disorder will be tolerated. These environmental signs can also promote greater levels of concern about crime from local residents, acting as an important symbol of the extent that the neighbourhood is in decline, and providing clear visual cues for residents that warn them of their potential risk (Ferraro 1995). The role of disorder has since formed the basis of Innes' (2004) 'signal crimes perspective', which explores how signs of disorder within a local community come to be defined as potentially dangerous and hence indicators of potential risk that lead to fear. Further work on the social perception of disorder in the UK (e.g. Jackson 2004; Farrall et al. 2009) suggests that the issue of crime is entangled in the public imagination with issues of cohesion, collective efficacy, social change and tension (Girling et al. 2000). Rather than being about an 'irrational' sense of crime, fear may express and distil lay diagnoses about neighbourhood breakdown and stability.

However, the link between low level signs of disorder and subsequent levels of crime and concerns about crime has not been universally accepted. In a landmark study using observational data systematically collected from Chicago, Sampson and Raudenbush (1999) found that this association was largely spurious, except perhaps when considering robbery. Harcourt (2001) also argues that the link between disorder and crime may have been exaggerated, criticising the findings from empirical treatments of the disorder perspective. Yet more recent, empirically robust, experimental evidence from Holland shows that the likelihood of low level crimes (such as littering) being influenced by visible signs of low level disorder such as graffiti may be low (Keizer et al. 2008). More interestingly, in a later quasi-experimental study, Keizer and colleagues (2011) also demonstrate that signs prohibiting norm violation (e.g. 'no littering' signs) may actually encourage such acts if there are obvious examples of norm violation nearby.

## *Defensible Space*

Variations in crime have also been linked directly to the physical structure of the built environment in local areas (Newman 1978). Highlighting the importance of accommodation type and the effect of property design, Newman and Franck (1982) argue that a community's capacity for social control is directly influenced by the physical design of the neighbourhood. Drawing on ideas of 'territoriality', they argue that the design of the local area can either foster increased opportunities for informal surveillance and a more proprietary attitude towards the immediate neighbourhood, or promote restrictions on informal social control with the creation of isolated 'out of sight' areas that are difficult to oversee. Critical to the sense of ownership that the local area engenders is how the space is marked out and bounded, with a clear demarcation between private and public areas encouraging local residents to have a stake in the local area; caring for it, policing it, and reporting strangers and others who have no apparent good purpose to be there (Rock 2007). The impact of the physical environment on levels of crime has since greatly influenced the growing interest in crime science (Smith and Tilley 2005), as well as the recent focus on situational crime prevention strategies (Bullock et al. 2010).

A number of researchers have also demonstrated direct links between the built environment and fear, with Newman and Franck (1982) demonstrating that fear was higher amongst residents of larger housing blocks. Similarly, Taylor and colleagues (1984) report that the presence of surveillance opportunities and physical barriers that restricted access to parts of the local area, were associated with lower levels of fear. Other physical elements of the built environment have also been linked with reductions in fear of crime, with the increased use of surveillance cameras (Gill and Spriggs 2005) and improved street lighting (Vrij and Winkel 1991) featuring in research (see also Schweitzer et al. 1999). However, Merry (1981b) highlights areas that qualify as architecturally defensible, yet which nevertheless go undefended by local residents. She therefore argues that the social processes involved in whether residents informally protect their local environment may be more important than physical attributes of the area.

## Improving the Estimation of Neighbourhood Effects

There is now an increasing body of evidence in support of the existence of neighbourhood effects on crime and perceptions of crime. However, neighbourhood effects research has not been without criticism (for clear reviews see Mayer and Jencks 1989; Wikström and Sampson 2003; Oakes 2004). These have primarily centred on three main limitations with early studies. First is the inability to consider individual and neighbourhood effects simultaneously without associated losses of information, with neighbourhood effects measured at the individual level overestimating levels of precision, or between resident variability sacrificed in favour of accurate measurement of between neighbourhood differences. Second is the potential existence of selection effects, with the non-random sorting of individuals within areas limiting the extent that causal effects of neighbourhoods can be identified. Third is the use of inadequate neighbourhood geographies to represent local areas, with the majority of studies relying on administrative boundaries that bear little resemblance to the lived realities of local residents.

Significant inroads have now been made on the first problem, with more recent studies adopting a multilevel modelling framework to accurately incorporate multiple levels of influence simultaneously (see for example Sampson et al. 1997; Wilcox-Rountree and Land 2000; Taylor 2001; Krivo et al. 2009). This has also prompted an increasing emphasis on so-called 'cross-level interactions' between individual and neighbourhood effects enabling researchers to more directly examine the effect of contextual influences on individual residents (Wikström and Sampson 2003).

The second problem is selection bias. More specifically, if people select into or are constrained to live in certain areas – as is sometimes the case – then any observed associations between neighbourhood and crime may be bound up also with associations between neighbourhood and individual characteristics that are related to selective mobility. Attempts have been made to adjust for the non-random allocation of individuals to areas, with the Moving To Opportunity study in the US the most well cited example of a experimental design enabling casual neighbourhood effects to be identified (Kling et al. 2004).[1] Work using instrumental variables to adjust estimates also offers the potential to provide more refined assessments of neighbourhood

---

[1] Although even in experimental conditions where such moves are observed and the process is randomised, there is no guarantee that the effects noted are due to the change in setting. Sampson et al. (2002, p.466) state that '[t]he clear tendency has been to interpret MTO results in terms of the effects of changing [from] concentrated poverty, but...such an assertion is arbitrary – any number of changes in social processes associated with poverty may account for the result'.

effects (e.g. MacDonald et al. 2012). But more criminological work is needed on treating, in Sampson's (2012, p.67) words, 'selection as a social problem not a statistical nuisance.' Such an approach would embed selection and mobility into a broader conception of neighbourhood effects and social context.

The third main problem revolves around the ability to identify adequate neighbourhood units for analyses. Studies are still limited by the availability of suitable area boundaries to represent local areas, and as a result administrative boundaries that have relatively little correspondence to individuals' conceptions of their local neighbourhood have often been used, for example postcode sectors or electoral wards. There is now an increasing understanding of the potential impact that the use of different area geographies can have on results, with studies placing greater attention on the sensitivity of effects to the choice of neighbourhood boundary (e.g. Wikström and Sampson 2003; Manley et al. 2006; Hipp 2007; Oberwittler and Wikström 2009; Weisburd et al. 2009). But the use of these administrative boundaries is also at odds with the more sophisticated treatments of neighbourhood found in community studies (Lupton 2003), failing to fully capture the contingent nature of neighbourhood for individual residents. As such, comparatively little consideration has been given to the sorts of anchoring characteristics like physical barriers, landmarks, and roads, or the social characteristic of areas that inform individuals' definitions of their neighbourhood.

One way forward to better incorporate this understanding of the physical and social character of neighbourhoods in empirical studies can be found in the new super output area (SOA) geographies introduced with the advent of the 2001 census in England and Wales for the dissemination of population statistics. In particular, the middle layer of this geography (MSOA) maps more closely onto the neighbourhood definitions suggested by community studies than other administrative boundaries. These are composed of an average of 2,500 households grouped together based on spatial proximity, and homogeneity of dwelling type and tenure. Importantly, during their construction they also included a consultation stage with local authorities to ensure that they represented meaningful geographic areas, and attempts were made to ensure they did not cross clear physical boundaries.

Community studies also view neighbourhood boundaries as flexible and permeable, with residents drawing on more than their immediate surroundings. This emphasises the relations between places when individuals define their local neighbourhood, with individual definitions of their neighbourhood partially a reflection of comparisons with the areas that surround them, and their beliefs about how the neighbourhood is perceived by others (Chaskin 1998). The contingent nature of neighbourhood boundaries can be considered as a form of spatial autocorrelation, with residents from neighbourhoods that are in closer proximity to one another sharing similar experiences and neighbourhood effects 'spilling over' neighbourhood boundaries (see for example Morenoff et al. 2001). Measures of spatial autocorrelation are typically used as an indicator of social processes that are operating at a different level of influence to the chosen area geography (see Manley et al. 2006). But in combination with more detailed neighbourhood boundary data, they can also provide us with an indication of the competing spheres of influence on

**Table 4.1** Worry about victimisation across neighbourhoods – adjusting for spatial autocorrelation

|  | Model I | | Model II | |
|---|---|---|---|---|
| Neighbourhood fixed effects | | | | |
| Neighbourhood disadvantage | 0.01 | | 0.01 | |
| Urbanicity | 0.06 | ** | 0.05 | ** |
| Population mobility | 0 | | 0 | |
| Age profile | 0.01 | * | 0.01 | * |
| Housing structure | −0.02 | ** | −0.02 | ** |
| Ethnic diversity | 0.27 | ** | 0.2 | ** |
| BCS interviewer rating of disorder | 0.06 | ** | 0.06 | ** |
| Recorded crime (IMD 2004) | 0.07 | ** | 0.05 | ** |
| Neighbouring area effects | | | | |
| BCS interviewer rating of disorder | | | 0.06 | ** |
| Recorded crime (IMD 2004) | | | 0.04 | * |
| Spatial autocorrelation | 0.027 | ** | 0.026 | ** |
| Neighbourhood variance | 0.016 | ** | 0.015 | ** |
| Individual variance | 0.811 | ** | 0.811 | ** |
| Base sample size | 102,133 | | 102,133 | |

**P<0.01; *P<0.05

individuals. By incorporating information about those neighbourhoods that surround each resident's own local neighbourhood, we can assess the extent of these shared influences, as well as the extent to which individuals base their judgements on the broader local area. This further clarifies the importance of the immediate neighbourhood, adjusting neighbourhood effects for any dependencies exhibited by neighbourhoods in close proximity.

To illustrate how improved neighbourhood boundaries and adjustments for spatial autocorrelation can extend our understanding of the complex impact of neighbourhood characteristics on individual residents, we present here some brief results from an analysis of the neighbourhood influences on individual levels of worry about crime, based on 3 years (2002–2005) of survey data from the British Crime Survey covering a total of 102,133 residents grouped in 5,196 local neighbourhoods (MSOA). In addition to the specific area of each resident, we also identify all of the areas whose boundaries touch each of our sampled neighbourhoods, with each area surrounded by between one and 18 neighbouring MSOAs, and an average of five neighbours.[2] Full details and methodology can be found in Brunton-Smith and Sturgis (2011) and Brunton-Smith and Jackson (2012).

To examine the influence of neighbourhoods on people's worries about falling victim crime – and to assess the possibility that individuals draw on environmental cues from beyond their own neighbourhood boundaries – Table 4.1 includes details from two multilevel models. Model 1 includes a range of measures of the immediate neighbourhood of each resident, covering the level of socio-economic disadvantage,

---

[2] Analyses in Brunton-Smith and Jackson (2012) were restricted to urban neighbourhoods, however here we extend the focus to also incorporate data on rural locations.

urbanicity, population mobility, age and housing structure, and level of ethnic diversity, along with measures of the levels of crime and observed signs of disorder within the area. Individual covariates are also included in the model (not reported here), adjusting for differences in worry based on gender, age, ethnicity, education social class, newspaper readership, marital status, length of residence in the area, and personal experience of victimisation in the previous year (distinguishing between household and personal crime, and single and repeat victimisation).

This model confirms the importance of neighbourhood characteristics for explaining variations in worry, with the crime rate, extent of visible disorder, and the social and organisational structure of the neighbourhood all exerting direct and independent effects on the expressed worry of otherwise similar people living in otherwise similar neighbourhoods. Residents of more ethnically diverse neighbourhoods are also significantly more worried about crime. Importantly, there is also residual spatial autocorrelation indicating that – for worry about crime at least – residents look beyond their own neighbourhood when assessing their potential risks. Model 2 adds measures of the levels of crime and disorder in surrounding areas, with higher levels of worry amongst residents from areas that are also surrounded by greater levels of crime and disorder. That this operates in addition to the main effects observed in the previous model suggests that residents are drawing on multiple spheres of influence simultaneously when forming judgments about their own potential risks of crime. Allowing for the impact of neighbouring areas therefore seems to be a potentially useful way forward in understanding the complex ways that individual outcomes are influenced by their local surroundings.

## The Growth of Neighbourhood Crime Reduction Policies

In tandem with growth in academic interest in the possible explanatory power of neighbourhood effects, there has been concomitant policy emphasis on the importance of neighbourhoods, driven by the recognition that crime and other social outcomes vary significantly across neighbourhood contexts. In the US this has included a greater emphasis on community policing (Skogan 2003), and zero tolerance strategies designed to tackle low-level signs of disorder (Dennis 1997). In the UK a number of policy initiatives have been introduced since the early 1990s including neighbourhood policing, neighbourhood wardens, the new deal for communities, and crime and disorder reduction partnerships (Hughes and Rowe 2007). These have placed community centre stage as potential solution to crime problems (Johnston and Mooney 2007), and have also emphasised the importance of reducing public concerns about crime alongside crime reduction strategies (Innes and Roberts 2008).

The importance of neighbourhood level strategies as an effective element of crime reduction was first seen with the widespread implementation of neighbourhood watch. Established in 1982, the number of neighbourhood watch schemes grew dramatically, with British Crime Survey data suggesting as many as 27 % of households were members at its peak in 2000. Although this has since fallen, with recent estimates placing the figure at around 13 % of households reported as members (Scribbins et al. 2011). With support from the police, neighbourhood watch represented a clear attempt to directly involve community members in local crime reduction efforts, whilst also aiming to reassure members of the public and lower levels of fear (Laycock and Tilley 1995; Scribbins et al. 2011).

In general terms, neighbourhood watch schemes promote the direct involvement of community members in the prevention of criminal activities, both by acting as an additional means of surveillance in support of the police, and by directly involving residents in activities to reduce their own risks of crime (Laycock and Tilley 1995). At the same time, neighbourhood watch strategies also strengthen the informal social controls available to reduce crime by enhancing a sense of community cohesion amongst residents. However, in practice the actual operation of neighbourhood watch schemes varies considerably across the country, with a range of different activities undertaken. In reviewing existing studies of neighbourhood watch, Laycock and Tilley (1995) find some evidence that these schemes can be helpful in reducing crime, and enhancing levels of community cohesion amongst members. Importantly, they also note that evidence has consistently shown that neighbourhood watch has been most successfully implemented in middle class, low crime rate areas (a finding also shown in Scribbins et al. 2011). This points to potential difficulties in generating informal social control in the local areas that need it most, with more directed intervention strategies needed to support these communities.

Perhaps the clearest example of neighbourhood policy being informed by neighbourhood effects research was the introduction of the Reassurance Policing Programme (Innes et al. 2004, 2005; Fielding and Innes 2006) and by extension the Neighbourhood Policing Programme that followed the initial trials (Home Office 2005; Hughes and Rowe 2007). Initially piloted by Surrey police, this was in part guided by the signal crimes perspective outlined by Innes (2004), emphasising the symbolic function of the police in dealing with particular crimes and low level disorder within communities that held a particular resonance with local residents. Recognising the potential reassurance role served by the police, this therefore prioritised those activities that were of most concern to the community, not necessarily those that were most serious. The focus, therefore, was on locally identified priorities, and more citizen-focused policing (Home Office 2005; Millie 2007), with the resulting neighbourhood policing initiative intended to "re-establish connections between the police and local communities that were lost as the Unit Beat Policing reorganised and professionalised the service in the 1960s and 1970s" (Hughes and Rowe 2007, p.329). The work of Sampson and colleagues (1997) on collective efficacy can also be seen in the aims of neighbourhood policing, with the focus on local priorities intended to promote greater levels of community cohesion and reinforce informal social control mechanisms operating at the neighbourhood level (Quinton and Tuffin 2007). In addition, the initiative intended to raise public confidence in policing, improving the familiarity and visibility of the police amongst local residents (Mason 2009).

There are now approximately 3,700 neighbourhood policing teams operating in England and Wales, all placing specific focus on the importance of addressing local crime and disorder issues that resonate with residents. This includes an emphasis on direct communication with residents to identify priorities for action, provision of information to residents about the activities that have been undertaken, and engagement with local agencies and community members to respond effectively to neighbourhood problems. Initial pilot evaluations of across 16 areas found increases in confidence in police, perceptions of the levels of crime, and resident feelings of safety, with falls in self reported victimization and perceptions of anti-social behavior when compared to areas that had not implemented neighbourhood policing (Quinton and Tuffin 2007). Evidence of positive effects from the national implementation of neighourhood policing are less clear, with a 2 year evaluation by Mason (2009) finding no consistent pattern of reduced crime or improvements to victim satisfaction. This may be partially explained by insufficient engagement with the local community, with recent evidence from the British Crime Survey suggesting that only two-fifths of people were aware that a neighbourhood policing team was operating in their local area and 31 % had seen, heard or read details about them (Scribbins et al. 2011). This study also found that only 24 % of residents had any form of direct contact with their local police (the police had knocked on their door, they had approached the police on the street, or spoken to the police at an event or public meeting), and that less than a third of these individuals had been asked about problems in their area during the contact. Of course, this is based on early implementation of the scheme, with the possibility for improved outcomes as police forces adopt neighbourhood policing strategies more fully.

Before the introduction of neighbourhood policing, similar ideas about a resident facing approach to community problems underpinned the establishment of neighbourhood warden schemes, originating out of the National Strategy for Neighbourhood Renewal in 1999 (NRU 2004). Focused primarily in deprived urban areas, these neighbourhood officials were intended to act as a direct link between community members and a range of formal agencies including, but not restricted to, the police. Because these wardens were community based it ensured they were accessible to local residents, with wardens taking part in a varied range of activities to reduce crime and fear of crime. Evaluation of the 84 neighbourhood warden schemes that were initially funded pointed to a number of positive outcomes from this scheme, reducing fear of crime, raising confidence in the police, and improving the quality of life of residents (NRU 2004). There was also evidence of improved resident perceptions of the extent of neighbourhood problems, including low level ASB.

These local wardens have since been rolled out nationally under the label of Police Community Support Officers as part of the Neighbourhood Policing Programme, acting as a visible form of policing in local neighbourhoods, with a more direct remit to form a link between police and communities (Hughes and Rowe 2007; Millie 2010). Established via the 2003 Police Reform Act, they hold only limited legal sanctions to tackle crime, with their primary function to further promote confidence amongst local neighbourhood residents by tackling issues of importance to local residents that regular offices were unable to prioritise and acting as a visible presence in the local area (Crawford and Lister 2004). However, despite bolstering police numbers within local communities, there adoption has not been uniformly successful, with the lack of clear guidelines defining the scope of their role leading to varied implementations across local areas (Paskell 2007).

A recognition of the importance of neighbourhoods and the potential impact of a strategy focused on local communities can also be seen in the Communities and Local Government New Deal for Communities (NDC) initiative, a 10 year safer neighbourhoods strategy initiated in 1998 that proposed to focus on the most deprived neighbourhoods in England and Wales (Batty et al. 2010c). This had a broad remit focusing on strengthening community, improving housing and the physical environment, raising education levels, improving health, and reducing worklessness (Batty et al. 2010c). As such, crime reduction and public confidence formed part of a larger programme of neighbourhood renewal work, with the aim of recognising the importance of local community interactions (Parkinson et al. 2006).

A total of 39 NDC areas (ten in London, and the remainder spread across England) were selected that were suffering high levels of deprivation. Each received a range of tailored neighbourhood interventions, with some designed specifically to address local crime issues and reduce fear of crime. This included the introduction of specific target hardening measures in problem areas, improvements to security for homes and businesses, improvements to community facilities, the creation of 'boundary markers', and various landscaping activities (Batty et al. 2010a). This also included specific initiatives to respond directly to low level signs of disorder within the local area. NDC areas also invested heavily in neighbourhood wardens (with a similar remit to PCSOs) to promote community safety, as well as placing a further emphasis on closer links between community, the police, and other local service providers (Batty et al. 2010a). This included close working with the police to identify crime 'hotspots' for targeted intervention, and focused work on reducing young peoples' involvement in anti-social behaviour. Here, the intention was to foster stronger bonds of informal control amongst local residents, enhancing their ability to intervene to reduce crime and disorder.

As with the neighbourhood policing strategy, the NDC has seen some clear successes. The main inroads have been made in the public confidence sphere, with large reductions in fear of crime and higher levels of public confidence in the police. Survey evidence also points to moderate success at reducing levels of crime, with NDC areas experiencing significantly lower levels of criminal damage and overall victimisation than comparison neighbourhoods selected with similar deprivation profiles (Batty et al. 2010a). However, when compared against the national average NDC areas typically improved at a slower rate, highlighting remaining inequalities between local areas. Evaluation of the NDC areas also points to evidence of a 'diffusion of benefits' from the specific neighbourhoods included in the initiative to surrounding local areas as policing and community safety strategies 'spill-over' neighbourhood boundaries, demonstrating how local neighbourhood areas within close proximity can be inextricably linked with one another.

However, despite improvements being made to community cohesion across the duration of the initiative, these improvements were mirrored in other local areas not part of the NDC scheme. Batty and colleagues (2010b) report relatively low levels of participation from local residents in specific NDC activities, with less than half of residents involved at any time, and considerably fewer regularly participating. Therefore, whilst improvements are evident to a range of outcomes, there is no clear evidence of an increasing role for community cohesion.

The other principle area that crime reduction policy has been influenced by the potential existence of neighbourhood effects has been in the development of Crime and Disorder Reduction Partnerships (CDRP). Established via the crime and disorder act of 1998, these are multiagency groups operating within local areas that hold a core objective of reducing crime and fear of crime (Hughes and Rowe 2007). Although primarily designed to co-ordinate crime and disorder reduction strategies at a broader spatial scale, implicit in their development was the same recognition of the importance of a locally specific focus that been at the heart of research on neighbourhood effects on crime and perceptions of crime. To respond effectively to local problems CDRPs draw on evidence about crime and perceptions of crime at the local level, with a focus on partnership working between various agencies including the emergency services, local authorities, and voluntary organisations. This includes crime mapping to identify problem areas or crime hotspots, surveys of local communities, and direct consultation with local residents, providing a clear evidence base to tailor crime reduction strategies (Hough and Tilley 1998). This also directly acknowledges the variations in crime and perceptions that exist between local neighbourhoods, with CDRP encouraged to direct resources at those areas representing the highest risk.

## Final Considerations

An expanding body of research demonstrating clear spatial patterning of crime and public reactions to crime has now accumulated in criminology. This, in turn, has prompted the development of a range of mechanisms intended to explain these patterns. Most promisingly, research is now beginning to explore in detail the links between individual, psychological, explanations, and these broader ecological explanations (Sampson et al. 2002; Wikström and Sampson 2003; Jackson et al. 2012). This recognition of the interaction between individuals and their communities is a crucial step forward, with developments in multilevel modelling enabling researchers to empirically test these associations.

Yet there remain a number of methodological challenges, which combine to limit the impact of this body of work. To our minds, one important stumbling block is the relative lack of attention that has been given to adequate conceptions of neighbourhood. Many studies rely on administrative boundaries, treating neighbourhoods simply as an empirical unit to identify clusters of individuals. In the previous demonstration we have aimed to provide a more detailed conception of neighbourhood that simultaneously stays within the confines of readily available administrative data sources, using spatial autocorrelation models to bridge the gap between community studies and empirical treatments of neighbourhood effects. This recognises that each local resident may have multiple neighbourhood definitions that inform their daily activities, and that, as a result may be looking beyond their immediate surroundings when forming judgments of the area.

Initial empirical findings looking specifically at fear of crime confirm that individuals are influenced by more than just their immediate surroundings, with the crime and disorder profile of the broader area also important. Despite including a range of neighbourhood measures, this also revealed the existence of considerable remaining variability between local neighbourhoods (and resulting from similarities between neighbourhoods in close proximity to one another), pointing to the existence of further important neighbourhood effects. These likely include the sorts of community level processes not easily captured by available administrative data. The use of ecometrics – neighbourhood measures constructed from individual survey data – may provide a way forward in filling this conceptual gap, with considerable progress now being made in improving the estimation of these contextual effects (Raudenbush and Sampson 1999).

But focusing solely on geographical units has its limitations, no matter how small these are, and how closely they map on to individual conceptions of their own neighbourhood. Questions still remain over the extent that these can successfully capture the lived realities of individual residents and the complex of different communities that individuals belong to. For example, Wikström (2002) found that 14–15 year olds in Peterborough spent only 20 % of their waking hours in their own 'neighbourhood', with 28 % in school, 14 % outside of the neighbourhood (and not in school), and 38 % at home suggesting that the exposure and impact of single physical environments may vary considerably between individuals. What may be more pertinent for understanding the relationship between environment and behaviour is gaining a better understanding of which people use spaces where and when. This type of work has been the focus of time-use studies for some years (e.g. Gershuny and Sullivan 1998), and presents a potentially illuminating way forward for future studies. For example, Wikström and colleagues (2011) have recently developed a space-time-budget (STB) methodology that captures both the where, when, and with whom. This moves us away from relying solely on neighbourhoods as the focus of analysis in the conventional sense because one needs to capture all the environments that individuals interact with, not just where they live.

The idea that neighbourhood effects influence crime levels and perceptions of crime can be seen in a range of crime prevention initiatives over the last quarter of a century, including the establishment of neighbourhood policing teams and the widespread use of Police Community Support Officers to act as an informal link between communities and law enforcement agencies. Neighbourhood regeneration schemes have also been implemented, including the NDC initiative, which have focused specifically on those neighbourhoods identified as the most at risk of disadvantage and crime. These initiatives have been informed, at least in part, by the theoretical developments from neighbourhood effects studies, with an emphasis on neighbourhood cohesion, and a recognition of the need to tackle low level problems in local areas that are of most significance to residents, as well as strategies that have focused specifically on improving the built environment.

However, the extent to which these initiatives have been successful is variable, with some consistent evidence that they have served to reduce fear of crime and raise public confidence, but less clear evidence that they have had a parallel impact on crime rates. Of most difficulty seems to be the promotion of community cohesion to strengthen the levels of informal social control available to communities to respond directly to crime and disorder. Practitioners also need to be aware of the effect of the wider geographical area beyond the immediate neighbourhood. Attributes of adjacent localities can have just as much of an effect as immediate social and physical conditions. Therefore tackling crime and disorder hot-spots may not just have an influence locally, but also in neighbouring areas. Conversely, policies focused at the neighbourhood level should also consider the broader context in which neighbourhoods are situated, with strategies designed to incorporate the different spheres of influence on resident's daily lives.

# References

Adams, R., & Serpe, R. (2000). Social integration, fear of crime, and life satisfaction. *Sociological Perspectives, 43*, 605–629.

Baldwin, J., & Bottoms, A. E. (1976). *The urban criminal*. London: Tavistock.

Batty, E., Beatty, C., Foden, M., Lawless, P., Pearson, S., & Wilson, I. (2010a). Involving local people in regeneration: Evidence from the New Deal for Communities Programme. *The New Deal for Communities National Evaluation: Final report* (Vol. 2). Department for Communities and Local Government.

Batty, E., Beatty, C., Foden, M., Lawless, P., Pearson, S., & Wilson, I. (2010b). Making deprived areas better places to live: Evidence from the New Deal for Communities Programme. *The New Deal for Communities National Evaluation: Final report* (Vol. 3). Department for Communities and Local Government.

Batty, E., Beatty, C., Foden, M., Lawless, P., Pearson, S., & Wilson, I. (2010c). The New Deal for Communities Experience: A final assessment. *The New Deal for Communities Evaluation: Final report* (Vol. 7). Department for Communities and Local Government.

Brantingham, P. J., & Brantingham, P. L. (1981). *Environmental criminology*. Beverly Hills: Sage.

Brunton-Smith, I., & Jackson, J. (2012). Urban fear and its roots in place. In V. Ceccato (Ed.), *Urban fabric of crime and fear* (pp. 55–82). Dordrecht: Springer.

Brunton-Smith, I., & Sturgis, P. (2011). Do neighbourhoods generate fear of crime? An empirical test using the British Crime Survey. *Criminology, 49*(2), 331–369.

Bullock, K., Clarke, R., & Tilley, N. (Eds.). (2010). *Situational prevention of organised crime*. Devon: Willan.

Bursik, R. J. (1988). Social disorganisation and theories of crime and delinquency: Problems and prospects. *Criminology, 26*(4), 519–551.

Bursik, R. J., & Grasmick, H. G. (1993). *Neighbourhoods and crime: The dimensions of effective community control*. New York: Lexington Books.

Carr, P. (2005). *Clean streets: Controlling crime, maintaining order and building community activism*. New York: New York University Press.

Chaskin, R. (1998). *Defining neighbourhoods*. Paper presented at American Planning Association Casey Symposium on Indicators. Chicago, Illinois.

Covington, J., & Taylor, R. B. (1991). Fear of crime in urban residential neighbourhoods: Implications of between- and within-neighbourhood sources for current models. *The Sociological Quarterly, 32*(2), 231–249.

Crawford, A., & Lister, S. (2004). *The extended policing family: Visible patrols in residential areas*. York: Joseph Rowntree Foundation.

Dennis, N. (1997). *Zero tolerance: Policing a free society*. London: Institute of Economic Affairs.

Farrall, S., Jackson, J., & Gray, E. (2009). *Social order and the fear of crime in contemporary times*. Oxford: Oxford University Press, Clarendon Studies in Criminology.

Ferraro, K. F. (1995). *Fear of crime: Interpreting victimization risk*. Albany: State University of New York Press.

Fielding, N., & Innes, M. (2006). Reassurance policing, community policing and measuring police performance. *Policing and Society, 16*(2), 127–145.

Flatley, J., Kershaw, C., Smith, K., Chaplin, R., & Moon, D. (2010). Crime in England and Wales 2009/10: Findings from the British Crime Survey and Police Recorded Crime (3rd ed.). *Home Office Statistical Bulletin*. London: Home Office.

Gershuny, J., & Sullivan, O. (1998). The sociological uses of time-use diary analysis. *European Sociological Review, 14*(1), 69–85.

Gill, M., & Spriggs, A. (2005). *Assessing the impact of CCTV* (Rep. No. 292). London: Home Office Research, Development and Statistics Directorate.

Girling, E., Loader, I., & Sparks, R. (2000). *Crime and social order in Middle England: Questions of order in an English town*. London: Routledge.

Hale, C. (1996). Fear of crime: A review of the literature. *International Review of Victimology, 4*(2), 79–150.

Harcourt, B. (2001). *Illusion of order: The false promise of broken windows policing*. Cambridge: Harvard University Press.

Herbert, S. (2005). *Citizens, cops and power*. Chicago: University of Chicago Press.

Hipp, J. R. (2007). Block, tract, and levels of aggregation: Neighborhood structure and crime and disorder as a case in point. *American Sociological Review, 72*(5), 659–680.

Home Office. (2005). *Neighbourhood policing: Your police, your community, Our commitment*. London: Home Office.

Hough, M., & Tilley, N. (1998). *Auditing crime and disorder: Guidance for local partnerships. Crime detection and prevention series: Paper 91*. London: Home Office Police Research Group.

Hughes, G., & Rowe, M. (2007). Neighbourhood policing and community safety: Researching the instabilities of the local governance of crime, disorder and security in contemporary UK. *Criminology and Criminal Justice, 7*(4), 317–346.

Innes, M. (2004). Signal crimes and signal disorders: Notes on deviance as communicative action. *The British Journal of Sociology, 55*(3), 335–355.

Innes, M., & Roberts, C. (2008). Reassurance policing, community intelligence and the co-production of neighbourhood order. In T. Williamson (Ed.), *The handbook of knowledge based policing*. Chichester: Wiley.

Innes, M., Hayden, S., Lowe, T., Mackenzie, H., Roberts, C., & Twyman, L. (2004). *Signal crimes and reassurance policing volumes 1 and 2*. Guildford: University of Surrey.

Innes, M., Hayden, S., Lowe, T., & Roberts, C. (2005). *Signal crimes and reassurance policing volume 3*. Guildford: University of Surrey.

Jackson, J. (2004). Experience and expression: Social and cultural significance in the fear of crime. *British Journal of Criminology, 44*(6), 946–966.

Jackson, J., Bradford, B., Stanko, E. A., & Hohl, K. (2012). *Just authority? Trust in the police in England and Wales*. Oxon: Routledge.

Johnston, C., & Mooney, G. (2007). 'Problem' people, 'problem' places? New labour and council estates. In R. Atkinson & G. Helms (Eds.), *Securing an urban rennaisance: Crime, community and British urban policy*. Bristol: The Policy Press.

Keizer, K., Lindenberg, S., & Steg, L. (2008). The spreading of disorder. *Science, 322*(5908), 1681–1685.

Keizer, K., Lindenberg, S., & Steg, L. (2011). The reversal effect of prohibition signs. *Group Processes and Intergroup Relations, 14*(5), 681–688.

Kling, J. B., Liebman, J. B., Katz, L. F., & Sanbonmatsu, L. (2004). *Moving to opportunity and tranquillity: Neighbourhood effects on adult economic self-sufficiency and health from a randomized housing voucher experiment Princeton*. Princeton: Princeton University.

Krannich, R. S., Helen Berry, E., & Greider, T. (1989). Fear of crime in rapidly changing rural communities: A longitudinal analysis. *Rural Sociology, 54*(2), 195–212.

Krivo, L. J., Peterson, R. D., & Kuhl, D. C. (2009). Seggregation, racial structure, and neigborhood violent crime. *The American Journal of Sociology, 114*(6), 1765–1802.

Laycock, G., & Tilley, N. (1995). *Policing and neighbourhood watch: Strategic issues. Police research group. Crime detection and prevention series: Paper 60*. London: Home Office Police Department.

Lupton, R. (2003). *Neighbourhood effects: Can we measure them and does it matter?* (Rep. No. 73). London: London School of Economics.

MacDonald, J. M., Hipp, J. R., & Gill, C. (2012). The effects of immigrant concentration on changes in neighborhood crime rates. *Journal of Quantitative Criminology*. Online first, 2 June 2012.

Manley, D., Flowerdew, R., & Steel, D. (2006). Scales, levels and processes: Studying spatial patterns of British census variables. *Computers, Environment and Urban Systems, 30*(2), 143–160.

Mason, M. (2009). *Findings from the second year of the neighbourhood policing programme evaluation* (Home Office Research Report 14). London: Home Office.

Mayer, S. E., & Jencks, C. (1989). Growing up in poor neighbourhoods: How much does it matter? *Science, 243*(4897), 1441–1445.

Merry, S. E. (1981a). *Urban danger: Life in a neighbourhood of strangers*. Philadelphia: Temple University Press.

Merry, S. E. (1981b). Defensible space undefended: Social factors in crime control through environmental design. *Urban Affairs Review, 16*(4), 397–422.

Millie, A. (2007). Tackling anti-social behaviour and regenerating neighbourhoods. In R. Atkinson & G. Helms (Eds.), *Securing an urban rennaisance: Crime, community and British urban policy*. Bristol: The Policy Press.

Millie, A. (2010). Whatever happened to reassurance policing? *Policing: A Journal of Policy and Practice, 4*(3), 225–232.

Morenoff, J. D., Sampson, R. J., & Raudenbush, S. W. (2001). Neighborhood inequality, collective efficacy, and the spatial dynamics of urban violence. *Criminology, 39*(3), 517–560.

Morris, J. (2006). *The national reassurance policing programme: A ten site evaluation*. London: Home Office. Findings 273.

Neighbourhood Renewal Unit. (2004). *Neighbourhood warden scheme evaluation* (Research Report 8). Wetherby: Office of the Deputy Prime Minister.

Newman, O. (1978). *Defensible space: Crime prevention through urban design*. New York: McMillan.

Newman, O., & Franck, K. A. (1982). The effects of building size on personal crime and fear of crime. *Population and Environment, 5*(4), 203–220.

Oakes, J. M. (2004). The (Mis)estimation of neighbourhood effects: Causal inference for practicable social epidemiology. *Social Science & Medicine, 58*(10), 1929–1952.

Oberwittler, D., & Wikström, P. O. H. (2009). Why small is better. Advancing the study of the role of behavioural contexts in crime causation. In D. Weisburd, W. Bernasco, & G. Bruinsma (Eds.), *Utting crime in its place: Units of analysis in spatial crime research* (pp. 35–59). New York: Springer.

Park, R., & Burgess, E. (1924). *Introduction to the science of sociology*. Chicago: University of Chicago Press.

Parkinson, M., Champion, T., Turok, I., Crookston, M., Davies, L., Katz, Y. B., Park, A., Berube, A., Coombes, M., Dorling, D., Evans, R., Glass, N., Hutchins, M., Kearns, A., Martin, R., & Wood, P. (2006). *State of English Cities: A research study* (Vol. 1). London: Office of the Deputy Prime Minister.

Paskell, C. (2007). Community police relations: Support officers in low-income neighbourhoods. In R. Atkinson & G. Helms (Eds.), *Securing an urban rennaisance: Crime, community and British urban policy*. Bristol: The Policy Press.

Putnam, R. D. (2000). *Bowling alone – The collapse and revival of American community*. New York: Simon and Schuster.

Putnam, R. D. (2007). E pluribus unum: Diversity and community in the twenty-first century. The 2006 Johan Skytte Prize Lecture. *Scandinavian Political Studies, 30*(2), 137–174.

Quinton, P., & Tuffin, R. (2007). Neighbourhood change: The impact of the national reassurance policing programme. *Policing, 1*(2), 149–160.

Raudenbush, S. W., & Sampson, R. J. (1999). 'Ecometrics': Toward a science of assessing ecological settings, with application to the systematic social observation of neighborhoods. *Sociological Methodology, 29*(1), 1–41.

Rock, P. (2007). Sociological theories of crime. In M. Maguire, R. Morgan, & R. Reiner (Eds.), *The oxford handbook of criminology* (4th ed., pp. 3–42). Oxford: Oxford University Press.

Sampson, R. J. (2012). *Great American City: Chicago and the enduring neighbourhood effect*. Chicago/London: The University of Chicago Press.

Sampson, R. J., & Groves, W. B. (1989). Community structure and crime: Testing social disorganisation theory. *The American Journal of Sociology, 94*(4), 774–802.

Sampson, R. J., & Raudenbush, S. W. (1999). Systematic social observation of public spaces: A new look at disorder in urban neighborhoods. *The American Journal of Sociology, 105*(3), 603–651.

Sampson, R. J., Raudenbush, S. W., & Earls, F. (1997). Neighbourhoods and violent crime: A multilevel study of collective efficacy. *Science, 277*(5328), 918–924.

Sampson, R. J., Morenoff, J. D., & Earls, F. (1999). Beyond social capital: Spatial dynamics of collective efficacy for children. *American Sociological Review, 64*(5), 633–660.

Sampson, R. J., Morenoff, J. D., & Gannon-Rowley, T. (2002). Assessing "neighborhood effects": Social processes and new directions in research. *Annual Review of Sociology, 28*, 443–478.

Schweitzer, J. H., Woo Kim, J., & Mackin, J. R. (1999). The impact of the built environment on crime and fear of crime in urban neighbourhoods. *Journal of Urban Technology, 6*(3), 59–73.

Scribbins, M., Flatley, M., Parfrement-Hopkins, J., & Hall, P. (2011). Public perceptions of policing, engagement with the police and victimisation: Findings from the 2009/10 British crime survey. *Home Office Statistical Bulletin, 19/10*, 1–61.

Shaw, C. R., & McKay, H. D. (1942). *Juvenile delinquency and urban areas*. Chicago: University of Chicago Press.

Singer, L. (2004). *Reassurance policing: An evaluation of the local management of community safety* (Home Office Research Study, 288). London: Home Office.

Skogan, W. (1986). Fear of crime and neighbourhood change. In A. S. Reiss Jr & M. Tonry (Eds.), *Communities and crime* (pp. 203–229). Chicago: University of Chicago Press.

Skogan, W. (1990). *Disorder and decline: Crime and the spiral of decay in American neighborhoods*. Berkeley: University of California Press.

Skogan, W. (2003). *Community policing: Can it work?* Oxford: Wadsworth Publishing.

Skogan, W., & Maxfield, M. G. (1981). *Coping with crime: Individual and neighborhood reactions*. Beverly Hills: Sage.

Smith, M., & Tilley, N. (Eds.). (2005). *Crime science: New approaches to preventing and detecting crime*. London: Willan Publishing.

Taylor, R. B. (2001). *Breaking away from broken windows: Baltimore neighbourhoods and the Nationwide fight against crime, grime, fear, and decline*. Boulder: Westview Press.

Taylor, R. B., & Covington, J. (1993). Community structural change and fear of crime. *Social Problems, 40*(3), 374–397.

Taylor, R. B., Gottfredson, S. D., & Brower, S. (1984). Block crime and fear: Defensible space, local social ties and territorial functioning. *Journal of Research in Crime and Delinquency, 21*(4), 303–331.

Thrasher, F. (1927). *The gang*. Chicago: University of Chicago Press.

van Ham, M., Manley, D., Bailey, N., Simpson, L., & Maclennan, D. (2012). In M. van Ham, D. Manley, N. Bailey, L. Simpson, & D. Maclennan (Eds.), *Neighbourhood effects research: New perspectives* (pp. 1–22). Dordrecht: Springer.

van Ham, M., Manley, D., Bailey, N., Simpson, L., & Maclennan, D. (2013). Understanding neighbourhood dynamics: New insights for neighbourhood effects research. In M. van Ham, D. Manley, N. Bailey, L. Simpson, & D. Maclennan (Eds.), *Understanding neighbourhood dynamics: New insights for neighbourhood effects research* (pp. 1–22). Dordrecht: Springer.

Vrij, A., & Winkel, F. W. (1991). Characteristics of the built environment and fear of crime: A research note on interventions in unsafe locations. *Deviant Behaviour, 12*(2), 203–215.

Weisburd, D., Bruinsma, G., & Bernasco, W. (2009). Units of analysis in geographic criminology: Historical development, critical issues, and open questions. In D. Weisburd, W. Bernasco, & G. Bruinsma (Eds.), *Putting crime in its place: Units of analysis in spatial crime research* (pp. 35–59). New York: Springer.

Wikström, P.-O. H. (1991). *Urban crime, criminals and victims*. New York: Springer.

Wikström, P.-O. H. (2002). *Adolescent crime in context* (Report to the Home Office). Cambridge: Institute of Criminology.

Wikström, P.-O. H., & Sampson, R. J. (2003). Social mechanisms of community influences on crime and pathways in criminality. In B. G. Lahey, T. E. Moffitt, & A. Caspi (Eds.), *Causes of conduct disorder and juvenile delinquency*. New York: The Guildford Press.

Wikström, P.-O. H., Treiber, K., & Hardie, B. (2011). Examining the role of the environment in crime causation: small area community surveys and space-time budgets. In D. Gadd, S. Karstedt, & S. F. Messner (Eds.), *The sage handbook of criminological research methods*. London: Sage.

Wilcox-Rountree, P., & Land, K. C. (2000). The generalizability of multilevel models of burglary victimization: a cross-city comparison. *Social Science Research, 29*(2), 284–305.

Wilson, J. Q., & Kelling, G. L. (1982). Broken windows. *Atlantic Monthly, 249*(3), 29–38.

Woldoff, R. A. (2006). Emphasizing fear of crime in models of neighbourhood social disorganisation. *Crime Prevention and Community Safety, 8*, 228–247.

Chapter 5
# An Environmental Justice Framework for Understanding Neighbourhood Inequalities in Health and Well-Being

Jamie Pearce

## Introduction

Neighbourhood based explanations for inequalities in health and well-being have recently had considerable traction in academic and policy circles. Place-based determinants of health including the local urban infrastructure, features of the physical environment, and community social capital have been viewed as an intuitive explanation for the stark inequalities across neighbourhoods only short distances apart. It has been posited that in addition to who you are, where you live, work and play matters for your health, and that 'place' explains a component of the socio-spatial arrangement in health documented in many countries. Researchers have devoted considerable energy towards distinguishing 'contextual' from compositional' accounts, often relying on statistical modelling to partition variance at difference levels (e.g. individuals, households and neighbourhoods that are fixed in space). It is argued that place exerts an influence on a range of health outcomes (e.g. mortality, cancer incidence) and related behaviours (e.g. smoking, nutrition and alcohol consumption).

Whilst this line of investigation has been instructive in developing socio-ecological explanations of health and behaviours (i.e. that health is affected by a multitude of factors operating in different contexts and at various levels), it is increasingly recognised that it has provided only a partial account for the geographical inequalities in health. Rather, as has long been recognised in human geography and elsewhere, neighbourhoods are fluid, non-bounded and their makeup partially reflect broader macro-level social and economic processes that have accumulated over many years and decades. As Wacquant (2008) notes:

---

J. Pearce (✉)
Centre for Research on Environment, Society and Health (CRESH),
Institute of Geography, School of Geosciences, The University of Edinburgh,
Drummond Street, Edinburgh EH8 9XP, UK
e-mail: jamie.pearce@ed.ac.uk

> To forget that urban space is a *historical and political construction* in the strong sense of the term is to risk (mis)taking for 'neighbourhood effects' what is nothing more than the spatial retranslation of economic and social differences (p. 9)

Yet few studies have attempted to develop a coherent picture of the multi-scalar processes operating in and through neighbourhoods to track the historical development of places and then consider the implications for health and well-being of people who occupy these spaces. Similarly, there has been little work that has shed light on the ways in which properties of local neighbourhoods can mediate the associations between health and the corresponding social, political and environmental determinants. These omissions have been an important impediment to developing robust accounts for the socio-spatial patterning of health across neighbourhoods and perhaps why geographical inequalities in health continue to rise.

This chapter considers the ways in which the adoption of an 'environmental justice' framework might contribute to work in the field of spatial health inequalities. From an academic perspective, environmental justice is concerned with how environmental and social differences are interconnected. For some socio-demographic groups and individuals the environment is a positive element of their well-being, whereas for others the environment is a risky place with a deficiency the availability of beneficial attributes (Walker 2011). The (mal)distribution of environmental 'goods' and 'bads', the implications of this arrangement, as well as the social, economic and political structures explaining environmental injustices have received attention. Here it is contended that by drawing on the theoretical arguments developed by environmental justice scholars, new insights into the socio-spatial patterning of health and well-being are likely. An environmental justice framing also offers a radical departure for those working in the field of spatial inequalities in health from the current vogue for identifying 'neighbourhood effects'. This alternative approach offers novel opportunities including identifying the unequal availability of environmental pathogens and resources, the social and political processes underlying this arrangement including historical accounts, and the implications for health and well-being of the unjust organisation of environmental goods.

The chapter is arranged in four substantive sections. First, the historical documentation of health inequalities at the local level, and a brief assessment of the posited explanations are provided. In the second section, discussion focuses on the ways in which an environmental justice framework has been incorporated into work on health inequalities. Importantly, the commentary includes a consideration of the restricted application of this framing which has to date hindered insights into environmental factors and their social, political and economic antecedents in understanding inequalities in health. Some broad suggestions for future research in the field of environmental justice and health inequalities are offered in the third part of this chapter before a final section draws some conclusions.

## Neighbourhoods and Health Inequalities

Palpable differences in health outcomes and experiences across neighbourhoods have been recognised and documented for at least 200 years (Chadwick 1843; Pearce and Dorling 2009). Socially and economically disadvantaged neighbourhoods routinely

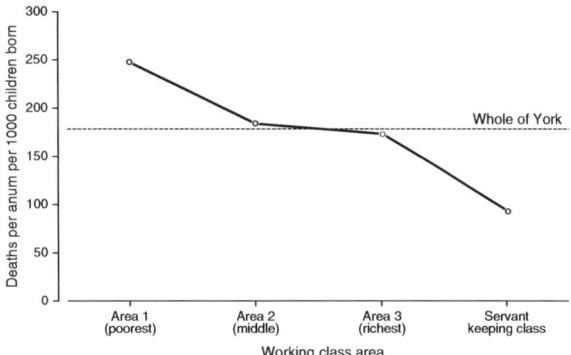

Fig. 5.1 Infant mortality rates (under 12 months) in York in 1898 (Adapted from data in: Davey Smith et al. (2001))

have relatively worse health than more advantaged places. For example, in England during the mid-19th century the pioneering work of Edwin Chadwick, Friedrich Engels, and others revealed the strength of the health gradient across neighbourhoods of newly industrialised cities (Davey Smith et al. 2002). In his 1899 survey of poverty in York, Benjamin Seebohm Rowntree documented a firm social gradient in infant mortality across three working-class districts of the city that were arranged by occupation type and income (Rowntree 1901). There was a gradient in infant mortality across area type with the worst health outcomes in the poorest neighbourhood (see Fig. 5.1). The infant mortality rate in this area of the city was over 2.5 times that of the 'servant-keeping' classes. For the first time, these studies unambiguously disclosed that health was causally linked to socio-economic position, which itself was rooted in the social structures around which society was constructed.

More than 100 years later, health inequalities in England and most other countries remain ubiquitous. There is for example irrefutable evidence that health varies markedly across small distances such as between neighbourhoods in the same city. For example a recent World Health Organization (WHO) report noted 28-year difference in life expectancy between two Glaswegian children living only a few suburbs apart (WHO Commission on Social Determinants of Health 2008). Crucially, disparities in health are not restricted to dichotomous gaps (e.g. poorest neighbourhood versus the rest) but are evidenced across the social spectrum. For instance, when neighbourhoods are stratified by measures of poverty or socio-economic status, health incrementally improves from the least to most advantaged area. Whilst the immediate postwar period (until the 1970s) was characterised by less social differentials in health, from the 1980s health inequalities have risen rapidly. Further, since the 1970s, relative inequalities in health across areas have tended to widen, and in some countries this increase has been rapid (Shaw et al. 2005). Over the past 30 years in nation-states such as the United Kingdom, the United States, and New Zealand, geographical inequalities in life expectancy have sharply increased by between 50 and 60 % (Pearce and Dorling 2006; Shaw et al. 2005; Singh and Siahpush 2006). Remarkably, in the UK at least, small area mortality measures from a century ago are strong predictors of the contemporary geography of mortality (Dorling et al. 2000) which suggests that area-level poor health stubbornly persists.

## *What Explains Geographical Inequalities in Health?*

Whilst researchers have become adept at monitoring and describing inequalities in health, explanations for this changing socio-spatial profile remains elusive. Nonetheless, it is almost certainly no coincidence that the increase in health inequalities since the 1970s has coincided with the adoption and implementation of a market-oriented neoliberal political agenda. This approach, which has been enthusiastically adopted in many countries, has privileged the needs of the market in determining the political and economic priorities of the state. The associated deregulation of the labour market, the retrenchment of welfare state including constraints on social security have resulted in widening inequalities in social position. Even in a nation-state such as New Zealand which has previously been considered relatively egalitarian there has been a discernible shift in governmental support away from the welfare state. Successive New Zealand governments have adopted a more market-oriented policy agenda, which has eroded the long accepted assumptions of a universal and freely accessible public health system. This is despite New Zealand's tradition of progressive social policy which included being the first nation in the world to implement a universal healthcare system and lay the foundations for a welfare state following the enactment of the Social Security Act in 1938. Since the landmark election of the radical 1984 administration, successive governments (of various political persuasions) have enthusiastically placed market needs at the centre of policy prioritisation. These changes have led some commentators to claim that the reforms in New Zealand resulted in social and economic changes that were more rapid than in any other nation. Recent evidence suggests that by the end of the twentieth century, socioeconomic inequalities in health in New Zealand had reached extremely high levels by OECD standards (O'Dea and Howden-Chapman 2000).

The implications of this major shift toward market-oriented policy agenda in New Zealand and elsewhere has been profound. National-level income inequality, a marker of social stratification, is a key driver of a nation's health and well-being (Dorling et al. 2007). For example, among richer nations, countries that have maintained high levels of income inequality in recent years have the highest prevalence of a host of measures of poor health including mental illness (Wilkinson and Pickett 2007). As commentators such as Richard Wilkinson, Kate Pickett and Danny Dorling have argued, the uneven allocation of resources in non-egalitarian societies is not only disadvantageous for more socially deprived groups, but is also harmful to more affluent groups (Dorling 2010; Wilkinson and Pickett 2010). Income inequality, a marker of social stratification, is also a key causal factor in explaining health and well-being at the national-level (Dorling et al. 2007). Hence, the macro-level factors that establish and maintain global and national social and economic inequality are harmful to all groups across the social spectrum. Further, from a geographical perspective, given what has been garnered from the development of the 'social determinants' and 'socio-ecological' models of health and well-being, it is unsurprising that greater social and economic inequality is equated with a divergence in health outcomes within society. In tandem with the increasingly unequal power relations in society, rising social and economic inequalities are likely to affect the socio-spatial arrangement of environmental and neighbourhoods resources. Disempowered and increasingly marginalised communities operating in a neoliberalised setting are less able to affect decision making and offset the potent forces of the vested interests among powerful groups. More socially advantaged communities are hence better positioned to advocate for health promoting neighbourhood resources (e.g. health care provision or green spaces) and are also less likely to contest the positioning of health damaging disamenities (e.g. busy roads or polluting facilities). Predictably, geographical, social and ethnic inequalities in health have risen accordingly.

## *'Strange Geographies' of Neighbourhoods and Health*

Given that neighbourhood differences in health are well documented, it is not surprising that many researchers have adopted the seemingly common sense notion that differences in health related resources across residential neighbourhoods are likely influence the health status of local residents. The premise of research here is that where you live matters for your health as well as who you are. Often using datasets on individuals combined with information about their neighbourhoods, and statistical methods such as multilevel modelling, researchers have sought to partition variation at the individual and neighbourhood levels. This has enabled the identification of 'contextual effects' that operate independently of the composition of the resident population. As we shall see, this conceptualisation is proving problematic.

A number of pathways linking neighbourhoods to the health of local residents have been posited. Commonly assessed ecological attributes include physical (e.g. air pollution or aspects of the built environment green space), social (e.g. social capital) and cultural constructs (e.g. local perceptions of crime). A multitude of health outcomes, behaviours and experiences with biologically-plausible associations have been studied. Diez Roux (2003) for example develops a conceptual model for exploring the role of residential environments in explaining cardiovascular risk. It is suggested that the residential environment can influence cardiovascular disease through proximate biological factors such as blood pressure, body mass index and blood lipds. As Fig. 5.2 depicts, these proximate factors are in turn influenced by features of the physical and social environments. The physical environment might exert an effect through aspects of the urban infrastructure (e.g. the availability of food stores, cycle lanes and street connectivity which in turn can affect key factors in the aetiology of cardiovascular disease) or through ambient concerns such as air or noise pollution. The social environment may be pressing because place-based social norms, social support and cohesion as well as safety and violence might be considered to influence psychosocial, stress and behavioural factors.

However, recent commentaries from health geographers and others have expounded that such conceptualisations of how neighbourhoods get 'under the skin' to affect health and well-being are restricted on a number of grounds. First, it is

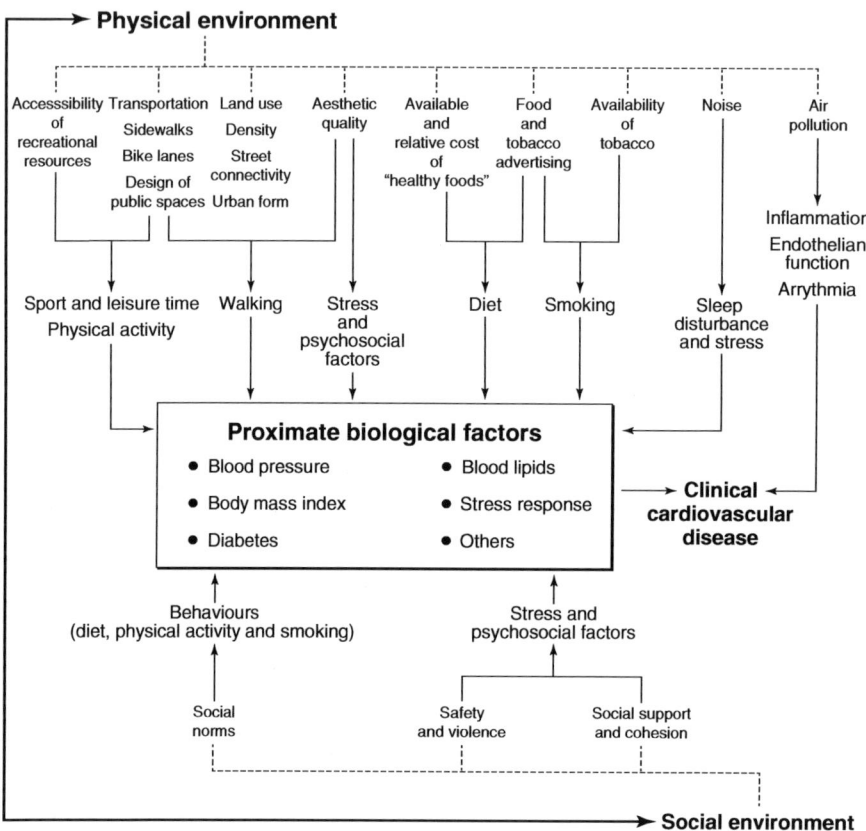

**Fig. 5.2** Schematic representation of possible pathways linking residential environments to cardiovascular risk (*Source*: Diez Roux 2003, p. 572)

argued that the distinction between context and composition is an insufficient conceptualisation of the relationships between health and place; contextual (relating to the places in which people live, work and play) and compositional (attributes of the individuals residing in a particular place) factors are rarely mutually exclusive (Macintyre et al. 2002). For example, unemployment is commonly conceived as a compositional attribute, yet employment status cannot be separated from a range of contextual factors including local, regional and global markets and economies which are likely to be considered a contextual construct. Further, contextual factors are rarely articulated around theoretical considerations but rather utilise readily available off-the-shelf measures such as census-based deprivation indices. Smith and Easterlow (2005) also note this false dualism but go further to suggest that geographical accounts of health inequalities have become 'locked into context' or what they describe as a 'strange geographies'. These authors advocate for more theoretically informed compositional accounts of spatial inequalities in health. This limitation has resulted in a partial account of how a suite of place-based factors are significant in understanding geographical inequalities in health. A second criticism is that many conceptualisations of neighbourhood processes adopted in the literature rely on static and often tightly bounded notions of space that are artificially constrained by administrative boundaries (e.g. census units) and that this 'local trap' fails to account for the different geographical settings in which individuals with varying health trajectories live their lives (Cummins 2007). Crucially, the multi-scalar nature of place-based factors (from the global to local) are under explored in health research. Useful exceptions to this assertion include Schulz and colleagues' conceptual model for understanding cardiovascular disease which include 'fundamental', 'intermediate' and 'proximate' geographical constructs potentially explain disparities in this health concern (Schulz et al. 2005). The implicated factors range from residential segregation at the 'fundamental' level to social support at the 'proximate' level. Related, Pearce and Witten (2010) develop a multi-scalar account of the emerging obesity 'epidemic'. This approach recognises that health outcomes such as obesity are affected by a complex web of environmental factors ranging from broad societal shifts at the national or global level, to various processes and practices that can be viewed within residential neighbourhoods. Third, missing also is a longitudinal perspective that considers the accumulation of environmental risk over the life course. Finally, surprisingly little attention has been provided to evaluating the extent to which neighbourhood factors can operate in different ways for diverse socio-demographic groups. Hence the role of neighbourhoods as mediators of health inequalities deserves further investigation.

In this Chapter, the utility of an 'environmental justice' framework for understanding the role of neighbourhoods in affecting health is examined. It is argued that although there is clear evidence that some (often socially advantaged) neighbourhoods accumulate health-promoting (and indeed other beneficial) resources, the attribution of neighbourhood effects on health may be misguided. The central claim here is that 'neighbourhood effects' are an unsatisfactory conceptualisation of geographical health inequalities, and this notion implies that deeply entrenched social concerns with complex multiscalar explanations are locally-situated quandaries that belong to the local community and can be addressed simply through local-level interventions such as modifications to the local infrastructure or urban design initiatives affected through planning decisions. There is a danger that assigning social problems to neighbourhoods is akin to the 'victim blaming' ethos that has dominated political discourses around health and healthcare. Whilst places are important for understanding health, behaviours and practices, it has to be recognised that neighbourhood are largely a manifestation of socially and politically embedded issues such as the underinvestment in (socially disadvantaged) areas, lack of empowerment, the entrapment and forced migration amongst many other socio-political concerns. Our conceptualisation of the relationship between neighbourhoods and health can be enhanced by adopting an environmental justice framing because of the attention this provides to: the (mal)distribution of environmental good and bad; the social, economic and political processes leading to this socio-spatial arrangement; why particular population groups inhabit those spaces; and to the processes leading to broader notions of health and well-being. It is argued that it is important for researchers to turn their attention to examining the processes that lead to socially disadvantaged and vulnerable populations occupying spaces that are prejudicial for a full and healthful life.

## Environmental Justice and Health Inequalities

### An Environmental Justice Framework

The principles of environmental justice incorporate the protection from environmental deprivation, including adverse health impacts, irrespective of individual or area-level socioeconomic status (Cutter 1995). There has been an implicit application of an environmental justice framework in public health research to investigate disparities in the neighbourhood availability of potentially health promoting resources (e.g. green spaces) and exposure to pathogenic environmental characteristics (e.g. air pollution). With a foundation in the civil rights movement in the United States, work in this national context and beyond is increasingly demonstrating that minority, socially disadvantaged groups and places suffer the from the double jeopardy of environmental and social disadvantage.

Some theorists have critiqued the scope of this early framing suggesting that environmental justice research remains entrenched in the distributional facets of environmental harms (Reed and George 2011), grounded in a deficit model of health. Drawing on developments in justice theory, two further lines of enquiry have been identified. First, whilst recognising that the distribution of environmental goods in society remains a significant line of research endeavour, it is also prudent to scrutinise the social and political processes that create and maintain environmental maldistributions. For example, accounts offered from political ecology of the complex socio-political processes that operate at a range of geographical scales have been instructive. Second, and related, there needs to be greater attention applied to the position of those affected by the distribution of environmental goods. Schlosberg (2007) for example draws on 'capabilities theory' that has been developed by commentators such as Amartya Sen and Martha Nussbaum to broaden the concept of environmental justice. The capabilities approach was provided as an alternative to traditional welfare economics and emphasises what people are *effectively* able to do and be (i.e. the ability to achieve a given functioning) (Abel and Frohlich 2012). The effective opportunities to complete an individual's desired actions and activities, and the distinction between the realisable and realised functions are important (Sen 1985). Nussbaum uses this framing to develop a 'capabilities set' to include aspects such as living an active life, bodily health, emotions, working, playing, etc. (Nussbaum 2000). Schlosberg suggests that it is not only important consider the distribution of environmental goods but also "how those goods are transformed into the flourishing of individuals and communities" (p. 4). A conception of environmental justice that incorporates notions of distribution, recognition, participation and capabilities offers obvious potential for scholarship in the field of neighbourhood-level health inequalities. To date there are few empirical studies that have embraced broader conceptualisations of environmental justice to frame questions relating to health inequalities.

The remainder of this chapter provides an overview of work on this broad theme, and argues that researchers in the field of health and place could fruitfully widen their theoretical framework to incorporate perspectives from the field of environmental justice. In particular, it is contended that applying an environmental justice framework in this way can help to deepen our understanding social and economic inequality and how such processes impact life chances at the local level, which in turn lends insights into the relationship between place and health.

## *Socio-Spatial Distribution of Neighbourhood Pathogens and Salutogens*

Using distributional aspects of the environmental justice framework, the accumulating international evidence suggests environmental pathogens (disease causing) and salutogens (health supporting) tend to be unequally distributed across neighbourhoods. The evidence base is particularly extensive in North America where low income, African-American and other disadvantaged populations are often exposed to disproportionately high levels of environmental 'bads' and low levels of access to health-promoting environmental 'goods'. A recent review of the environmental disparities literature incorporating an extensive range of environmental matters noted that 'the poor and especially the non-white poor bear a disproportionate burden of exposure to suboptimal, unhealthy environmental conditions in the United States' (Evans and Kantrowitz 2002, p. 323). The earliest work to adopt an environmental justice framework to consider the distribution of environmental disamenities tended to focus on the sitting of hazardous waste and noxious facilities, predominantly located in the vicinity of African-American communities (Bullard 1983; United Church of Christ 1987; US General Accounting Office 1983). For example, Bullard's (1983) seminal work in Houston, Texas revealed that solid-waste sites were predominantly found in the vicinity of neighbourhoods and schools with large African-American populations, which he attributed to institutionalised racism. Since this early work, the focus of environmental justice research has expanded beyond considerations of 'environmental racism' to examine a broader range of 'vulnerable' populations, a wider set of environmental concerns, and locations outside of the United States.

Whilst race continues to have salience for environmental justice scholars, an increasing number of researchers have evaluated the environmental justice concerns of other vulnerable social and demographic groups such as low-income, socially deprived, elderly or young populations. The facet of the environment that has probably received the most attention is ambient air pollution. Here, the research findings overwhelmingly demonstrate that neighbourhoods with a high proportion of low income and/or ethnic minority populations are often exposed to higher levels of a range of air pollutants within many urban areas (Perlin et al. 2001). For example, studies in New Zealand have demonstrated that levels of particulate pollution tend to be significantly higher in more socially deprived and low-income neighbourhoods across the country (Kingham et al. 2007; Pearce and Kingham 2008; Pearce et al. 2006a). These findings were consistent with work in the UK, where exposure to nitrogen dioxide pollution was found to be most heavily concentrated in the poorest neighbourhoods, communities and places with a younger population (Mitchell and Dorling 2003). A study of the distribution of carbon monoxide and nitrogen dioxide in the urban area of Birmingham, UK similarly found that exposure was strongly related to ethnicity and poverty (Brainard et al. 2002).

The extension of environmental justice concerns beyond the realms of hazardous facilities and air pollution has assisted in developing a more nuanced assessment of the social and economic dimensions of environmental issues. There is a burgeoning literature on the social and spatial dimensions of neighbourhood environmental characteristics such as green space availability, water quality, noise and transport as well as environmental 'events' such as heat waves or 'natural' disasters (e.g. Hurricane Katrina). Whilst the literature from political ecology (and elsewhere) consistently shows that 'vulnerable' groups (e.g. low income, ethnic minority) endure the most during environmental disasters (Peet et al. 2010), no consistent picture has emerged on the relationship between the socio-economic characteristics of areas and their community resources and environmental pathogens/salutogens. In some cities socioeconomically poorer areas, or areas that are ethnically segregated (particularly in the United States), have been found to be relatively poorly endowed with community resources (Pacione 1989; Sooman and Macintyre 1992). However, research in other urban areas has shown no clear association between socioeconomic deprivation and community resource access, or areas of higher deprivation have been found to have higher levels of community resource access (Field et al. 2004; Knox 1982; Lineberry 1977; McLafferty 1982).

The application of an environmental justice framework has been fruitful in helping to understand the role of the political economy in explicating the distribution of resources and some of the related effects of the adoption of neoliberalism, as well as to reveal the social fault lines in the resources available to different communities and the long- and short-term effects and responses to 'natural' disasters. Over the past 30 years there has been a long tradition in the discipline of human geography in examining the socio-spatial distribution of neighbourhood community resources including facilities with public health implications. Neomaterialist theorising posits a systematic underinvestment in community infrastructure in poor areas (Lynch 2000). Marxist interpretation draws attention to the flight of capital (or disinvestment) from certain urban neighbourhoods in response to the urban growth cycle (Harvey 1973; Smith 1984) and the litany of social and infrastructural problems that arise. The evidence base for this assertion is better developed for some types of resources than others. For example, studies of geographical access to health care provision have noted that in deprived areas access to secondary care such as specialist physicians (Mansfield et al. 1999) and pharmacies (Dokmeci and Ozus 2004) has been found to be worse than in non-deprived areas in the United States and Turkey; one component of the so called 'inverse care law' (Hart 1971). On the other hand studies of inequalities in proximity to primary health care provision are equivocal. British studies have found that access to a GP is better in more deprived neighbourhoods across the country (Adams and White 2005), Similar results have been noted in Perth, Australia, where average distances to the closest GP were shown to be significantly lower in poorer areas of the city (Hyndman and Holman 2001). However, the opposite trend has been noted in the United States, where, for example, more deprived areas have been found to have poorer access to primary care (Guagliardo et al. 2004). There is evidence that there is a social gradient in access to recreational resources (Giles-Corti and Donovan 2002). For example, research in Scotland has found an inequitable distribution in recreational facilities in the favour of high-income neighbourhoods (Macintyre et al. 1993). A link has been suggested between poor access to safe facilities in deprived areas, levels of exercise that are sufficiently low to endanger health, and a high level of exercise-related problems such as obesity (Kavanagh et al. 2005; Oliver and Hayes 2005). Work from New Zealand found that at the national level, socially disadvantaged neighbourhoods had a lower availability of total green space (although these neighbourhoods had marginally more *usable* green space) (Richardson et al. 2010b).

There has been a remarkable amount of work examining whether socially deprived neighbourhoods have lower availability of retailers selling high-quality and nutritious food, or, in other words, whether there is a presence of what has become termed a 'food desert' (Clarke et al. 2002). The clearest evidence for food deserts is in the USA where it has been suggested that supermarkets are relocating away from poorer inner-city areas, increasing the likelihood of food deserts developing (Alwitt and Donley 1997; Zenk et al. 2005). For instance, a study in four areas of the USA found a larger proportion of supermarkets and gas stations with convenient stores located in wealthier and white-dominated neighbourhoods compared with the poorest and black neighbourhoods (Morland et al. 2002). Outside of North America, the evidence for food deserts is mixed (Cummins and Macintyre 2006). Early work in Glasgow supported the existence of food deserts (Ellaway and Macintyre 2000; Sooman and Macintyre 1992) but more recent work has not (Cummins and Macintyre 1999, 2002b; Cummins et al. 2005). On the other hand, results of work in Leeds have generally been supportive of food deserts (Clarke et al. 2002; Whelan et al. 2002). Other work in Britain (Pearson et al. 2005) and in Australia (Winkler et al. 2006) failed to find evidence of food deserts.

A clear limitation of this research in terms of understanding the role of neighbourhoods in understanding health outcomes, behaviours and practices has been the restricted focus on single environmental attributes. Missing from earlier work has been a systematic assessment of a range of community resources. Further, very few studies have considered whether access to community resources varies between deprived and non-deprived neighbourhoods at a local or regional scale, and even fewer have examined trends at a national level. These omissions have been significant as it has resulted in a piecemeal understanding of urban disinvestment. In turn, this has been instrumental in academic and policy thinking that emphasises intervention at the neighbourhood level rather than recognising the antecedent macro-level explanations. In other words, the neighbourhood itself has been viewed as the fundamental problem that requires consideration, diverting attention from the processes that led to the establishment and perpetuation of environmental concerns in the first place, as well as the reasons for why particular populations occupy such spaces. Recent work in New Zealand, a country with stark geographical inequalities in health across a range of measures (Pearce and Dorling 2006), has started to address this research gap. The authors collated data on numerous health promoting resources (including supermarkets, health facilities, schools and recreational facilities) for the whole country and examined the distribution of these resources by neighbourhood social deprivation (Pearce et al. 2006b). The findings from the study were patent: there was a strong linear association between each of the community resources and neighbourhood deprivation (Pearce et al. 2007b, 2008c). However, and contrary to many of the results in other countries, a pro-equity distribution was observed; that is more socially deprived neighbourhoods tended to have greater availability of the various types of community resources. These findings call into question some of the assumptions of the neo-material model of health inequalities. However, subsequent work by the same team has found that facilities which may be considered health damaging (e.g. fast food, tobacco and alcohol outlets) show the same association with area-level deprivation (i.e. greater availability in high deprivation neighbourhoods) (Pearce et al. 2007a, 2008a, 2009b). An inspection of the cross-sectional associations between many of these community resources and health outcomes with a biologically plausible link (e.g. neighbourhood access to fast food and individual-level measures of diet) have tended not to find that there is a dose response and/or significant association (Hiscock et al. 2008; Pearce et al. 2008, 2009a, b; Witten et al. 2008). This evidence suggests that in New Zealand, neighbourhood access to community resources do not tend to be associated with the health outcomes and behaviours of local residents.

The criticism that researchers have been content to consider only single environmental attributes extends beyond studies of community resources and applies to work on neighbourhood physical environmental features. As Evans and Kantrowitz (2002) argue:

> We suspect that the potential of environmental exposure to account for the link between SES [socioeconomic status] and health derives from the multiple exposures to a plethora of suboptimal environmental conditions. That is, we would argue that a particularly important and salient aspect of reduced income is exposure to a confluence of multiple, suboptimal environmental conditions. (p. 304)

This issue is a concern because a variety of environmental factors are likely to simultaneously (and potentially multiplicatively) affect health outcomes. Hence it is plausible that residents of socially disadvantaged areas are exposed to a wide array of low-quality environmental features. Recent work by a team of researchers at the Universities of Glasgow and Edinburgh have attempted to extend the parameters of previous environmental justice research that has tended to consider single environmental features by assessing the socio-spatial distribution of multiple dimensions of the physical environment across the UK. The authors developed the *Multiple Environmental Deprivation Index* (MEDIx)[1] which was a small area measure of a range of environmental characteristics that have significance for health and wellbeing (Richardson et al. 2010a). Environmental characteristics were selected on the basis of the strength of the epidemiological evidence and the availability of environmental data at the national level. Components of MEDIx included measures of air pollution, climate, UV radiation and green space. By comparing the index to an area measure of income deprivation, the authors found that, at the national level, multiple environmental deprivation increased as the degree of income deprivation rose (Pearce et al. 2010). Further, disaggregated analyses of the constituent environmental variables provided consistent results. The authors tested the applicability and replicability of these methods in a different international context and found that in New Zealand the findings were broadly consistent with the UK (Pearce et al. 2011). These findings are revealing because they demonstrate that multiple dimensions of the physical environment (in the UK and New Zealand at least) are systematically distributed to the disadvantage of low income, vulnerable and less healthy populations. Further work on the social distribution of multiple dimensions of the (health-related) environment in other international settings is warranted.

## *Capabilities, Neighbourhoods and Health*

Whilst there is mounting evidence that across neighbourhoods salutogenic and pathogenic features are unequally distributed, it is perhaps surprising that few studies have examined the implications for inequalities in health outcomes. As Brulle and Pellow (2006) note, the fields of environmental justice and health inequalities remain principally distinct domains. This is despite the theoretical and policy-related potential of enhancing our understanding of the differential impacts of neighbourhood attributes for different socio-demographic groups (for example see Pearce and Maddison (2011) for a review in the field of inequalities in physical activity). It is feasible that if socially disadvantaged communities are exposed to raised levels of environmental pathogens and have fewer community assets, then due to the additional effects of material deprivation and psychosocial stress they are likely to be more susceptible to the health effects of these environmental contributors (O'Neill et al. 2003) and that socioeconomic factors may modify the relationship between the environment and health outcomes (Samet and White 2004). Fewer studies still have simultaneous considered multiple dimensions of the local environment. As Pearce and colleagues (2010) note, it has been widely assumed but largely untested that the disproportionate burden of poorer quality environments among socially disadvantaged groups foreshadows adverse health effects. Exposure to environmental 'goods' and 'bads' may have dissimilar

---

[1] MEDIx data for the UK and New Zealand is can be downloaded at www.cresh.org.uk.

effects on communities differentiated in terms of their socio-economic profile. Socioeconomic disadvantage could act to compound the influence of environmental deprivation on health status.

A notable exception to this assertion is the body of work examining the differential effects of air pollution on the health of various socioeconomic groups. There is mounting evidence, particularly from the United States, that the socio-spatial distribution or air quality explains a component of the pollution-related health gradient that is observed between and across North American cities. For example, extensive work in Hamilton, Canada demonstrated that socio-economic status modified the relationship between air pollution exposure and mortality (Jerrett et al. 2004). The largest health effects attributed to air pollution exposure occurred in lower socioeconomic areas. In the same city, a separate study noted that differences in exposure to air pollution accounted for some of the socio-economic differences in circulatory disease (cardiovascular and stroke) mortality (Finkelstein et al. 2005). The findings from Canada are consistent with research elsewhere including the United States (Zeka et al. 2006), and Norway (Naess et al. 2007). This work has also helped to establish that socio-economic status modifies the relationship between air pollution and respiratory health. However, other studies, including some work in these same countries, have often arrived at different conclusions. A study of three large Latin American cities found that educational level did not modify the relationship between particulate air pollution exposure and mortality (O'Neill et al. 2008). Findings in France (Laurent et al. 2008) and the United States (Schwartz 2000) also established that socio-economic status did not affect the association between pollution exposure and health. Most surprisingly, a study in China found that elderly residents living in areas with a higher gross domestic product (GDP) were more susceptible to the effects of air pollution than those living in low GDP areas (Sun and Gu 2008). A possible explanation for these equivocal findings is the variations in the methods adopted between studies, particularly the geographical scale at which socioeconomic characteristics are captured. Finer measures of socio-economic status (e.g. individual-level or small geographical areas) have tended to find that socio-economic characteristics modify the relationship between air pollution and mortality (Laurent et al. 2008). Further, comparisons across studies and between countries are problematic due to the different types of air pollution studied, as well as the diversity of methods used to capture pollution exposure.

An additional criticism levelled at this earlier research has been the reductionist nature of the analysis and the failure to recognise the holistic nature of the environment as it relates to health. Similar to the work on socio-spatial patterning of environmental pathogens and salutogens (see above), most studies have considered only a single environmental attribute (e.g. ambient particulate pollution) with few empirical investigations considering a range of environmental features. The development of the *Multiple Environmental Deprivation Index* (see above for details of how the index was constructed) (Richardson et al. 2010a) has begun to address this concern for the UK. The aim of this work was to use this novel area-level measure coupled with routine vital statistics (mortality records) disaggregated by neighbourhoods to assess the implications of exposure to multiple environmental deprivation for health and health inequalities. The research team found that MEDIx had an independent association with health that remained after taking into account the age, sex and socio-economic profile of each area (Pearce et al. 2010). Area-level health progressively worsened as multiple environmental deprivation increased. However, this effect was most pronounced in the least income-deprived areas. The authors contend that the findings highlight the importance of the physical environment in affecting population-level health, and the importance of examining the social and political processes that have resulted in low income populations enduring a disproportionate burden of multiple environmental deprivation.

## New Directions in Environmental Justice and Health Inequalities Research

As we have seen, environmental justice scholarship has remained largely disconnected from field of health inequalities research. It has been argued that perspectives from the environmental justice literature offer some considerable opportunities to further our understanding of the socio-spatial patterning of health outcomes and behaviours. In this section, three possible (and non-exhaustive) directions for further work at the interface of these two fields are expounded.

First, current theoretical framing of neighbourhood 'effects' on health are limited. Neighbourhoods do not operate independently of broader societal factors and in the vacuum that the analytical design of many studies would suggest. As Cummins and colleagues (2007) note:

> The way that areas are delineated administratively, the distribution of services, infrastructure and linkages among places and the ways that places are represented are not seen as socially and politically neutral but as the outcome of dynamic social relations and power struggles between groups in society. (p. 1828)

Rather, neighbourhoods are to a significant extent the social and spatial expression of decades (if not centuries) of economic, social and political factors including structural adjustments affecting the employment base, capital (dis)investment in resources and selective migration streams amongst many other factors, as well as the well documented bottom up 'place making'. Similarly decisions taken at the macro-level (e.g. by multinational companies marketing health-related products such as tobacco or food) exert effects that are felt at the neighbourhood level (e.g. encouraging demand for such products through enhancing local availability and promotion). Disparities in these factors have become accentuated by the neoliberal agenda that has been adopted in most industrialised economies. Therefore, it is not often desirable (or in practice possible) to detach a 'neighbourhood effect' from the broader structural and political events. Rather than exerting independent effects, neighbourhoods can therefore be one canvas on which health disadvantage plays out. A more sophisticated analysis might involve an exploration of how "human and physical phenomena need to be understood as an outcome of interrelated processes which may operate simultaneously at various spatial scales" (Cummins et al. 2007, p. 1828). Future research can usefully move beyond relying on cross-sectional associations between neighbourhood properties and health outcomes to incorporate consideration of the historical trajectory of neighbourhood changes and how the reworking and reproduction of neighbourhoods over time impacts historical and contemporary geographies of health. This approach will also serve to deepen our understanding of the temporal course of environment risk. Further, and related, it is important to develop our understanding of the relationships between macro-level drivers (e.g. global, international and national-level factors) and neighbourhood-level expressions of these factors. Greater attention to these complex multi-scalar processes will provide enhanced theoretical insights into the factors driving local-level inequalities in health. The application of complexity theory to this concern may provide valuable insights (Gatrell 2005).

Second, as Walker (2009) argues, distributional justice constitutes only one facet of the discourse of environmental justice concerns but yet has been the dominant line of investigation in most environmental justice studies. Further work could usefully incorporate a capabilities perspective and consider the differential impact of neighbourhood inequalities in environmental resources on the health and behaviours of various socio-demographic groups. Specifying these effects amongst various 'vulnerable' socio-demographic groups (e.g. the elderly, ethnic minorities, children) will be insightful. This research agenda will be further enhanced by an exploration of a wider array of environmental concerns (e.g. beyond an almost exclusive focus on air quality) including more work that builds on the UK and New Zealand research on 'multiple environmental deprivation'. It is feasible that residents of socially disadvantaged areas are exposed to a wide selection of low-quality environmental features. Further studies beyond the UK and New Zealand that clarify this assertion will provide fertile perspectives into the ubiquitous socio-economic gradient in health. Of relevance here is the concern expressed with the dominant 'deficit' model for evidence based public health. An 'assets' model that draws on the theory of salutogenesis to examine important health assets which support the formation of health rather than the prevention of disease offers a useful framework (Morgan and Ziglio 2007). This approach is likely to offer insights into health effects of the socio-spatial patterning of environmental resources and disamenities.

Third, and related to the above concern of engaging with a broader theorisation of environmental justice concerns, more attention should be paid to the social, economic and political processes that lead to the non-equitable distribution of environmental resources and disamenities as well as the processes that account for the geographical sorting of the population into environments that differ in terms of their capacity to support health. Further health research that considers mobility and migration as a substantive line of enquiry rather that a technical nuisance to be 'controlled' for is particularly welcome. Such research endeavour requires a longitudinal perspective that involves consideration of a multitude of interrelated and multi-scalar factors including for example institutional decision making and planning, employment structures, community empowerment amongst many other concerns. Considering these questions will provide perspectives on the accumulation of environmental (dis) advantage over the life course.

## Conclusion

> Social injustice is killing people on a grand scale.
>
> (WHO 2008, p. 26)

An enhanced understanding of the explanations for widening geographical inequalities in health that have been observed in most country is an intellectual and policy priority. As Woodward and Kawachi argue, health inequalities are not only unfair and avoidable but the 'spillover' effects associated with the increased differentials in health such as crime, violence, the spread of infectious disease, as well as alcohol and drug use, affect all of society (Woodward and Kawachi 2000). However, as we have seen, current conceptualisations of the role of place (particularly neighbourhoods) in explaining spatial health inequalities are limited.

This chapter has argued that an environmental justice framework provides a useful means for understanding the observed socio-spatial polarisation in health. The adoption of this alternative conceptualisation offers the potential for a radical shift in the current framing used in the 'neighbourhoods and health' literature. In particular a greater engagement with the environmental justice literature, as well as related fields such as political ecology and urban theory, offers the chance to reconsider our conceptualisation of 'place effects' on health. An environmental justice approach presents the opportunity to recognise the multifaceted nature of the (health-related) environment. Deficit models of health are unlikely to be sufficient to understand and address health inequalities, and theories of salutogenesis offer promising insights into explaining geographical inequalities in health. In particular, an environmental justice framing encourages researchers to systematically consider the simultaneous 'triple jeopardy' of social, health and environmental inequalities. In addition, and critically, the current focus of neighbourhoods and health research has been to isolate environmental characteristics from broader structural factors. Neighbourhoods tend to be viewed as being the problem that requires mending at the local level, an assertion that is akin to the 'victim blaming' ethos that remains customary in much UK public health policy making. An environmental justice framing on the other hand encourages researchers to evaluate the macro-level process that account for environmental disparities (e.g. unequal investment in local infrastructure) and the forces that affect the migration and mobility patterns which explain why certain (often low income) groups inhabit places that are disadvantageous for health. This advance in conceptual thinking is important because it provides the opportunities to connect the neighbourhoods literature with broader concerns relating to the establishment and perpetuation of economic and social inequality.

This discussion of the complex relationships between neighbourhoods and health and the integration of an environmental justice framework are timely. The current austerity agenda that has been prompted by the crisis in the financial sector in the UK and elsewhere has potentially significant implications for socio-spatial assemblages of health and related concerns. As argued elsewhere, the spatial consequences of economic retrenchment are unlikely to be evenly shared (Pearce 2013). Some places are bound to be affected more than others. Austerity measures such as the reduction in welfare payments and local council budgets are likely to undermine the social determinants of health model through higher unemployment, job insecurity, insufficient resources to improve local infrastructure, disruption of local community networks, and a host of other factors relating to the social geography of the country. There is also likely to be a significant resorting of the population as housing benefit payments are reduced and many low income residents seek alternative accommodation in areas with lower rent. These and other factors may result in a further widening in geographical inequalities in health and other social outcomes.

At the same time as the austerity agenda is being pursued, the UK Government are seeking to implement a 'localism agenda' (including its flagship 'Big Society' project) which seeks to empower local people and communities in order to strengthen local accountability and decision making. Critics of this approach describe the localism agenda as a thinly disguised attempt to scale back state intervention and pass responsibility to communities who are often not well positioned to self manage. It is highly likely that the response will be non-equitable whereby some communities are well placed to enthusiastically take-up the government's challenge, and others are not. It is feasible that the implementation of the localism agenda will have profound implications for the multitude of factors affecting health inequalities that are mediated through neighbourhoods. In these times of economic austerity and increasingly fluid political accountability, is clear that the research community have a particular responsibility to understand and foresee the (often unanticipated) effects of policy decisions on disadvantaged populations and, in particular, to contest decisions taken by political elites that are likely to result in wider social, environmental and health inequalities.

## References

Abel, T., & Frohlich, K. L. (2012). Capitals and capabilities: Linking structure and agency to reduce health inequalities. *Social Science & Medicine, 74*(2), 236–244.

Adams, J., & White, M. (2005). Socio-economic deprivation is associated with increased proximity to general practices in England: An ecological analysis. *Journal of Public Health, 27*(1), 80–81.

Alwitt, L. F., & Donley, T. D. (1997). Retail stores in poor urban neighborhoods. *Journal of Consumer Affairs, 31*(1), 139–164.

Brainard, J. S., Jones, A. P., Bateman, I. J., & Lovett, A. A. (2002). Modelling environmental equity: Access to air quality in Birmingham, England. *Environment and Planning A, 34*, 695–716.

Brulle, R. J., & Pellow, D. N. (2006). Environmental justice: Human health and environmental inequalities. *Annual Review of Public Health, 27*, 103–124.

Bullard, R. (1983). Solid waste sites and the Houston Black community. *Sociological Inquiry, 53*(2–3), 273–288.

Chadwick, E. (1843). *Report on the sanitary condition of the labouring population of Great Britain. A supplementary report on the results of a special inquiry into the practice of interment in towns*. London: W Clowes and Sons for HMSO.

Clarke, G., Eyre, H., & Guy, C. (2002). Deriving indicators of access to food retail provision in British cities: Studies of Cardiff, Leeds and Bradford. *Urban Studies, 39*(11), 2041–2060.

Cummins, S. (2007). Commentary: Investigating neighbourhood effects on health – avoiding the 'local trap'. *International Journal of Epidemiology, 36*(2), 355–357.

Cummins, S., & Macintyre, S. (2002). Food 'deserts' – evidence and assumption in health policy making. *British Medical Journal, 325*, 436–438.

Cummins, S., & Macintyre, S. (2006). Food environments and obesity – neighbourhood or nation? *International Journal of Epidemiology, 35*, 100–104.

Cummins, S., & Macintyre, S. (1999). The location of food stores in urban areas: A case study in Glasgow. *British Food Journal, 101*, 545–553.

Cummins, S., Petticrew, M., Higgins, C., Findlay, A., & Sparks, L. (2005). Large scale food retailing as an intervention for diet and health: Quasi-experimental evaluation of a natural experiment. *Journal of Epidemiology and Community Health, 59*, 1035–1040.

Cummins, S., Curtis, S., Diez-Roux, A. V., & Macintyre, S. (2007). Understanding and representing 'place' in health research: A relational approach. *Social Science & Medicine, 65*(9), 1825–1838.

Cutter, S. L. (1995). Race, class and environmental justice. *Progress in Human Geography, 19*, 111–122.

Davey Smith, G., Dorling, D., & Shaw, M. (2001). *Poverty, inequality and health in Britain, 1800–2000: A reader*. Bristol: Policy Press.

Davey Smith, G., Dorling, D., Mitchell, R., & Shaw, M. (2002). Health inequalities in Britain: Continuing increases up to the end of the 20th century. *Journal of Epidemiology and Community Health, 56*, 434–435.

Diez Roux, A. V. (2003). Residential environments and cardiovascular risk. *Journal of Urban Health, 80*(4), 569–589.

Dokmeci, V., & Ozus, E. (2004). Spatial analysis of urban pharmacies in Istanbul. *European Planning Studies, 12*, 585–594.

Dorling, D. (2010). *Injustice: Why social inequality persists*. Portland/Bristol: Policy Press.

Dorling, D., Mitchell, R., Shaw, M., Orford, S., & Smith, G. D. (2000). The ghost of Christmas past: Health effects of poverty in London in 1896 and 1991. *British Medical Journal, 321*, 1547–1551.

Dorling, D., Mitchell, R., & Pearce, J. (2007). The global impact of income inequality on health by age: an observational study. *British Medical Journal, 335*, 873.

Ellaway, A., & Macintyre, S. (2000). Shopping for food in socially contrasting localities. *British Food Journal., 102*(1), 52–59.

Evans, G. W., & Kantrowitz, E. (2002). Socioeconomic status and health: The potential role of environmental risk exposure. *Annual Review of Public Health, 23*, 303–331.

Field, A., Witten, K., Robinson, E., & Pledger, M. (2004). Who gets to what? Access to community resources in two New Zealand cities. *Urban Policy and Research, 22*(2), 189–205.

Finkelstein, E. A., Ruhm, C. J., & Kosa, K. M. (2005). Economic causes and consequences of obesity. *Annual Review of Public Health, 26*(1), 239–257.

Gatrell, A. C. (2005). Complexity theory and geographies of health: a critical assessment. *Social Science & Medicine, 60*(12), 2661–2671.

General Accounting Office, U. S. (1983). *Siting of hazardous waste landfills and their correlation with racial and economic status of surrounding communitie*. Washington, D.C.: US General Accounting Office.

Giles-Corti, B., & Donovan, R. J. (2002). The relative influence of individual, social and physical environment determinants of physical activity. *Social Science & Medicine, 54*(12), 1793–1812.

Guagliardo, M. F., Ronzio, C. R., Cheung, I., Chacko, E., & Joseph, J. G. (2004). Physician accessibility: An urban case study of pediatric providers. *Health & Place, 10*(3), 273–283.

Hart, J. (1971). The inverse care law. *Lancet, 1*(7696), 405–412.

Harvey, D. (1973). *Social justice and the city*. London: Edward Arnold.
Hiscock, R., Pearce, J., Blakely, T., & Witten, K. (2008). Is neighborhood access to health care provision associated with individual-level utilization and satisfaction? *Health Services Research, 43*(6), 2183–2200.
Hyndman, J. C. G., & Holman, C. D. J. (2001). Accessibility and spatial distribution of general practice services in an Australian city by levels of social disadvantage. *Social Science & Medicine, 53*(12), 1599–1609.
Jerrett, M., Burnett, R. T., Brook, J., Kanaroglou, P., Giovisl, C., Finkelstein, N., & Hutchison, B. (2004). Do socioeconomic characteristics modify the short term association between air pollution and mortality? Evidence from a zonal time series in Hamilton, Canada. *Journal of Epidemiology and Community Health, 58*(1), 31–40.
Kavanagh, A. M., Goller, J. L., King, T., Jolley, D., Crawford, D., & Turrell, G. (2005). Urban area disadvantage and physical activity: A multilevel study in Melbourne, Australia. *Journal of Epidemiology and Community Health, 59*, 934–940.
Kingham, S., Pearce, J., & Zawar-Reza, P. (2007). Driven to injustice? Environmental justice and vehicle pollution in Christchurch, New Zealand. *Transportation Research Part D: Transport and Environment, 12*(4), 254–263.
Knox, P. L. (1982). *Urban social geography: An introduction*. New York/London: Longman.
Laurent, O., Pedrono, G., Segala, C., Filleul, L., Havard, S., Deguen, S., Schillinger, C., Riviere, E., & Bard, D. (2008). Air pollution, asthma attacks, and socioeconomic deprivation: A small-area case-crossover study. *American Journal of Epidemiology, 168*, 58–65.
Lineberry, R. L. (1977). *Equality and urban policy: The distribution of municipal public services*. Beverly Hills: Sage.
Lynch, J. (2000). Income inequality and health: Expanding the debate. *Social Science & Medicine, 51*, 1001–1005.
Macintyre, S., Maciver, S., & Sooman, A. (1993). Area, class and health – should we be focusing on places or people. *Journal of Social Policy, 22*(2), 213–234.
Macintyre, S., Ellaway, A., & Cummins, S. (2002). Place effects on health: How can we conceptualise, operationalise and measure them? *Social Science & Medicine, 55*(1), 125–139.
Mansfield, C. J., Wilson, J. L., Kobrinski, E. J., & Mitchell, J. (1999). Premature mortality in the United States: The roles of geographic area, socioeconomic status, household type, and availability of medical care. *American Journal of Public Health, 89*(6), 893–898.
McLafferty, S. (1982). Urban structure and geographical access to public services. *Annals of the Association of American Geographers, 72*(3), 347–354.
Mitchell, G., & Dorling, D. (2003). An environmental justice analysis of British air quality. *Environment and Planning A, 35*, 909–929.
Morgan, A., & Ziglio, E. (2007). Revitalising the evidence base for public health: An assets model. *Promotion & Education, 14*(2), 17–22.
Morland, K., Wing, S., & Diez Roux, A. (2002). The contextual effect of the local food environment on residents' diets: The Atherosclerosis Risk in Communities Study. *American Journal of Public Health, 92*(11), 1761–1767.
Naess, O., Nafstad, P., Aamodt, G., Claussen, B., & Rosland, P. (2007). Relation between concentration of air pollution and cause-specific mortality: Four-year exposures to nitrogen dioxide and particulate matter pollutants in 470 neighborhoods in Oslo. Norway. *American Journal of Epidemiology, 165*, 435–443.
Nussbaum, M. C. (2000). *Women and human development: the capabilities approach*. New York/Cambridge: Cambridge University Press.
O'Dea, D., & Howden-Chapman, P. (2000). Income and income inequality and health. In M. Tobias & P. Howden-Chapman (Eds.), *Social inequalities in health, New Zealand 1999: A summary*. Wellington: Ministry of Health.
Oliver, L., & Hayes, M. (2005). Neighbourhood socio-economic status and the prevalence of overweight Candian children and youth. *Canadian Journal of Public Health, 96*(6), 415–420.
O'Neill, M. S., Jerrett, M., Kawachi, L., Levy, J. L., Cohen, A. J., Gouveia, N., Wilkinson, P., Fletcher, T., Cifuentes, L., & Schwartz, J. (2003). Health, wealth, and air pollution: Advancing theory and methods. *Environmental Health Perspectives, 111*(16), 1861–1870.

O'Neill, M. S., Bell, M. L., Ranjit, N., Cifuentes, L. A., Loomis, D., Gouveia, N., & Borja-Aburtof, V. H. (2008). Air pollution and mortality in Latin America: The role of education. *Epidemiology, 19*(6), 810–819.

Pacione, M. (1989). Access to urban services – the case of secondary-schools in Glasgow. *Scottish Geographical Magazine, 105*(1), 12–18.

Pearce, J. (2013). Financial crisis, austerity policies and geographical inequalities in health. Environment and Planning A. In Press.

Pearce, J., & Dorling, D. (2006). Increasing geographical inequalities in health in New Zealand, 1980–2001. *International Journal of Epidemiology, 35*(3), 597–603.

Pearce, J., & Dorling, D. (2009). Tackling global health inequalities: Closing the health gap in a generation. *Environment and Planning A, 41*, 1–6.

Pearce, J., & Kingham, S. (2008). Environmental inequalities in New Zealand: A national study of air pollution and environmental justice. *Geoforum, 39*(2), 980–993.

Pearce, J. R., & Maddison, R. (2011). Do enhancements to the urban built environment improve physical activity levels among socially disadvantaged populations? *International Journal for Equity in Health, 10*(28).

Pearce, J., & Witten, K. (2010). *Geographies of obesity: Environmental understandings of the obesity epidemic*. Burlington/Surrey: Ashgate.

Pearce, J., Kingham, S., & Zawar-Reza, P. (2006a). Every breath you take? Environmental justice and air pollution in Christchurch, New Zealand. *Environment and Planning A, 38*, 919–938.

Pearce, J., Witten, K., & Bartie, P. (2006b). Neighbourhoods and health: A GIS approach to measuring community resource accessibility. *Journal of Epidemiology and Community Health, 60*, 389–395.

Pearce, J., Blakely, T., Witten, K., & Bartie, P. (2007a). Neighborhood deprivation and access to fast food retailing: A national study. *American Journal of Preventive Medicine, 32*(5), 375–382.

Pearce, J., Witten, K., Hiscock, R., & Blakely, T. (2007b). Are socially disadvantaged neighbourhoods deprived of health-related community resources? *International Journal of Epidemiology, 36*(2), 348–355.

Pearce, J., Day, P., & Witten, K. (2008a). Neighbourhood provision of food and alcohol retailing and social deprivation in urban New Zealand. *Urban Policy and Research, 26*(2), 213–227.

Pearce, J., Hiscock, R., Blakely, T., & Witten, K. (2008b). The contextual effects of neighbourhood access to supermarkets and convenience stores on individual fruit and vegetable consumption. *Journal of Epidemiology and Community Health, 62*(3), 198–201.

Pearce, J., Witten, K., Hiscock, R., & Blakely, T. (2008c). Regional and urban-rural variations in the association of neighbourhood deprivation with community resource access: A national study. *Environment and Planning A, 40*, 2469–2489.

Pearce, J., Hiscock, R., Blakely, T., & Witten, K. (2009a). A national study of the association between neighbourhood access to fast-food outlets and the diet and weight of local residents. *Health & Place, 15*(1), 193–197.

Pearce, J., Hiscock, R., Moon, G., & Barnett, R. (2009b). The neighbourhood effects of geographical access to tobacco retailers on individual smoking behaviour. *Journal of Epidemiology and Community Health, 63*, 69–77.

Pearce, J., Richardson, E., Mitchell, R., & Shortt, N. (2010). Environmental justice and health: The implications of the socio-spatial distribution of multiple environmental deprivation for health inequalities in the United Kingdom. *Transactions of the Institute of British Geographers, 35*(4), 522–539.

Pearce, J. R., Richardson, E. A., Mitchell, R. J., & Shortt, N. K. (2011). Environmental justice and health: a study of multiple environmental deprivation and geographical inequalities in health in New Zealand. *Social Science & Medicine, 73*(3), 410–420.

Pearson, T., Russell, J., Campbell, M. J., & Barker, M. E. (2005). Do 'food deserts' influence fruit and vegetable consumption? A cross-sectional study. *Appetite, 45*, 195–197.

Peet, R., Robbins, P., & Watts, M. (2010). *Global political ecology*. New York/Oxon: Routledge.

Perlin, S. A., Wong, D., & Sexton, K. (2001). Residential proximity to industrial sources of air pollution: Interrelationships among race, poverty, and age. *Journal of the Air & Waste Management Association, 51*(3), 406–421.

Reed, M. G., & George, C. (2011). Where in the world is envrionmental justice? *Progress in Human Geography, 35*(6), 835–842.

Richardson, E., Mitchell, R., Shortt, N., Pearce, J., & Dawson, T. (2010a). Developing summary measures of health-related multiple physical environmental deprivation for epidemiological research. *Environment and Planning A, 42*, 1650–1668.

Richardson, E., Pearce, J., Mitchell, R., Day, P., & Kingham, S. (2010b). The association between green space and cause-specific mortality in urban New Zealand: An ecological analysis of green space utility. *BMC Public Health, 10*, 240.

Rowntree, B. S. (1901). *Poverty, a study of town life*. New York/London.

Samet, J. M., & White, R. H. (2004). Urban air pollution, health, and equity. *Journal of Epidemiology and Community Health, 58*(3–5).

Schlosberg, D. (2007). *Defining environmental justice: Theories, movements, and nature*. New York/Oxford: Oxford University Press.

Schulz, A. J., Kannan, S., Dvonch, J. T., Israel, B. A., Allen, A., 3rd, James, S. A., House, J. S., & Lepkowski, J. (2005). Social and physical environments and disparities in risk for cardiovascular disease: The healthy environments partnership conceptual model. *Environmental Health Perspectives, 113*(12), 1817–1825.

Schwartz, J. (2000). Assessing confounding, effect modification, and thresholds in the association between ambient particles and daily deaths. *Environmental Health Perspectives, 108*(6), 563–568.

Sen, A. (1985). *Commodities and capabilities*. Amsterdam: North-Holland.

Shaw, M., Davey Smith, G., & Dorling, D. (2005). Health inequalities and New Labour: How the promises compare with real progress. *British Medical Journal, 330*, 1016–1021.

Singh, G. K., & Siahpush, M. (2006). Widening socioeconomic inequalities in US life expectancy, 1980–2000. *International Journal of Epidemiology, 35*(4), 969–979.

Smith, N. (1984). *Uneven development: Nature, capital, and the production of space*. New York: Blackwell.

Smith, S. J., & Easterlow, D. (2005). The strange geography of health inequalities. *Transactions of the Institute of British Geographers, 30*(2), 173–190.

Sooman, A., & Macintyre, S. (1992). Scotland's health – a more difficult challenge for some? The price and availability of healthy foods in socially contrasting localities in the West of Scotland. *Health Bulletin, 51*(5), 276–284.

Sun, R., & Gu, D. (2008). Air pollution, economic development of communities, and health status among the elderly in Urban China. *American Journal of Epidemiology, 168*(11), 1311–1318.

United Church of Christ. (1987). *Toxic wastes and race in the United States: A national report on the racial and socio-economic characteristics of communities surrounding hazardous waste sites*. New York: United Church of Christ Commission for Racial Justice.

Wacquant, L. J. D. (2008). *Urban outcasts: a comparative sociology of advanced marginality*. Cambridge: Polity Press.

Walker, G. (2009). Beyond distribution and proximity: exploring the multiple spatialities of environmental justice. *Antipode, 41*(4), 614–636.

Walker, G. (2011). *Environmental justice*. New York/Oxon: Routledge.

Whelan, A., Wrigley, N., Warm, D., & Cannings, E., (2002). Life in a 'food desert'. *Urban Studies, 39*, 2083–2100.

WHO Commission on Social Determinants of Health. (2008). *Closing the gap in a generation: health equity through action on the social determinants of health. Final report of the Commission on Social Determinants of Health*. Geneva: WHO.

Wilkinson, R. G., & Pickett, K. E. (2007). The problems of relative deprivation: Why some societies do better than others. *Social Science & Medicine, 65*(9), 1965–1978.

Wilkinson, R. G., & Pickett, K. (2010). *The spirit level: Why greater equality makes societies stronger*. New York: Bloomsbury Press.

Winkler, E., Turrell, G., Patterson, C. (2006). Does living in a disadvantaged area mean fewer opportunities to purchase fresh fruit and vegetables in the area? Findings from the Brisbane food study. *Health and Place, 12*, 306–319.

Witten, K., Hiscock, R., Pearce, J., & Blakely, T. (2008). Neighbourhood access to open spaces and the physical activity of residents: A national study. *Preventive Medicine, 47*(3), 299–303.

Woodward, A., & Kawachi, I. (2000). Why reduce health inequalities? *Journal of Epidemiology and Community Health, 54,* 923–929.

Zeka, A., Sullivan, J. R., Vokonas, P. S., Sparrow, D., Spiro, A., III, Cantone, L., Kubzansky, L., & Schwartz, J. (2006). Inflammatory markers and particulate air pollution: Characterizing the pathway to disease. *International Journal of Epidemiology, 35*(5), 1347–1354.

Zenk, S. N., Schulz, A. J., Israel, B. A., James, S. A., Bao, S. M., Wilson, M. L. (2005). Neighborhood racial composition, neighborhood poverty, and the spatial accessibility of supermarkets in metropolitan Detroit. *American Journal of Public Health, 95,* 660–667.

# Chapter 6
# Capitalist Urbanization Affects Your Life Chances: Exorcising the Ghosts of 'Neighbourhood Effects'

**Tom Slater**

> It is not even necessary to engage reality to reveal the fundamental practical difficulty posed in existing market economies with respect to social justice, within the restricted terms of the free-market model itself. This is the dependence of the distribution of life chances being generated on the pre-existing distribution of income, wealth and other resources. Those with most money will have the greatest power to influence what is produced, while those who happen to own (or otherwise control) land and its natural resources or have capital to invest can exert an influence denied those with only their labour to sell. Very simply, the claim to generate social justice depends on the justice of the distribution that already drives the system.
>
> David M. Smith (1994, p.281).

## Neighbourhoods Cause Poverty?

'Neighbourhood effects' stems from an understanding of society that adheres to one overarching assumption, that 'where you live affects your life chances'. It's seductively simple, and on the surface, very convincing. Somebody growing up in, say, a seven-bedroom mock-Tudor mansion in a leafy residential suburb surrounded by golf courses in the stockbroker belt of Surrey, England, will have far more 'chances' in life than somebody growing up in a stigmatised social housing estate less than 30 miles away in Tower Hamlets (for decades one of the most multiply deprived urban areas in England, with high levels of unemployment, struggling schools, poor health outcomes and little green space). Who could argue against that? The striking simplicity and inherent 'fait accompli' of this line of thinking in a complex world has led to the emergence of *analytic hegemony* in urban studies: neighbourhoods matter and

---

T. Slater (✉)
Institute of Geography, School of GeoSciences, The University
of Edinburgh, Drummond Street, Edinburgh EH8 9XP, UK
e-mail: tom.slater@ed.ac.uk

shape the fate of their residents (and their young residents most acutely), therefore, urban policies must be geared towards poor neighbourhoods, seen as incubators of social dysfunction. A belief in *causal* neighbourhood effects is now the dominant paradigm amongst policy elites, mainstream urban scholars, journalists, and think tank researchers.

In cities of advanced societies, an acceptance of the neighbourhood effects thesis is not something confined to journal pages and conference discussions. It has shaped flagship urban policies and has had dramatic and at times devastating consequences for not just the appearance of cities but for their poorest citizens. One (in)famous illustration is the HOPE (Housing Opportunities for People Everywhere) VI program in the United States, where nearly all public housing projects classified in a 1993 federal audit as 'severely distressed' have now been demolished, and their very low-income, predominantly African-American residents subjected to 'dispersal'. HOPE VI had many motives, and addressing the terrible legacy of racial segregation via public housing was a most welcome development, but the view that "concentrated poverty"[1] was detrimental to the life chances of those living in projects classified as severely distressed, and that the land upon which the projects stood could be put to 'higher and better use', arguably took over the initial racial desegregation imperative. An equally large-scale but less researched policy was Housing Market Renewal (HMR) in England, which gained momentum in 2001 when the Labour government launched its *National Strategy for Neighbourhood Renewal*, with one central and highly ambitious goal: that "nobody should be seriously disadvantaged by where they live in 20 years." HMR - designed by British housing scholars - designated large swathes of urban land in northern England as having poorly performing housing markets (reflected in low house prices relative to regional and national averages) due to the existence of "obsolete" terraced housing (Cole and Nevin 2004). The view amongst the policy's architects was that low house prices were stifling economic growth, and therefore having detrimental effects on the life chances of those living in the areas targeted by HMR. Substantial funds were released by the then Labour government to allow municipal authorities to use Compulsory Purchase Orders to acquire land and housing from owner occupiers so that the "obsolete" dwellings could be demolished in favour of newly built housing developments aimed at wealthier prospective buyers. In both HOPE VI and HMR, the human costs have been truly abysmal – studies of persons displaced under these programmes document a litany of social harms, among them broken social networks, continued housing problems, inadequate counselling services, health deterioration and various forms of psychosocial stress, particularly acute among the elderly, where reactions to the demolition of their long-time homes can only be classified as grief (Goetz 2003; Fullilove 2004; Popkin et al. 2005; Allen 2008; Crookes 2011).

It would be inaccurate and certainly unfair to state that anyone pressing the neighbourhood effects view of cities is intent on causing misery amongst the poor. Many

---

[1] A swashbuckling analytical indictment of "concentrated poverty" has been advanced by Stephen Steinberg (2010b).

of the leading voices in the debate are well-intentioned scholars who are driven by a strong sense of social justice. Their research agendas are guided by an honest and entirely admirable wish to determine the extent to which neighbourhoods matter in people's lives in many different respects, in the hope that their research findings will offer the evidence necessary for policy remediation vis-à-vis entrenched poverty and stunted life chances (in respect of health, education, crime, employment, and so on). In February 2010 I travelled to St. Andrews to attend an ESRC-funded seminar on neighbourhood effects, the first of three seminars that have led to this and two other edited volumes on the subject (van Ham et al. 2012, 2013). The seminar was entitled "Neighbourhood Effects: Theory and Evidence" and it featured a truly international cast of speakers, drawing an admirably large and diverse audience consisting of academics from several disciplines, urban planners and policy makers. I arrived with an open mind, not knowing very much about the neighbourhood effects literature, and hoping to be enlightened theoretically, conceptually, methodologically, perhaps politically. But although the event took place in an excellent, collegial spirit, and the levels of scholarly accomplishment and analytic perspicacity were most impressive, I left the event shaking my head in sheer frustration, bewildered and bemused by the failure of every speaker to engage in matters of political economy, to acknowledge the structural causes of poverty, to pay much attention to the role of the state and of the institutional arrangements that would seem impossible to ignore in discussions of life chances in poor neighbourhoods.

This chapter, then, is primarily concerned with an absolutely fundamental *structural* question that is rarely, if ever, tabled at virtual or actual gatherings of those concerned with 'neighbourhood effects'. That question is: *why do people live where they do in cities*? If *where* any given individual lives affects their life chances as deeply as neighbourhood effects proponents believe, it seems crucial to understand *why* that individual is living there in the first place. It has long been a scientific fact and predicament that in most cities of the world there are neighbourhoods of astounding affluence and neighbourhoods of persistent (and often deepening) poverty, often side-by-side. Life chances will of course be very different for residents of these very different neighbourhoods, but stating the obvious and 'controlling' for various externalities (especially popular amongst statistically-oriented urban sociologists) does not explain why such urban inequality exists. My contention in this essay is that neighbourhood effects, when viewed through *explanatory* rather than descriptive analytic lenses, take the appearance of phantoms. I therefore attempt to exorcise them by examining the structural factors that give rise to differential life chances and the inequalities they produce. If we invert the neighbourhood effects thesis to: *your life chances affect where you live*, then the problem becomes one of understanding life chances via a theory of capital accumulation and class struggle in cities. Such a theory provides an understanding of the injustices inherent in letting the market (buttressed by the state) be the force that determines the cost of housing, and correspondingly, the major determinant of where people live. The ecological determinism practiced by neighbourhood effects believers stands on very shaky ground when placed in the context of well over a century of theoretical advances in respect of how differential life chances are created in cities.

## Private Property Rights, and Locational See-Saws

The question of why people live where they do in cities is not new, and numerous competing explanations have been advanced in a substantial theoretical and empirical literature. As Eric Clark (1987) explained:

> Suburbanization, urban renewal, ex-urbanization and gentrification are complex processes which have been researched and explained from an array of perspectives focusing on different aspects. Some of the foci of explanations found in the literature are: demand for space and accessibility, demographic change, cultural identity, lifestyle and preference shifts, housing supply and demand, employment structure, land rent, class conflict, managerial institutions, and the structure of capitalist economies. (p.5)

In one book chapter it is impossible to do justice to all the urban applications of (often centuries-old) theories of residential differentiation, not to mention the rich intellectual traditions from which they are drawn and the conflicts between those traditions. Clark, citing Andrew Sayer, also notes that "given the complexity of processes which comprise the broad notion of urban change, it should not surprise us to find that 'characteristically explanations are relatively incomplete, approximate and contestable.'" (ibid. p.5). But for our purposes it is a useful starting point to remind ourselves of the 1845 lessons provided by an impressionable 24 year-old who 3 years earlier had been despatched by his industrialist father from his native Germany to Manchester, England, in order to learn the practices of sound factory management, and in particular, how to extract maximum value from the proletariat. It's safe to say that the outcome of this parental decision was not what was intended.

In 1840s Manchester, the cradle of the English cotton industry and undergoing astonishingly rapid urbanization during the first half of the nineteenth century (to the extent that historians refer to it today as the archetypal 'shock city'), Engels was so horrified by what he saw that his destiny as a cotton lord was arrested and the seeds of communist theory were sown.[2] As is very well known, the abysmal living conditions of the working class labourers of the cotton mills were documented by Engels with poignant eloquence. But was it working people's quarters per se that *created* these conditions? Or, put another way, was it the insalubrious neighbourhoods to which workers were confined that stunted their life chances? Far from it:

> Everything which here arouses horror and indignation is of recent origin, belongs to the industrial epoch. The couple of hundred houses, which belong to old Manchester, have been long since abandoned by their original inhabitants; the industrial epoch alone has crammed into them the swarms of workers whom they now shelter; the industrial epoch alone has built up every spot between these old houses to win a covering for the masses whom it has conjured hither from the agricultural districts and from Ireland; the industrial epoch alone enables the owners of these cattlesheds to rent them for high prices to human beings, to plunder the poverty of the workers, to undermine the health of thousands, in order that they alone, the owners, may grow rich. In the industrial epoch alone has it become possible that

---

[2] According to radical historian Jonathan Schofield, "Without Manchester there would have been no Soviet Union. And the history of the twentieth century would have been very different." (quoted in Jeffries 2006).

> the worker scarcely freed from feudal servitude could be used as mere material, a mere chattel; that he must let himself be crowded into a dwelling too bad for every other, which he for his hard-earned wages buys the right to let go utterly to ruin. This manufacture has achieved, which, without these workers, this poverty, this slavery could not have lived. True, the original construction of this quarter was bad, little good could have been made out of it; but, have the landowners, has the municipality done anything to improve it when rebuilding? On the contrary, wherever a nook or corner was free, a house has been run up; where a superfluous passage remained, it has been built up; the value of land rose with the blossoming out of manufacture, and the more it rose, the more madly was the work of building up carried on, without reference to the health or comfort of the inhabitants, with sole reference to the highest possible profit on the principle that no hole is so bad but that some poor creature must take it who can pay for nothing better.

Engels' explanation for the grim life and hopeless life chances of Manchester's industrial proletariat could not be clearer. In this passage, we learn that it is *capitalist urbanization* that condemns workers to social suffering on an epic scale. The dense *concentration* of a particular (poor) category of urban dwellers in certain neighbourhoods of Manchester and the *effects* generated by that concentration is not at all the central issue to be addressed; in fact, such a concern appears ludicrous. By contrast, the villain is the capitalist quest for profit, both from industrial expansion and from the valuable land upon which workers dwell. In a seminal essay published in 1843, Engels identified *private property rights* as the chief institutional arrangement that made capitalist urban expansion possible, for it "encompassed all the myriad features of political economy – wages, trade, value, price, money – that he had seen at work in Manchester" (Hunt 2009, p.98). It was private property rights that had created not just grinding urban poverty but also a profoundly unequal city, captured in words that are as relevant to countless cities worldwide today as they were to Manchester in 1845:

> The town itself is peculiarly built, so that a person may live in it for years, and go in and out daily without coming into contact with a working-people's quarter or even with workers, that is, so long as he confines himself to his business or to pleasure walks. This arises chiefly from the fact, that by unconscious tacit agreement, as well as with outspoken conscious determination, the working-people's quarters are sharply separated from the sections of the city reserved for the middle-class.... I have never seen so systematic a shutting out of the working-class from the thoroughfares, so tender a concealment of everything which might affront the eye and the nerves of the bourgeoisie, as in Manchester.

For Engels, the grotesque social divides of the 'cottonopolis' and the unequal life chances within it were wrought by the system ('systematic') - one sustained by private property rights. They were not wrought by the 'effects' on life chances of different categories of resident living in starkly contrasting residential districts. The lasting analytic contribution of all Engels writings is that we come to understand the value of urban land as a *collective social creation*, and therefore something that should always be under social ownership. If a tiny parcel of land located in the heart of a large, vibrant and growing city is valued, it is because centrality and accessibility are necessary in that society, and because collective *social* investments over time (in the form of labour) produced that large, vibrant, growing city. So private property rights throw a spanner in these social arrangements - they allow a select few landowners to capture most, if not all, of this social investment in the form of ground rent (which is simply the charge that owners are able to demand for the rights to use their land). Engels noted that the rents paid by the working classes can rarely or never increase above a certain maximum, so the preferred strategy of the landowners is to banish people from valuable land in the interests of maximising profit from that land and putting it to its "highest and best use", to use the dismal planning language. The outcome is, of course, an unequal city segregated by social class. But we must remember that Engels was writing at a certain time and a certain place, laying the foundations for important analyses that followed many years later that can help us understand why people live where they do in cities. In this respect it is helpful to elucidate the analytical tradition from which the Marxist critique of capitalist urbanization emerged in the mid-late twentieth century, to which I now turn.

At the time Engels was penning his indictments of private property rights, 'urban studies' was not even a fledgling field of social science inquiry - the field as we know it was arguably born in 1920s Chicago as a consequence of the prolific writings of Robert Park and his colleagues based at the University of Chicago's sociology department. There was little place for Marxist reasoning amidst all their land use models, ethnographic accounts of Chicago life, appeals to natural science metaphors, and interpretations of the city as social 'laboratory'. By the 1960s, urban studies had crystallized into a hegemonic blend of the social and spatial theories of the 'Chicago School' of sociology, infused with the methods and assumptions of neoclassical economics. Morphological analyses, with, suspiciously, 'half the city submerged under Lake Michigan' (Smith 1992, p.110), portrayed the suburbanization of middle-class and wealthy households as *the* driving force of urban growth, suburban expansion, and overall metropolitan housing market change. Among the numerous legacies of the Chicago School (some constructive, others obstructive) arguably the most enduring was the idea that the urban environment tends towards 'equilibrium' much as an organism does, with individuals and groups sorting themselves into 'natural areas' that constituted a city symbiotically balanced between cooperation and conflict (Metzger 2000). This logic - an attempt to account for why certain population categories lived in certain districts of the city - laid the foundation for ideas of *spatial equilibrium* and economic competition that were used to develop neoclassical models of urban land markets in the late 1950s and early 1960s (Alonso 1964; Muth 1969). These models explained suburbanization in terms of an overriding *consumer preference* for space, combined with differences in the ability of high- and low-income households to engage in locational trade-offs between access to centralized employment and the cheaper land prices available on the lower-density urban fringe. The neoclassical models seemed to account for the spatial paradox of the U.S. city: middle-class and wealthy households living on cheap suburban land, poor and working-class households forced to crowd into dense apartment blocks on expensive, centrally-located inner-city land.

In the course of creating sophisticated mathematical models of land use, however, the neoclassical urbanists had built everything on the shaky foundation of *consumer sovereignty*. Viewed through neoclassical analytic lenses, the form and function of the city is always and everywhere attributable to the result of *choices* made by individual consumers of land and housing. Each consumer 'rationally chooses' amongst available options in order to maximize their 'utility,' subject to the constraints of their available resources.[3] Institutions then compete against each other to serve the needs of these utility-maximizing consumers. In respect of neighborhoods and housing, the resulting market will produce the spatial tradeoffs between space and accessibility that structure different residential patterns. All that remains to complete the calculus are the 'optimal' political-economic conditions for the operation of such a competitive market; if it is allowed to operate free of any constrictive regulations enforced by the state, the incentives for both producers and consumers to make rational and economically sound decisions will push the urban environment towards an equilibrium while yielding the maximum amount of utility for the maximum number of people.

Neoclassical theories continue to dominate urban theory and especially urban policy, and, not surprisingly, they constitute the analytic foundation of proponents of neighbourhood effects, who leave political-economic structures unquestioned just as they assume that people living where they do is entirely a matter of individual choices made under resource constraints.[4] But it is the very conceptual simplicity of neoclassical urban thought (always geared to the consumption preferences of suburbanizing consumers) that leaves it wide open to stinging critique. If we are interested in 'resource constraints', surely we need to know how those constraints come about? Surely the more urgent task is to consider the *limits* on individual choice, the *boundaries* set by ever-present inequalities of wealth and power? What about the limited choices available to the poor and working classes? Surely there is far more to the question of neighbourhood change than descriptive accounts of who moves in and who moves out? These questions were at the forefront of a radical (literally) shift in how we understand cities that began in the early 1970s, and continue to shape a critical imagination, where normative concerns for a more socially just spatial arrangement of urban places drives intellectual inquiry. It is unfortunate that much of the literature reflecting such concerns appears to have bypassed the research questions of those locked within the neighbourhood effects paradigm.

In 1969, David Harvey decamped from his first teaching post at the University of Bristol, England and arrived in Baltimore, a city with districts hit hard by grotesque racial injustices, systematic disinvestment, and rioting in the wake of the assassination of Martin Luther King Jr. The previous year he had submitted the manuscript of *Explanation in Geography* (a landmark text in the quantitative/positivist geographical tradition), and felt politically irresponsible:

> I turned in the manuscript in the summer of 1968 with near revolutions going on in Paris, Berlin, Mexico City, Bangkok, Chicago and San Francisco. I had hardly noticed what was happening. I felt sort of idiotic. It seemed absurd to be writing when the world was collapsing in chaos around me and cities were going up in flames. (Harvey 2006, p.187).

---

[3] Neighbourhood effects proponents are usually in the neoclassical tradition, and in light of these resource constraints they usually ask if adequate 'controlling' for them has taken place during statistical calculation.

[4] In respect of the segregating tendencies of social housing allocation, the paradigm of consumer choice has been called into question in a revealing critique of the New Labour policy of 'Choice Based Letting' in the UK (Manley and Van Ham 2010).

To understand the origins of inequality and injustice in the city in which he now resided, Harvey quickly became the leading force of a new analytic framework that returned to the roots of contemporary neoclassical theory - the classical political economy debates between Smith, Ricardo, Malthus, and Marx. Harvey's (1973) *Social Justice and the City* was the manifesto of this new urban studies, which sought to understand how cities

> ...are founded upon the exploitation of the many by the few. An urbanism founded on exploitation is a legacy of history. A genuinely humanizing urbanism has yet to be brought into being. It remains for revolutionary theory to chart the path.... (Harvey 1973, p.314).

Harvey offered a panoramic view of urbanism and society, and in later work he outlined a comprehensive analysis of economic, urban, and cultural change (Harvey 1982, 1985, 1989, 2000, 2003). But his 1973 attack on the dominant neoclassical explanation of inner-city decline and ghetto formation is crucial for any critique of 'neighbourhood effects'. He took dead aim at the models of urban structure that Alonso (1964) and Muth (1969) had built using the principles of agricultural land-use patterns that had been devised by a Prussian landowner, Johann Heinrich von Thünen (1793–1850):

> After an analytic presentation of the theory, Muth seeks to evaluate the empirical relevance of the theory by testing it against the existing structure of residential land use in Chicago. His tests indicate that the theory is broadly correct, with, however, certain deviations explicable by such things as racial discrimination in the housing market. We may thus infer that the theory is a true theory. This truth, arrived at by classical positivist means, can be used to help us identify the problem. What for Muth was a successful test of a social theory becomes an indicator of what the problem is. The theory predicts that poor groups must, of necessity, live where they can least afford to live.
>
> Our objective is to eliminate ghettos. Therefore, the only valid policy ... is to eliminate the conditions which give rise to the truth of the theory. In other words, we wish the von Thünen theory of the urban land market to become not true. The simplest approach here is to eliminate those mechanisms which serve to generate the theory. The mechanism in this case is very simple - competitive bidding for the use of the land. (Harvey 1973, p.137, emphasis added).

This critique is acutely relevant today, when neoclassical assumptions have been revitalized and appropriated by the political triumphs of neoliberalism, where cities "have become the incubators for many of the major political and ideological strategies through which the dominance of neoliberalism is being maintained" (Brenner and Theodore 2002, p.375–6). Municipal administrations now act less as regulators of markets to protect marginalized residents and more as *entrepreneurial* agents of market processes and capital accumulation (Harvey 1989; Peck 2005), resulting in spectacular wealth inequalities within and between cities.

Any scholar working at the forefront of exciting theoretical advances is bound to attract strong graduate students, and so it was that Harvey was joined in the late 1970s by Neil Smith, whose work on gentrification and uneven development was profoundly intertwined with the deep engagement with Marx (and Marxist thought) exhibited by his advisor. Both scholars added a geographical, spatial dimension to something that had fascinated Marx – the powerful contradictions of capital investment. Investments are required to create the places that must exist in order for profits to be made - offices, factories, shops, homes, and all the rest of the infrastructure that constitutes a city. Yet once these investments are committed to a certain place, capital cannot be quickly or easily shifted to newer, more profitable opportunities elsewhere. This is because capitalists are always forced to choose between investing to maintain the viability of previous capital commitments (or exploiting new opportunities), and neglecting or abandoning the old. Therefore capital investment is always animated by a geographical tension: between the need to *equalise* conditions and seek out new markets in new places, versus the need for *differentiation* (and particularly a division of labour that is matched to various places' comparative advantage). The result is a dynamic "see-saw" of investment and disinvestment over time and across space, in an ongoing process of *uneven geographical development* (Smith 1982, 1984; Harvey 1973, 1982, 2003). Capitalism is always creating new places, new environments designed for profit and accumulation, in the process devalorizing previous investments and landscapes:

> The logic behind uneven development is that the development of one area creates barriers to further development, thus leading to underdevelopment, and that the underdevelopment of that area creates opportunities for a new phase of development. Geographically this leads to the possibility of what we might call a "locational seesaw": the successive development, underdevelopment, and redevelopment of given areas as capital jumps from one place to another, then back again, both creating and destroying its own opportunities for development. (Smith 1982, p.151).

Capital investment in a particular land use will eventually face an unavoidable depreciation: buildings and other infrastructure age, and require ongoing labour and capital for maintenance and repair. As new urban growth adopts better construction and design technologies, land uses developed in previous generations become less competitive and less profitable. The resultant flight of capital away from certain areas of the city – depreciation and disinvestment – has devastating implications for people living at the base of the see-saw, the bottom of the urban class structure. To take one example from the domain of housing: landlords in poorer inner-city neighborhoods are often holding investments in buildings that represented what economists and urban planners call the 'highest and best use' over a century ago; spending money to maintain these assets as low-cost rental units becomes ever more difficult to justify with each passing year, since the investments will be difficult to recover from low-income tenants. It becomes rational and logical for landlords to "milk" the property, extracting rent from the tenants, spending the *absolute minimum* to maintain the structure, and waiting as potential ground rent increases in the hopes of eventually capturing a windfall through redevelopment and gentrification (Smith 1979). With the passage of time, the deferred maintenance becomes apparent: people with the money to do so will leave a neighborhood, and financial institutions "red-line" the neighborhood as too risky to make loans (Squires 1992). Neighborhood decline accelerates, and moderate-income residents and businesses moving away are replaced by successively poorer tenants who move in - they cannot access housing anywhere else. The lack of maintenance expenditure leads to tough housing conditions for those poorer tenants, amidst myriad other consequences of capital disinvestment such as high unemployment, poor schools, inadequate retail services, dismal health outcomes, and so on. Crucially, and in sharp contrast to much popular and intellectual perception, such areas usually see social networks and community ties within them *strengthen* as a coping mechanism for the withdrawal of capital. Residents living in disinvested parts of cities fall back on what they know and what they have - each other.

What is the relevance of this account of uneven development to a critique of neighbourhood effects? Let us recall the textbook definition: that where you live affects your life chances. Applied to poor people, it infers that the influences of what surrounds you explain your poverty - it's all about negative role models wallowing in a culture of concentrated poverty that stop people rising up and finding a better life and escaping their neighbourhood. As Steinberg (2010a) has argued, "Stripped of its prosaic veneer, the 'neighborhood effects' model assumes that poverty feeds on itself, that it metastasizes and is a cancer on the body politic." So, in any society where class inequality is present, or diffracted through a racial-ethnic prism (or through any other aspect of socio-cultural polarization), residential turnover leading to entrenched disinvestment almost invariably unleashes an all-encompassing, discriminatory and stigmatizing argument: that *the clustering of poor people causes neighborhood decline.* But the Marxist view of capitalist urbanization explains that poorer residents and businesses can only afford to move in *after* a neighborhood has been devalorized - *after* capital disinvestment and the departure of the wealthy and middle classes. The system, therefore, is causing neighbourhood disinvestment and truncating the life chances of the poor, who become stuck in place due to the exclusive nature of a city's highly competitive housing market.

In capitalist property markets, the decisive 'consumer preference', to take the neoclassical language, is *the desire to achieve a reasonable rate of return on a sound financial investment* (Smith 1979). Private property rights feed those desires, but there is a price to pay to access those 'rights'. Contrary to the neoclassical paradigm, there is nothing natural or optimal about such a situation. The clear injustice is that the owners of capital tend to see spectacular wealth gains at the expense of those residing in neighbourhoods robbed (often quite literally) of adequate investment. When the see-saw tips again and investment does arrive, it is seldom geared to the interests of the poor; on the contrary, tenants are evicted and displaced as 'rent gaps' (Smith 1979; Clark 1987) are exploited and gentrification begins. Although the term 'gentrification' was not coined until 1964, Engels recognized the systemic injustices upon which it thrives as early as 1872:

> The expansion of the big modern cities gives the land in certain sections of them, particularly in those which are centrally situated, an artificial and often enormously increasing value; the buildings erected in these areas depress this value, instead of increasing it, because they no longer correspond to the changed circumstances. They are pulled down and replaced by others. This takes place above all with centrally located workers' houses, whose rents, even with the greatest overcrowding, can never, or only very slowly, increase above a certain maximum. They are pulled down and in their stead shops, warehouses, and public buildings are erected. ... The result is that the workers are forced out of the center of the towns towards the outskirts ...

Capitalist property markets, in short, favour the creation of urban environments to serve the needs of capital accumulation:

> Capitalist development has therefore to negotiate a knife-edge path between preserving the exchange values of past capital investments in the built environment and destroying the value of these investments in order to open up fresh room for accumulation. Under capitalism there is, then, a perpetual struggle in which capital builds a physical landscape appropriate to its own condition at a particular moment in time, only to have to destroy it, usually in the course of a crises, at a subsequent point in time. (Harvey 1978, p.124).

Harvey was quick to show that the urban process under capitalism was far more than a matter of capital flows: it was about class inequality, the formation of an exploited and alienated urban working class and "the violence which the capitalist form of accumulation inevitable inflicts upon it" (p.124), and the possibilities the contradictions of capitalism create for resistance by the working class (class struggle). On that final point, Engels identified only one 'concentration effect' in 1840s Manchester: the intensity of working class clustering and oppression in particular districts meant that "the spatial configuration of the city only accelerated the nurturing of class consciousness" (Hunt 2009, p.109). This is seldom something associated with poor districts today by researchers working on neighbourhood effects – such effects are almost always seen as deeply negative, destructive phenomena. So, where Pierre Bourdieu (1993) may have been correct in offering a view of 'site effects' whereby "bringing together on a single site a population homogeneous in its dispossession strengthens that dispossession" (p.129), the history of class struggle and social movements teaches us that such dispossession can strengthen the possibilities for *repossession* of a 'right to the city' (Lefebvre 1970).

## "When We Control For Everything, We Lose Control"

> In short, a policy predicated on the claim that the demolition of their homes will advance the interests of the very people whose homes are being destroyed is a preposterous sham. Stephen (2010b, p.222).

Having summarized (albeit rather crudely) the analytic tradition of Marxist theory towards addressing the urbanization question, and having offered a glimpse of the lessons it offers in respect of understanding why people live where they do, we turn now to some of the conclusions and arguments of proponents of the neighbourhood effects thesis, and particularly, a consideration of the *political* implications of such scholarship. As (Sampson et al. 2002) comprehensive review reveals, the neighbourhood effects literature was enormous even a decade ago, so I can only offer a taster and initial critique of some of the work that reflects the popular view that where you live affects your life chances at precisely the same time as it ignores systemic injustices. In doing so it becomes pertinent to draw upon some of the rather lonely yet crucial critiques that have been advanced in recent years to call the neighbourhood effects paradigm into question and, in particular, to elucidate its troubling political import.

As (Manley et al. 2011) explain, the initial stimulus to engage with neighbourhood effects was provided by Wilson (1987, 1991) in his influential attempts to wrench the 'underclass' term away from conservative researchers (among them Charles Murray) and give it an economic and spatial foundation, one faithful to the initial coinage of the term by Gunnar Myrdal in the 1960s. Wilson was particularly persistent with the research question of entrenched unemployment in neighbourhoods exhibiting high poverty; he attributed 'joblessness' (to use his preferred term) not only to the refusal of employers to hire residents from certain neighbourhoods with a negative reputation, but to the very *concentration* of residents experiencing long term unemployment, which led to "negative social dispositions, limited aspirations, and casual work habits" (Wilson 1991, quoted in Manley et al. 2011, p.153). Wilson's arguments influenced a generation of liberal scholars interested in far more than simply the labour market outcomes of 'concentrated poverty', as Manley et al. summarise:

> explanations of neighbourhood effects....include role model effects and peer group influences, social and physical disconnection from job-finding networks, a culture of poverty leading to dysfunctional values, discrimination by employers and other gatekeepers, access to low quality public services, and high exposure to criminal behaviour. (ibid.)

We are dealing, then, with *ecological thematics*: for the liberal proponents of thesis, it is the *neighbourhood* that is the problem to be addressed by policy, over and above the personal characteristics of its residents (the exclusive focus of conservative scholars). Bauder (2002) captures these thematics succinctly:

> The idea of neighbourhood effects suggests that the demographic context of poor neighbourhoods instills 'dysfunctional' norms, values and behaviours into individuals and triggers a cycle of social pathology and poverty that few residents escape. ...[It] implies that the residents of the so-called ghettos, barrios and slums are ultimately responsible for their own social and economic situation. (p.85)

'Neighbourhood effects' is therefore more than just a concept – it is an *instrument of accusation*, a veiled form of class antagonism that conveniently has no place for any concern over what happens *outside* the very neighbourhoods under scrutiny. Take, for instance, the conclusions of a logit regression analysis of 'school dropout rates' among Australian teenagers from various socioeconomic categories (Overman 2002). After stating that "living in an area where the immediate neighbourhood has low socioeconomic status has a negative effect on dropout propensities", the author offers the following policy recommendation:

> government policies placing small clusters of low SES [socio-economic status] families in better [sic] neighbourhoods may have little significant impact on dropout rates. 'Forced' mixing through government housing programmes may need to ensure that low SES families are well dispersed throughout more affluent neighbourhoods, rather than concentrated in 'sink' estates. (p.128)

This article is completely silent on the general quality of the education system, the relevance of the school curriculum to the hopes and dreams of young people, the availability of inspiring teachers, the educational opportunities beyond school corridors (such as in apprenticeships and mentoring schemes), the possibility that leaving school early to find work might be an economic necessity (part of a household survival strategy), the possibility that teenagers may have to quit school to care for a frail relative. These issues are ignored in favour of recommending that the Australian government disperse poor teenagers as widely as possible, because when clustered together, they feed off each other in a shameful school dropout culture.

Were these conclusions not dressed up in scientific language and legitimized by numbing tables of parameter values, they would be highly controversial. Unequal educational attainment needs to be considered as an offshoot of the unequal provision of public goods and unequal treatment by the state of the different areas. The degree of inequality between neighbourhoods with bad schools and good schools is not a property of the neighbourhood, but a property of the school system. However, Overman's analysis offers considerable ammunition for an embrace of the neighbourhood effects thesis: concentrations of poverty in 'low status' neighbourhoods in Australia apparently harm life chances, so reducing those concentrations by scattering poor young Australians among richer young Australians (prospective educational 'role models', presumably) apparently solves the problem.

Another example comes from a recent study of 'neighbourhood income mix' in respect of the earnings of adults in Sweden. Deploying regression analysis of government data on income, education, labour market, and population, Galster and colleagues (2008) comment that their "robust results" are

> consistent with the view that, for males who are not fully employed, low-income neighbors provide negative role models and middle-income (but not high-income) neighbors provide access to networks with valuable employment-related information. For those already fully employed, high-income neighbors probably are valuable because they provide access to networks with information about opportunities for more lucrative employment. (p.868)

Most troubling of all about these words is that *not a single Swede of any income category was interviewed for the study*. Therefore, the authors of this study could not possibly offer detailed analytical insights about social networks vis-à-vis employment opportunities in different urban districts of Sweden (it is not a social networks analysis, for a start). Their conclusions also contradict ethnographic inquiries that offer evidence that if there is one thing common to the experience of living in a poor neighbourhood anywhere, it is precisely the "valuable employment-related information" that is passed around as a collective economic survival strategy (e.g. Venkatesh 2006). As Cheshire (2006) has quite correctly reminded us, "people derive welfare from living near to other complementary – usually similar – households" (p.1240). Furthermore, rather than serve as role models for those worse off, middle-income neighbours are in fact far more likely to socialize and share information *among themselves*, as has been documented in countless sociological analyses of the middle-classes (e.g. Butler and Robson 2003). In addition, to men who are "not fully employed" low-income neighbours offer solidarity, empathy, informal social care, community and kinship. They may not provide full employment, but that does not make them "negative role models".[5]

---

[5] On the question of "negative role models", aside from the underclass caricatures invoked by such language, it speaks volumes of the neighbourhood effects literature that *not a single scholar of the genre* ever asks the burning question: what turns someone into a negative role model? What leads someone to impress on others the view that everyone is against them, that there is no hope in their neighbourhood, so they might as well give up on finding a job and join a gang and immerse themselves into a world of drugs, crime, etc? The answers can be found in terrain never explored by analysts of a neoclassical persuasion: a hostile entry-level labour market, the lack of a living wage or basic income, the absolute indignity of living in a stigmatized territory, the expansion of the penal fist of the state, the compassion fatigue displayed by civic institutions – the tragedy is that there are simply so many structural factors that condemn so many to poverty and social suffering (Bourgois 2003).

The assumptions of the Galster et al. article are matched in a study of whether neighbourhoods matter in the 'transition from welfare to work' in the Dutch city of Rotterdam (van der Klaauw and van Ours 2003). Before any data are presented and analysed, here is what the authors assume, and what guides their empirical inquiry:

> On the one hand in a neighbourhood with high unemployment, there might be less (informal) information about jobs available, i.e. social networks in these neighbourhoods are less valuable when searching for a job. On the other hand, the attitude towards joblessness and the social norms concerning work may differ between low and high unemployment neighbourhoods. (p.961)

The language is revealing: "less valuable", "attitude towards joblessness" and "social norms" - the instrument of accusation at work. Another regression analysis later, here is the principal finding and the recommended policy package:

> Our empirical results show that the neighbourhood affects the individual transition rates from welfare to work of young Dutch welfare recipients. These transition rates are lower if the unemployment rate within the neighbourhood is higher. .... From a policy point of view this implies that when it comes to youth unemployment policy special attention should be given to young welfare recipients in high unemployment neighborhoods. ...[I]f high unemployment rates have a negative effect on individual transition rates from welfare to work because they cause a negative attitude towards work then a policy of strict monitoring is useful. (p.984)

The structural factors that give rise to a tough entry-level labour market for young people in Rotterdam are strategically, perhaps judiciously disregarded. In addition, the "strict monitoring" recommended is particularly worrisome – effectively the authors are suggesting that the Dutch authorities need to travel down the U.S. welfare-to-work program path, regardless of its widely documented devastation of poor communities across the U.S. The title of their essay suggests as much, as does their comment that "in most literature on neighbourhood effects in the US results similar to ours are found." (p.982).

Considered together, the essays I have discussed above provide a snapshot of the neighbourhood effects genre, where authors are as quick to make sweeping assertions about communities into which they rarely (if ever) set foot as they are to ignore the political implications of those assertions. Bauder (2002) has advanced a powerful critique:

> The direct causality implied by neighbourhood effects models presents a simple and 'straight-forward' explanation for the social and economic marginality of inner-city residents, which entices through its use of quantitative methods and its claim to be objective and value-free. Yet...this literature makes ideological assumptions that remain unacknowledged by many researchers. One of these assumptions is that suburban middle-class lifestyles are normal, and inner-city, minority lifestyles are pathological. (p.89)

He continues to outline what is at stake:

> Researchers should be particularly critical of neighbourhood effects because the concept lends itself as a political tool to blame inner-city communities for their own marginality... [and] provides scientific legitimacy to neighbourhood stereotypes among employers, educators and institutional staff, and justifies slum-clearance and acculturation policies. (p.90)

A note here on methodology becomes necessary. I have no wish to denigrate a particular tradition of quantitative inquiry developed over many years of urban inquiry. It is unwise politically to cast aside the radical potential of quantitative exploration, especially when considered in creative tension with qualitative accounts (Sheppard 2001; Wyly 2011). But there is a serious analytic booby trap that afflicts those working with regression techniques (which appear to be something of an obsession in the neighbourhood effects literature). Regressions appear to show that it is not just that poor people live in poor neighborhoods, but that the neighborhood effects *exceed* what would be predicted by poverty alone. But even if that is true, dispersing the poor to wealthier places, as is so often advocated, would only eliminate that incremental difference (the 'concentration effect'), without even pretending to address the institutional and structural arrangements driving poverty. The incremental "after controlling for" logic and discourse thus seems deeply misguided. It is underpinned by a 'ceteris paribus' argument that is necessarily false - statistically controlling for characteristics of entrants into different neighborhoods does not make these individuals equal because we know (via Marxist insights discussed earlier) that the processes of allocation through space are *not random*. As two skilled and astute practitioners of the quantitative craft working on housing foreclosures in the US have warned: "When we control for everything, we lose control." (Wyly and Ponder 2011, p.560).

The lack of attention to wider structures is not just an occupational hazard of the statistician. Those who spend time in poor communities can also fall into the trap becoming so immersed in their research context that they fail to consider in significant depth the broader institutional political economy that would shed light on their findings. For instance, in a Chicago study of racial differences in neighbourhood social networks, Mario Small (2007) claims that his "findings are most consistent with the work of [William Julius] Wilson" after arguing that

> the consistency of the neighborhood poverty effect across different types of outcomes makes it difficult to rule out a neighborhood effect…. At a minimum, it is certainly the case that individuals with identical observed characteristics face alarmingly higher rates of social isolation if they live in high poverty neighborhoods than if they live in low poverty neighborhoods.

But is social isolation in a high poverty neighborhood a 'neighborhood effect', or something much deeper? In an exhaustive review of area-based policies in advanced capitalist societies, Andersson and Musterd (2005) remarked that "we should keep in mind that problems in the neighbourhood are seldom problems of the neighbourhood…..an area focus cannot by itself tackle the broader structural problems, such as unemployment, that underlie the problems of small areas." (p.386) In a similar vein, Wacquant (2008) argues that neighbourhood effects convey a "falsely depoliticized vision of urban inequality" and are best understood as *the effects of the state inscribed into urban space*: "in reality they track the extent to which the state works of fails to equalize basic life conditions and strategies across places" (p.284). (Small et al. 2008) attempt to protect themselves from such a critique when they state that their study of organizational ties (specifically, child care provision) in New York

City neighbourhoods "points to the importance of reconsidering the state and the non-profit sector, especially under the current political economy". Yet in their paper we actually learn very little about the historical path to the state's renouncement (at various scales) of its regulatory and protective functions that have placed such major pressures on the providers of social care for the population living at the bottom of the class structure in that city.

## Demolition and Emplacement

A decade ago Robert Sampson and colleagues (Sampson et al. 2002) wryly observed that "the study of neighborhood effects, for better or worse, has become something of a cottage industry[6] in the social sciences." (p.444). They did not see this as a problem *per se* – more troubling to these authors was that "the bad news is that this recent spurt in quantity has not been equally matched in quality; much hard work remains to be done." (ibid.) In the decade that followed these remarks, the cottage industry has shown no signs of relenting, and beyond the measurement and 'controlling' debates the literature appears to serve primarily as ideological justification for the policy hubris of creating 'mixed income communities' via "poverty deconcentration". Cheshire (2006) has elaborated the ramifications of such policies:

> Forcing neighbourhoods to be mixed in social and economic terms is treating the symptoms of inequality: it is on a par with applying leeches to lower a fever. At the same time, if there are welfare benefits derived from living in specialised neighbourhoods with other complementary households, the policy is directly destroying a potential source of welfare and a portion of the consumption benefits cities are capable of delivering. (p.1241)

Whilst I have some sympathy with (Manley et al. 2011) argument that we need far better information on how individuals sort into neighbourhoods, I do not share their view that

> The future for quantitative neighbourhood effects studies lies in the use of more sophisticated and tailored data which allows detailed geocoding of individuals and allows the modelling of selection mechanisms into neighbourhoods. (p.168)

---

[6]It is fascinating how the meaning of 'cottage industry' has developed ambiguity over time. Originally it was associated with protoindustrialisation, particularly in association with west Yorkshire woolens, and referred to a geographically dispersed but nonetheless regulated system of production that did not involve 'machinofacture' or the intense concentration of labour (Houston and Snell 1984). In scholarly circles today it usually refers to the intense concentration of research activity and output on a specific theme or sub-theme of inquiry. I am grateful to Innes Keighren and Charlie Withers for the historical clarification. The prodigious output on neighbourhood effects documented by Sampson and co matches that of the earlier intellectual cottage industry on the so-called "underclass" (Wacquant 1996), to which the neighbourhood effects literature is closely related.

More sophisticated and tailored data are unlikely to be used for anything other than *decision-based evidence making* by policy elites on the hunt for scientific legitimacy for agendas that trample over the rights of the poor who, through a combination of bad luck and systemic injustices, are relegated to stigmatised neighbourhoods that become the problem rather than the expression of the problem to be addressed. 'Neighbourhood effects' is a field of intellectual inquiry that is now so divorced from the raw realities of capitalist urbanization that only one solution seems appropriate, given its failure to say much of substance about poverty and especially class inequality in cities. It requires the same *demolition* that many of its authors prescribe for housing in the communities they study.

Recall that one of the main weapons in the policy elite's arsenal is the activation of an intense stigma already attached to the parts of the city where poverty is high. When a city district becomes blemished by all kinds of derogatory terms and phrases, it makes the job of implementing drastic policies considerably easier for their architects:

> Once a place is publicly labelled as a 'lawless zone' or 'outlaw estate', outside the common norm, it is easy for the authorities to justify special measures, deviating from both law and custom, which can have the effect – if not the intention – of destabilizing and further marginalizing their occupants, subjecting them to the dictates of the deregulated labour market, and rendering them invisible or driving them out of a coveted space. (Wacquant 2007, p.69)[7]

Therefore, in order to consign the neighbourhood effects genre to the intellectual dustbin, it is of paramount importance for a new cottage industry to emerge in the form of a research agenda that compliments the theoretical and political insights of Marxist approaches to capitalist urbanization to *show exactly how and why spaces are coveted by those who stand to lose them*. Crookes (2011) has called this a need for 'emplacement' studies in urban research, geared towards resisting the displacement fetish among urban policy elites. Drawing on Fullilove's (2004) mandate that "we can't understand the losses unless we first appreciate what was there" (p.20), Crookes draws upon his fieldwork among displacees from a northern English city subjected to Housing Market Renewal (see the third paragraph of this chapter) to argue that place attachment is especially strong in lower income neighbourhoods, and in direct response to marginalistion and residents' psychosocial and material needs. He convincingly points out that scholars of all political stripes are guilty of referring to poorer neighbourhoods in terms (decayed, traumatised, etc.) that do little to correct the view that people within them do not lead meaningful or valuable lives, against the odds. In respect of gentrification he says that

---

[7] Maloutas (2009) has argued that when Wacquant writes of territorial stigmatisation he is "definitely arguing about a growing neighbourhood effect" (p.830). This misconstrues stigmatisation, which is not a property of the neighbourhood, but a gaze trained on it. Its effects are the effects of symbolic structures applied to the neighbourhood, not a neighborhood effect.

Beginning at 'home' helps us to think gentrification differently by shifting the focus of our attention from displacement to what is already there and what could be lost… From this perspective gentrification is no longer about the production of space for more affluent users but the violent dispossession of home for people who, for various reasons, may have a much stronger connection to home and place than those that do the taking.

This crucial perspective and research agenda must serve as a point of departure for a serious challenge to unjust policies that are targeted at 'dysfunctional' neighbourhoods, and to the literature from which those policies draw credibility. It is also essential, as urban scholars and social scientists, to reflect upon why there is such an absence of research on extremely rich neighborhoods, and correspondingly, of policies aimed at dispersing the rich when their concentration caused such grievous collective disasters as the 2008 financial crisis.

# References

Allen, C. (2008). *Housing market renewal and social class*. London: Routledge.
Alonso, W. (1964). *Location and land use*. Cambridge: Harvard University Press.
Andersson, R., & Musterd, S. (2005). Area-based policies: A critical appraisal. *Tijdschrift voor Economische en Sociale Geografie, 96*(4), 377–389.
Bauder, H. (2002). Neighbourhood effects and cultural exclusion. *Urban Studies, 39*(1), 85–93.
Bourdieu, P., et al. (1993). *The weight of the world: Social suffering in contemporary society*. Cambridge: Polity Press.
Bourgois, P. (2003). Crack and the political economy of social suffering. *Addiction Research and Theory, 11*(1), 31–37.
Brenner, N., & Theodore, N. (2002). Cities and the geographies of 'actually existing neoliberalism. *Antipode, 34*(3), 349–379.
Butler, T., & Robson, G. (2003). *London calling: The middle classes and the global city*. London: Berg.
Cheshire, P. (2006). Resurgent cities, urban myths and policy hubris: What we need to know. *Urban Studies, 43*(8), 1231–1246.
Clark, E. (1987). *The rent gap and urban change: Case studies in Malmo 1860–1985*. Lund: Lund University Press.
Cole, I., & Nevin, B. (2004). *The road to renewal*. York: York Publishing Services.
Crookes, L. (2011). *The making of space and the losing of place: A critical geography of gentrification-by-bulldozer in the North of England*. PhD dissertation, University of Sheffield.
Engels, F. (1843). *Outlines of a critique of political economy*. Deutsch-Französische Jahrbücher.
Engels, F. (1845). *The condition of the working class in England*. London: Penguin Classics.
Engels, F. (1872). *The housing question*. Available from: http://www.marxists.org/archive/marx/works/1872/housing-question/index.htm.
Fullilove, M. (2004). *Root shock: how tearing up city neighborhoods hurts America and what we can do about it*. New York: One World.
Galster, G., Andersson, R., Musterd, S., & Kauppinen, T. M. (2008). Does neighborhood income mix affect earnings of adults? New evidence from Sweden. *Journal of Urban Economics, 63*(3), 858–870.
Goetz, E. G. (2003). *Clearing the way: deconcentrating the poor in urban America*. Washington, D.C.: Urban Institute Press.
Harvey, D. (1973). *Social justice and the city*. London: Arnold.
Harvey, D. (1978). The urban process under capitalism: A framework for analysis. *International Journal of Urban and Regional Research, 2*(1–4), 101–131.

Harvey, D. (1982). *The limits to capital*. Oxford: Blackwell.
Harvey, D. (1985). *The urbanization of capital*. Baltimore: Johns Hopkins University Press.
Harvey, D. (1989). From managerialism to entrepreneurialism: The transformation of urban governance in late capitalism. *Geografiska Annaler B, 71*(1), 3–17.
Harvey, D. (2000). *Spaces of hope*. Edinburgh: Edinburgh University Press.
Harvey, D. (2003). *The new imperialism*. Oxford: Oxford University Press.
Harvey, D. (2006). Memories and desires. In S. Aitken & G. Valentine (Eds.), *Approaches to human geography*. London: Sage.
Houston, R., & Snell, K. D. M. (1984). Proto-industrialization? Cottage industry, social change, and industrial revolution. *The Historical Journal, 27*(2), 473–492.
Hunt, T. (2009). *Marx's general: The revolutionary life of Friedrich Engels*. London: Holt McDougal.
Jeffries, S. (2006). On your Marx. *The Guardian*. http://www.guardian.co.uk/news/2006/feb/04/mainsection.stuartjeffries. Accessed 4 Feb.
Lefebvre, H. (1970). *The urban revolution*. Minneapolis: University of Minnesota Press.
Maloutas, T. (2009). Urban outcasts: A contextualized outlook on advanced marginality. *International Journal of Urban and Regional Research, 33*(3), 828–834.
Manley, D., & van Ham, M. (2010). Choice-based letting, ethnicity and segregation in England. *Urban Studies, 48*(14), 3125–3143.
Manley, D., van Ham, M., & Doherty, J. (2011). Social mixing as a cure for negative neighbourhood effects: Evidence-based policy or urban myth? In G. Bridge, T. Butler, & L. Lees (Eds.), *Mixed communities: Gentrification by stealth?* (pp. 151–168). Bristol: Policy Press.
Metzger, J. (2000). Planned abandonment: The neighbourhood life-cycle theory and national urban policy. *Housing Policy Debate, 11*(1), 7–40.
Muth, R. (1969). *Cities and housing*. Chicago: University of Chicago Press.
Overman, H. G. (2002). Neighbourhood effects in large and small neighbourhoods. *Urban Studies, 39*(1), 117–130.
Peck, J. (2005). Struggling with the creative class. *International Journal of Urban and Regional Research, 29*(4), 740–770.
Popkin, S., et al. (2005). Public housing transformation and the hard-to-house. *Housing Policy Debate, 16*(1), 1–24.
Sampson, R., Morenoff, J. D., & Gannon-Rowley, T. (2002). Assessing 'neighbourhood effects': Social processes and new directions. *Annual Review of Sociology, 28*, 443–478.
Sheppard, E. (2001). Quantitative geography: Representations, practices, and possibilities. *Environment and Planning D: Society & Space, 19*, 535–554.
Small, M. (2007). Racial differences in networks: Do neighborhood conditions matter? *Social Science Quarterly, 88*(2), 320–343.
Small, M., Jacobs, E., & Massengil, R. (2008). Why organizational ties matter for neighborhood effects: A study of resource access through childcare centers. *Social Forces, 87*(1), 387–414.
Smith, N. (1979). Toward a theory of gentrification: A back to the city movement by capital, not people. *Journal of the American Planning Association, 45*(4), 538–548.
Smith, N. (1982). Gentrification and uneven development. *Economic Geography, 58*(2), 139–155.
Smith, N. (1984). *Uneven development: Nature, capital and the production of space*. Oxford: Blackwell.
Smith, N. (1992). Blind man's bluff, or Hamnett's philosophical individualism in search of gentrification? *Transactions of the Institute of British Geographers, 17*(1), 110–115.
Smith, D. M. (1994). *Geography and social justice*. Oxford: Blackwell.
Squires, G. (Ed.). (1992). *From redlining to reinvestment: Community responses to urban disinvestment*. Philadelphia: Temple University Press.
Steinberg, S. (2010a). Reply to post by Mario Luis Small. Posted on Comurb listserv, January 17th.
Steinberg, S. (2010b). The myth of concentrated poverty. In C. Hartman & G. Squires (Eds.), *The integration debate: Competing futures for American cities* (pp. 213–227). New York: Routledge.

van der Klaauw, B., & van Ours, J. (2003). From welfare to work: does the neighborhood matter? *Journal of Public Economics, 87*(5–6), 957–985.
van Ham, M., Manley, D., Bailey, N., Simpson, L., & Maclennan, D. (Eds.). (2012). *Neighbourhood effects research: New perspectives*. Dordrecht: Springer. doi:10.1007/978-94-007-2309-2.
van Ham, M., Manley, D., Bailey, N., Simpson, L., & Maclennan, D. (Eds.). (2013). *Understanding neighbourhood dynamics: New insights for neighbourhood effects research*. Dordrecht: Springer.
Venkatesh, S. (2006). *Off the books: The underground economy of the urban poor*. Cambridge: Harvard University Press.
Wacquant, L. (1996). L' 'underclass' urbaine dans l'imaginaire social et scientifique américain. In S. Paugam (Ed.), *L'Exclusion. L'état des Savoirs* (pp. 248–262). Paris: La Découverte.
Wacquant, L. (2007). Territorial stigmatization in the age of advanced marginality. *Thesis Eleven, 91*(1), 66–77.
Wacquant, L. (2008). *Urban outcasts: A comparative sociology of advanced marginality*. Cambridge: Polity Press.
Wilson, W. J. (1987). *The truly disadvantaged*. Chicago: University of Chicago Press.
Wilson, W. J. (1991). Another look at the truly disadvantaged. *Political Science Quarterly, 106*(4), 639–656.
Wyly, E. (2011). Positively radical. *International Journal of Urban and Regional Research, 35*(5), 889–912.
Wyly, E., & Ponder, C. S. (2011). Gender, age, and race in subprime America. *Housing Policy Debate, 21*(4), 529–564.

# Chapter 7
# Social Mix: International Policy Approaches

**Keith Kintrea**

## Introduction

The popularity of 'neighbourhood effects' as a research theme in urban studies has been mirrored by the international appeal of 'social mix' as a policy prescription for disadvantaged neighbourhoods (see also van Ham et al. 2012, 2013). Chapters in this volume discuss approaches in the USA, the UK, the Netherlands and Canada but policies of some kind have also been tried in Australia and in many other European countries. In the literature, policies are captured by terms such as 'social mix', 'mixed communities' 'living together' and 'desegregation' but they all have the broad aim of diversifying the social composition of urban neighbourhoods. Typically, social mix as a policy object is not precisely defined but policies are usually targeted at neighbourhoods where concentrated disadvantage is found; just occasionally they seek to promote mix across a wider range of urban areas. This chapter reviews the international experience of social mix policies. It provides background on how the academic ideas on neighbourhood effects helped to influence this policy approach, examines the political imperatives for mixed communities and the mechanisms employed, and makes an assessment of their record.

## Neighbourhood Policy and Neighbourhood Effects

Across the developed world, there is a clustering of poor people in poor neighbourhoods, although they rarely house the majority of the poor. There is plenty of evidence that neighbourhoods where poor people are concentrated lack the resources

---

K. Kintrea (✉)
Urban Studies, School of Social and Political Sciences, University of Glasgow,
25 Bute Gardens, Glasgow G12 8RS, Scotland, UK
e-mail: Keith.Kintrea@glasgow.ac.uk

and amenities that better off neighbourhoods provide, and that services are of poorer quality, so the opportunities they offer to their residents are more limited. Therefore, disadvantaged areas represent a challenge for social justice, and this is one of the most important rationales which has underlain spatially targeted social policy over the decades. And poor areas are a visible manifestation of poverty and therefore a political problem; for whatever views there are about the diagnosis of and prescription for poor areas, poverty concentrations are rarely welcome.

Andersson (2006) sets three basic policy approaches to counter the segregation of the poor in cities. First, policy makers might seek to decrease inequalities between rich and poor neighbourhoods, for example through taxing the rich and redistributing resources to the poor across the whole society, therefore improving the relative position of the poor in poor areas. Second, they might seek to ensure that inequality in the poorest neighbourhoods is not extreme, through targeted area policy. Third they might seek to reduce the concentration of the poor by promoting a social mix, either across all neighbourhoods, or just in poor ones.

The first policy route draws attention to the economic context for poor neighbourhoods. Where income disparities are less (in countries with a strong welfare orientation) the disadvantages associated with living in a poor area are also less (Musterd 2002; Atkinson and Kintrea 2001). Ready contrasts can be made for example between neighbourhood services, crime rates and standards of living in poor areas strong welfare states compared to those in liberal market economies. However, income redistribution is rarely seen as a neighbourhood policy per se. Instead neighbourhood policy has generally followed the second path, that is targeted regeneration of poor neighbourhoods. This approach has been around a long time, in many countries since the 1970s or even before. So Maclennan (2013) argues that a number of good reasons remain to target specific neighbourhoods, including social justice, resource redistribution and encouraging investor confidence, whether neighbourhood effects are important or not. Similarly Bradford (2013) outlines how Canadian policy is now coming to terms -late compared to most other developed countries- with accumulating evidence of concentrated poverty and is developing new approaches to improving service delivery in disadvantaged neighbourhoods. Even in England, where strategic attention to poor neighbourhoods has now vanished, the limited opportunities offered by disadvantaged neighbourhoods are still recognised as a problem by the Conservative-Liberal coalition in its social mobility strategy (UK Cabinet Office 2011).

But it is the third policy route, that is social mix, that has been prominent recently. It rests significantly on a complex set of 'neighbourhood effects' propositions which hold that concentration of disadvantage is not just a spatial expression of inequality, but somehow makes inequality worse through social processes that operate in areas where large numbers of poor people are clustered. 'Neighbourhood effects', then, brings to the surface both the spatial and the social dimensions of neighbourhood, and the challenge of understanding what the relationships are between them. The scale and complexity of this challenge is part of the reason why it has become such a popular field of research endeavour.

Policy towards poor neighbourhoods has been influenced by both individual -or household- perspectives as well as perceived societal impacts. For individual residents, the central argument is that living in a poor area actively depresses people's life chances. Since Wilson (1987), using Chicago evidence, suggested that to be poor in a poor neighbourhood compounded people's poverty, 'neighbourhood effects' has held a fascination for urban researchers. The core of the idea is that the locality is an important arena in which social relationships help to construct beliefs, attitudes and expectations, which have important impacts on people's life chances. While neighbourhood effects can emerge, in principle, in any kind of area and can be positive or negative for wellbeing, the key concern is the potential negative impacts that may arise in disadvantaged areas, and serve to compound disadvantage.

There is a set of interlocking propositions about how neighbourhood effects 'work' to accentuate disadvantage (e.g. Atkinson and Kintrea 2002). The first is a necessary condition: that people with low incomes tend to live in relative isolation in their own areas. The second proposition relates to the type of social capital that exists in poor areas. Social capital can be understood as the resources that people get from their relationship with each other, and it is often divided into two types. The proposal is that 'bonding' social capital predominates in poor areas and is constraining and inward-looking, while 'bridging' social capital is found in greater quantities elsewhere and is a key resource for social and economic advantage (Portes 2000; Forrest and Kearns 2001). So in poor areas residents may be deprived of the resources that those with wider social networks have access to. This is particularly relevant for those who spend a greater proportion of their time in the local neighbourhood, including women, children, the unemployed and older people. A wider social network -a product of 'bridging' social capital- might have the potential to open up opportunities for economic activity, advancement or education so providing a route out of poverty, particularly in cases where less advantaged residents develop relations with better positioned people.

The third proposition is that the social norms which develop in poor areas under conditions of isolation and 'bonding' (rather than 'bridging') social capital are a further barrier to opportunity. Being trapped in social and spatial circuits where the majority of people are poor may foster low aspirations and expectations. It may lead to a sense of futility, and contribute to resentment and alienation. They may develop a set of norms and values which further distance them from success in the labour markets and the education system. For example worklessness may be regarded as unexceptional or even expected. This is of particular relevance for children and young people growing up in poor neighbourhoods as they position themselves in relation to further and higher education and the labour market (Oberwittler 2007; St Clair and Benjamin 2011).

Fourth, many poor neighbourhoods are said to characterised by a lack of social control and collective monitoring (e.g. Sampson and Raudenbush 1999). The consequences might be to reduce residents' access to services and jobs, for example because they are worried about going out after dark or crossing through certain parts of the neighbourhood. But it might also promote physical decay, vandalism, violence, crime and other social problems in public space which may cause residents to further retreat into the privacy of their homes and they avoid public places and/or their neighbours (Venkatesh 2000), leading to a downward spiral of neighbourhood disorder. In some extreme cases, this goes as far as threatening the basic needs of material and physical well-being. Again these impacts are not even among residents of poor neighbourhoods and disproportionately affect women, young people (e.g. Kintrea et al. 2011).

At the societal level, poor residential neighbourhoods are often identified as a key arena where social cohesion can break down (Musterd and Ostendorf 2009; Rowlands et al. 2009). In many European countries concentrated disadvantage is said to put at risk the stability of social and political systems as it undermines the ties people feel that have with society and with each other. Dorling (2010) argues that the segregation of the disadvantaged from the better off has a recursive effect; the more the poor are segregated, the more that wealthy groups are fearful and find it attractive to live yet further away. Wide social disparities are accompanied by 'cocooning' by better off groups, who choose their neighbourhoods (and also their transport, consumption and recreation activities) in ways which avoid the poor (Atkinson 2006). This extends segregation and this perhaps makes it even less likely that political strategies can be developed to challenge inequality.

'Communities', especially in old inner cities and mass housing estates of the post war era are frequently seen as the most problematic; their residents are said to be cut off from the norms and values of 'mainstream' society; they lack a sense of values in common with those beyond their boundaries. This is often exacerbated by ethnic and cultural differences between neighbourhoods. Residents in different neighbourhoods may come from distinctive backgrounds (plus they may be economically dissimilar) and they often seem to live 'parallel lives' or live in 'parallel societies'. For example, children go to different schools, worshippers are separated by religious affiliation, consumers buy in different shops, social networks rarely overlap and different languages are spoken. The most obvious expression of a breakdown of cohesion are outbreaks of unrest among marginalised groups from disadvantaged areas (see Wacquant 2008), which present a threat to public safety and social order. For example in the 5 days of disturbances in English cities in 2011, there was found to be a strong connection between area disadvantage and riots; more than half of those taken to court for rioting offences lived in the 20 % most disadvantaged areas of England and of the 66 local authority areas where riots took place, 30 were in the most deprived quarter of local authority areas in England (Riots Communities and Victims Panel 2012).

## Social Mix as a Contested Policy Approach

While poor neighbourhoods may be seen as a locus of disadvantage and disorder, neighbourhoods also seem to offer appealing possibilities to policymakers for 'broken societies' to be mended. Advocacy of social mix is sometimes focused on wealth and income but also sometimes connected to anxiety about the development of 'parallel lives' (Uitermark 2003; Bolt et al. 2010). The underlying belief is that mixed communities will bring benefits to disadvantaged individuals and households and, in consequence, to society as a whole. Perhaps the theory of how social mix can address social cohesion can best be summed up in the four dimensions discussed by Joseph and colleagues (2007) which embrace social networks, social control, role modeling and 'the political economy of place'.

The *social networks* approach assumes that social interaction will occur in mixed neighbourhoods between residents who have different characteristics. In other words disadvantaged people will be socially connected to the better off. This will allow disadvantaged people and minorities to expand their networks to encompass access to resources, information and jobs, and therefore to improve their economic status and participation in society at large. This dimension is particularly associated with the ideas originated by Granovetter (1973).

*Role modeling* assumes that social behaviour is learned through human contact. The theory is that better off residents, when mixed with others, will introduce disadvantaged residents to new models of behaviour, which will be positive for disadvantaged people's life chances, such as higher educational aspirations or seeking regular work.

The *social control* dimension assumes that higher income and mainstream residents will bring benefits to all residents through raising the level of social organisation, that is promoting (positive) common values and social organisation. Like *social networks* it depends on social ties and local interdependencies and trust, however not necessarily on intimate relationships. This is posited to bring benefits to poorer residents because it can help to prevent and address local problems, such as incivility and crime.

*The political economy of place* rests on the idea that higher income people living in the same neighbourhood as the poor will also act as more effective political advocates because they are better able to organise and articulate their case politically. They will also bring more wealth into the neighbourhood which will help to support better local private services, such as shops, which will give benefits to all residents.

In addition to these four dimensions referred to by Joseph and colleagues (2007), socially mixed neighbourhoods are also thought to be successful in fostering social cohesion because of general perceptions and reducing *stigmatisation* of these neighbourhoods. If they become perceived as more diverse, the discourse on diversity will change, with ultimately impacts on social cohesion through reduced exclusion and easier integration.

However, while the argument that social mix can challenge negative neighbourhood effects which arise from concentrated poverty has strong appeal, there are several important counter arguments about why social mix should not be seen as a policy solution, for example as summarised by Lees and colleagues (2012).

First, there is the question of whether social mix really leads to social mixing; certainly many authors over a long period suggest it may not (Atkinson and Kintrea 2000; Davidson 2012). A famous French study (Chamboredon and Lemaire 1970) quoted by both Blanc (2010) and Bacqué and Fijalkow (2012), as well as many other French papers on *mixité* shows that in the context of the *grands emsembles* (large housing complexes) different classes living near each other did not result in interaction. Without interaction, most of the posited benefits disappear.

Second, instead creating more cohesive and better functioning communities, many suggest it will have negative impacts social impacts. Blokland (2003). Kleinhans (2004) and Joseph and colleagues (2007) find virtually no support for the notion that positive role modeling effects will occur in mixed communities, and they found only weak support for the idea that benefits will accrue from neighbourhood-based social interaction. Amin (2002) has similarly claimed that mixed neighbourhoods are 'communities without community', marked by multiple identities and affiliations and lacking common interests.

The third criticism questions the motivations behind social mix policy and suggest that it is part of a neo-liberal attack on the poor, which panders to the idea that working class culture is inherently problematic. In this critique, social mix is 'gentrification by stealth', a state-led approach to destroying working class neighbourhoods and transforming them into sites which can be exploited better as real estate. Critics often point out that it is disadvantaged neighbourhood where lack of mix is seen as a problem and never rich ones (e.g. Lees 2008).

So, there are clearly contrasting opinions about the basis for the policy, and there is no academic consensus about the potential impact of social mixing policies. A key problem is that it is hard to get adequate data to measure neighbourhood effects, in particular to separate out their influence from that of household and individual characteristics (Feinstein et al. 2008). It has not helped that researchers around the world have had to use different data sets with different dependent and independent variables, informed by different ideas of how neighbourhood effects might work. For example, in the UK, Buck (2001), McCulloch (2001) and Bolster and colleagues (2006) have all used the British Household Panel Survey, albeit in different ways, to make comparisons over the life cycle between people that have lived in deprived areas and those that have not. Buck found that people's chance of starting a job, or leaving poverty, were decreased by living in a poor area, while their chances of re-entering poverty were higher but Bolster et al., looking at income, found no evidence of an impact from neighbourhood disadvantage over time. Van Ham and Manley (2010) used a matched spatially referenced sample of the population from two census rounds to investigate the impact of 1991 conditions on 2001 outcomes. They suggest that there are differences in employment outcomes by neighbourhood deprivation but that these are the consequence of selection effects (unemployed people tend choose to live in more deprived neighbourhoods) rather than neighbourhood effects *per se* (see also Smith and Easterlow 2005).

Evidence from research from several European countries supports to some degree the existence of neighbourhood effects (Blasius et al. 2007). The US evidence seems to show that neighbourhood make a measurable difference to various outcomes connected to life chances but are not nearly do important as family background factors. Galster (2007a) reviewed the international evidence and concluded cautiously that adults' experiences in the labour market (an outcome on which many studies have focused) were damaged by extended residence in areas where a high proportion of neighbours were poor. He was also able to conclude, even more tentatively, that there were benefits to disadvantaged individuals to live among a wider range of social groups, provided that the social distances between the poor and non-poor were not too wide.

But there is a difference between recognising that there are neighbourhood effects exist and having an evidence base strong enough to support policy interventions. While the literature tends to the conclusion that neighbourhood effects exist, it is still far from clear what mechanisms in disadvantaged areas are most associated with the compounding of disadvantage. Therefore it is hard to know what the key characteristics of alternative, more socially sustainable neighbourhoods should be. Galster (2013) explains there are several 'thorny practical issues' to be addressed in policy design including the composition of the neighbourhood population in terms of economic and ethnic mix, the concentration of particular groups, and the scale of neighbourhoods for the purpose of mixing. None of these dimensions can be determined from the research evidence but this has not stood in the way of policy makers across three continents introducing social mix and desegregation policies.

## Why Social Mix Policies?

Social mix as a policy approach became popular in the early 2000s in a large number of countries. Although its roots go back to the nineteenth century (Ley 2012; Cole and Goodchild 2000; Sarkissian 1976) and it was embedded as an idea in the early days of town planning and in post 1945 housing development (Arthurson 2008), in European countries at the turn of the twenty-first century it received an important boost from three sources.

First, social exclusion as a concept entered mainstream social policy in Europe the 1990s (Room 2005; Levitas 2005). With its emphasis on the processes whereby poverty was embedded, and not just the existence of poverty, policy makers begun to be interested in the neighbourhoods in which people were socialised and the influence they might have on people's life chances. In particular, across a number of countries, there was a perception that physical renewal of housing areas was not sufficient to improve people's lives fundamentally. Durose and Rees (2012) characterise this new approach as government viewing neighbourhoods as a less of a specific 'site' where interventions take place but more of a 'space' where key agendas about citizenship and strengthening communities were actualised.

Second, in European cities there was increasing ethnic diversity through the migration following expansion of the EU as well as wider international inflows, and larger numbers of asylum seekers and refugees. This was accompanied by differential and concentrated patterns of settlement and led to concerns about the overall cohesion and stability of society (van Gent et al. 2009), particularly among those nations for whom building solidarity between citizens was a long term political project. Social cohesion was also given prominence at the European Union level. Within a broad context of a desire for social cohesion and sustainability, and shared rights and responsibilities (e.g. Council of the European Union 2010), the Charter of Fundamental Rights of the Europe Union argues against 'ghettos' and promotes a broad social mix (Ponce 2010).

Third, many researchers were willing supporters of social mix; there was widespread excitement about the theory of neighbourhood effects, which researchers believed was well worth investigating, even if adequate data and evidence was hard to get (Atkinson and Kintrea 2002; Manley et al. 2012). Neighbourhood effects was a middle range theory embracing social and structures but also leaving rooms for agency, which appealed to researchers with social democratic political outlooks. Wilson's influential research (Wilson 1987, 1997) originated from Chicago, the emblematic city in which traditions of urban level socio-spatial analysis had been born in the 1920s. Neighbourhood effects seemed to offer a promise to lift neighbourhood research from a theoretically and methodologically weak social policy tradition into a world of cross disciplinary, multi-method social science. It also offered an alternative to unpalatable underclass perspectives on poverty which laid the blame of the lifestyle choices of its victims (e.g. Murray 1996). The excitement of researchers in the background to widespread policy making about desegregating neighbourhoods, however, does not mean, though, that policy was evidence-based. Bond and colleagues (2010) for the UK are highly critical about the quality of the evidence provided by academic reviews of neighbourhood effects and accuse researchers of peddling 'comfortable spin' that supported interventions rather than the conveying the 'inconvenient truth' of the uncertainties that surrounded the evidence.

The range of different theories about neighbourhood effects, the kinds of impacts that were actually measurable in most cases, the paucity of data, the fact that most of the most convincing accounts came from research in other counties, and the difficulty of distilling the precise contribution of the neighbourhood per se to different aspects of disadvantage all made it extremely difficult for policy makers to assess the evidence that was there, even if they had wanted to. Tunstall (2013) also points out the wide academic field in which neighbourhood effects research was conducted, in education and health research for example, and the difficulty that policy makers would face is surveying the field. She argues that as academics' scepticism grew about neighbourhood effects in the 2000s in the face of growing evidence, policy makers ploughed on regardless.

So in the UK, for example, although social mix ideas had begun to emerge as a byproduct of a separate policy stream about 'tenure diversification' the 1990s, it grew to be a key theme of housing, planning and regeneration policy during the New Labour period (1997–2010). Social; mix held tremendous appeal to policy makers as it was consistent with New Labour's focus on communities as a locus for social inclusion (see Imrie and Raco 2003; Durose and Rees 2012). At its height, the UK housing minister urged regeneration professionals and housing developers 'to think harder and faster and more creatively about how we should make mixed income communities the norms rather that the exception', adding that 'there's no place for ghettoes in early twenty-first century Britain'.

In the Netherlands, where there is also significant history of neighbourhood based social policy, the 1990s saw an emergence of the idea that the social composition of neighbourhoods should be changed. The key force behind this was government concern about the fragmentation of Dutch society under pressure from non-Dutch immigrant groups (van Gent et al. 2009; Bolt and van Kempen 2013), which persisted through periods of centre-left and more right-leaning governments in the 2000s. Similarly Swedish urban policy, which is dated by Andersson and colleagues (2009) back to the perceived failure of the 'million programme' of housebuilding in the 1970s, became clearly intended to 'break segregation' in the 2000s as minorities became more visibly concentrated in key neighbourhoods. Segregation was seen as a political problem in country that had long prided itself on a strong welfare state and solidarity between citizens.

In France, Blanc (2010), answering his own question 'why do policy makers believe in the virtues of social mix?' traces back anti- segregationist policies to the 'Anti Ghetto' Act of 1991 and the comprehensive Housing Act of 2000. Both of these have their roots in the crisis in suburban social housing estates, accommodating concentrations of immigrant families and the desire of the state to use its influence in the housing system to promote more integrative alternatives.

In all of these countries a clear feature of policy making was the careful negotiation of the distinction between social mix and ethnic mix. Although the New Labour governments in the UK became concerned about ethnic segregation, particularly after violent disturbances in Bradford and Oldham in the north of England in 2001, a clear distinction was drawn between 'social mix', which was focused on income groups often delivered through the medium of housing tenure, and 'community cohesion', which focused on relations between ethnic and/or religious groups. 'Social mix' was seen as part of housing, planning and regeneration policy and the responsibility of the housing and planning ministry, latterly called Communities and Local Government. 'Community cohesion' was seen more as an element of national security and identity and was the responsibility of the Home Office (the interior ministry).

In the Netherlands, as Bolt and van Kempen (2013) explain, the debate about socio-spatial segregation started with a focus on socio- economic disparities, but gradually became one about the integration of minorities into Dutch society. However, because of anti discrimination legislation, the actions that followed remained focused on income mix, with low income acting as a rough proxy for high levels of non-Dutch inhabitants in neighbourhoods. In France Blanc (2010) illustrates the sensitivities around race, religion and ethnicity in the social mix debate by drawing attention to the fact that the collection of official data on race and ethnicity is outlawed in France on the grounds that its use could be discriminatory. Therefore although a key issue in social segregation in France is the separation of minority groups, mainly in outer HLM estates, policy proceeds mainly through housing actions.

Australia and the USA have rather different backgrounds to their desegregation polices. The concentrations of poverty in both countries that have commanded policy attention have similarities to Europe; social housing (in the USA 'public housing') estates or 'projects' characterised by deep disadvantage. However, the smaller scale of such housing and their very deep poverty are distinctive compared to most Western European countries. In Australia, though, the approach is couched in similar description to the UK, about creating 'inclusive' 'sustainable 'communities' (Arthurson 2008). Until 2009 it was driven more by the concerns of housing authorities who own the rental housing and some federal agencies and individual states (see Atkinson 2008) rather than a national concern about the state of society. But the election of a Labour government in 2009 further increased the greater strategic focus on breaking cycles of disadvantage associated with social housing area (Hulse et al. 2010).

In the USA the background is a neglect of the poor and often punitive polices towards them that 'have helped place so many people in dangerous high poverty neighbourhoods isolated from job growth, good schools, basic personal security, decent health care and … from political influence and functioning public institutions' (Briggs et al. 2010, p.26). Unlike the European countries discussed in this chapter, there have not been major national policy initiatives around social mix nor are they central to government urban policy, and programmes have been implemented only in certain parts of the USA (Galster 2013; Fraser et al. 2012). The main driver of policy has been distinctive too: desegregation has been prompted significantly by successful legal challenges to the role of public authorities in discriminating against black residents by concentrating them in majority black housing projects, most notable the Gautreaux cases (Goetz 2003; Briggs et al. 2010). The implementation of desegregation in Chicago following Gautreaux judgment in the US supreme court in 1996 showed that desegregation was not only legally required but a practical proposition (using housing vouchers) which seemed to deliver benefits to residents of poor areas.

## Social Mix Approaches in Practice

Social mix policies is fundamentally about incentivising or changing the residential locations decisions that that face households in order to try to achieve an outcome where poor and better off people live in the same neighbourhoods. In fact, a key aspect of social mix policy across all countries is that they have relied overwhelmingly on housing programmes. The concentration on housing seems to arise partly from the longstanding, accepted role of government in shaping a nation's housing, and therefore 'scripting' (or perhaps 'socially engineering') who consumes what housing and where. So across a number of countries, by bolting on mixed communities or desegregation as an objective, policy makers were able to justify a modification of familiar, tested, housing-led programmes to meet new ends, through demolition and construction activities, land use planning and housing allocations. There are five broad housing-led approaches which have been taken.

## Social Mix Through Regeneration

The first is an inward-looking approach acting on areas of concentrated poverty and which seeks to introduce higher income residents and therefore to create a 'social mix', where before there was a majority of poor people. A variation on the same theme is to try to retain upwardly mobile residents who would otherwise leave a poor area. Such approaches often involve substantial physical restructuring to allow new housing development of a kind which will attract middle income earners, who have choices in where they live. Introducing social mix as part of area regeneration has been common in the UK, for example as part of the New Deal for Communities programme which operated in 39 areas in England in the 2000s (Cole and Green 2011) as well as the Mixed Communities Pilots (Lupton et al. 2010). In these cases, low demand social housing is demolished and replaced by housing for sale in order to attract economically active residents. In France *désenclavement* (opening up existing residential areas) is similarly practiced through the in the 'Urban Sensitive Zones' (ZUS) that applied to 500 urban areas to 2015, covering 4 million people (LévyVroelant 2007). Demolition, typically of high rise blocks, leads on to replacement with low rise private housing. The same approach is also used in the Netherlands (Bolt and van Kempen 2013) which has probably demolished proportionately the most social housing of any country, almost all of it technically sound property with high quality amenities.

## Deconcentrating the Poor

The second approach looks out from the areas of poverty and seeks to remove the deepest concentrations by dispersing the poor into different neighbourhoods, either through incentives or coercion (Goetz 2003). The population-exporting neighbourhood can then either be redeveloped as a 'mixed community' or it can be disposed of for another use. In practice this approach may have elements in common with the inward looking approach which, as the gentrification critique of social mixing holds, also effectively displaces people from the area targeted. As Galster (2013) outlines, removing the poor from poor areas has been the predominant approach in the USA. There have in fact been several identifiable US programmes, notably HOPE VI and the 'Moving to Opportunity' experiment which operated in 5 major cities in the 1990s and 2000s (Briggs et al. 2010). The US approach always involves some kind of 'housing choice voucher' (a household-related portable subsidy) which enables residents of public housing to leave public housing projects and find housing in the private rented market. However there are some important differences between the programmes; HOPE VI forced tenants to move out, and the aim was to disperse poor minority groups, mainly black households, to less minority-dense areas. MTO was specifically informed by the 'neighbourhood effects thesis' and was set up as a controlled experiment which compared volunteer movers with similar households who stayed behind in the projects.

## *Remaking Social Housing*

The third approach involves acting on the processes that shape segregation in the first place: housing development outcomes and housing choice. A particularly important force in sponsoring segregation is the role of social housing, which in many countries disproportionately houses poor people, usually because it is intended to play a role in meeting the needs for people who struggle to find adequate housing in the market. Because of the way in which social housing was planned and built in 'estates' (variously 'schemes', 'projects' or *'banlieux'*) it serves to concentrate poor people.

There are two different approaches under this heading though. The first is to consider this social composition of social housing as a whole, which in the European context is probably most relevant in the UK where the concentration of low income groups especially in social housing is the greatest among major countries (Stephens 2008). Hills (2007) in a review of policy for the UK government considered various ways to reinvent social housing and make it more attractive to higher income groups, by creating greater equality between the benefits offered by private and social housing respectively, for example by offering 'market rent' options or equity-sharing tenures. This proposal was not taken up by government; the main problem is that there has been no change in the key role of social housing in meeting needs. Without spending more on increasing the supply of social housing far beyond existing commitments it is difficult to see how an extension to more affluent groups could be justified. It is also difficult to see how those with choices would be attracted to rent in low-status areas among low-income people, if there are other alternatives. Similarly, when France introduced measures to avoid concentrating the poor in social housing, it might have been effective at the sectoral level but Blanc (2010) reports that the poorest tenants are still concentrated in the stigmatized peripheral high rise estates, which are avoided by middle class residents of social housing. Meanwhile it is clear that policy in some countries which have had a greater spread of income groups in social housing in changing.

An alternative way to reform social housing from within is to change the allocation criteria. In Germany, after a debate about tipping points and critical thresholds, there was pressure to avoid the over concentration of 'problem households' in social housing, which was seen as destabilising of communities, as existing residents sought to leave when many poor minority households moved in, therefore sending stable neighbourhoods into decline (Busch-Geertsema 2007). So reforms were undertaken allowing landlords themselves to choose which households to favour among those qualifying for social housing. The outcome was that some households found themselves excluded from social housing altogether. So a policy which was designed to promote inclusion had the opposite effect, at least for some households.

A similar approach is discussed by Bolt and van Kempen (2013). Imitating a local initiative in Rotterdam the Dutch government gave local authorities the possibility to exclude households who depend on benefits and who have not lived locally in the previous years from certain deprived areas. Although this was very controversial as, like the Germans example, effectively it was aimed at immigrants under the cloak of applying to all low income groups, Bolt and van Kempen report it was ineffective in its original Rotterdam setting, as it impacted on very few households. No other cities have chosen to apply to implement the law.

The fourth desegregation approach is to act on the pattern of new housebuilding, even if in most countries it is adding quite slowly to the housing stock. Under market conditions, new housebuilding tends to sift residents spatially according to their incomes. So sites developed in areas which are already high value and high status tend to be aimed at the rich, while more down market private housing is constructed in less well favoured areas. Meanwhile social housing tends to get constructed to limited budgets, which usually means finding cheaper land in less favoured areas. So the outcome of the development system tends to be to the reproduction- or the intensification- of the socio-spatial patterns which are found in the existing housing system.

A possible element of this approach is where government has a role in funding social housing it can potentially steer its production towards wealthy areas so that relatively poorer social housing tenants live alongside richer private sector residents. In the UK Bramley and colleagues (2007) show that even at the time when official enthusiasm for social mix was at its height, social housing in England was much more likely to be built in disadvantaged areas than non-disadvantaged ones. It appears that there was no strategy to ensure a more even distribution of social housing and the impact, of course, was simply to reinforce existing patterns of poverty concentration.

In contrast, in France, as well as programmes to diversify existing '*grands ensembles*' through selective demolition and the introduction of middle class homes, there is a legal requirement (the so called 'anti-ghetto act) that every commune (the lowest level of local government), except the very smallest and those in rural areas, should have 20 % of its stock as social housing by 2020 (Blanc 2010). At face value this appears to be a strong, and unique, anti-segregation policy. But, although Blanc is does not provide a full evaluation of the impact of this policy, he notes that the rather wide definition of social housing in the law, effectively covering any housing gets some type of financial support, which is 75 % of the national stock. This means that it includes far more than just the poorest groups. After considering the local politics and delivery of this policy he concludes that 'social housing is irrelevant in France as an indicator of poverty and/or income stratification' (2010, p.268), so it can be concluded that the anti- ghetto act is quite ineffective as a desegregation measure.

Finally on the theme of distributing social housing more evenly across urban areas, 'scattered site' public housing (very small developments in existing residential areas) have been built in the US since the 1970s as an alternative to large housing projects. Reviewing their experience, Goetz (2003) points to the difficulties of promoting them due to public budgets quailing at high land costs, but also because of resistance from existing residents. Although officials generally view them as successful and most, when completed, are well integrated into their neighbourhoods, they still only represent a tiny proportion of an already very small public housing sector. As time has passed, especially since HOPE IV, the programme has been reduced to 'boutique status' (Galster 2013).

## *Land Use Planning*

In aiming for more socially integrated patterns of new development, policy makers also can try to influence the pattern of new development through the planning system. In Spain, as an adjunct to policies developed in the 2000s to promote affordable housing, and in the context of a constitution which declares Spain to be a 'social state' promoting social rights, social mix obligations were built into planning law. In provinces across Spain land use law promotes social mix in order to avoid socio- spatial segregation of 'protected', i.e. targeted, low income, housing (Ponce 2010). In Sweden new construction is the main way in which social mix is promoted through housing and planning actions. However, Andersson and colleagues (2009) are not very positive about its contribution. Although there is a national commitment, social mix policies are a local government issue, there is no national policy, and the extent to which it is prioritised is very variable between municipalities. Even where planners are keen to promote social mix, it proves difficult in practice. The municipal housing companies which once provided such an important means of steering housing production and setting standards (Barlow and Duncan 1994) are now hardly distinguishable in Sweden from private developers in their imperatives. This is a similar story to housing associations in the UK and the Netherlands. Moreover, social mix tends to be a second order objective, less important for example than providing for housing shortages. The upshot to all this in Sweden is a weak, hard to implement and generally ineffective policy.

There is also some experience of this approach in the UK. Since the 1990s there has been the possibility of municipalities to influence the extent to which new developments provide 'affordable housing' by specifying requirements in area-wide development plans, and then using these as a basis to formulate planning agreements with developers, using the planning acts (Crook et al. 2002). As developers almost always prefer to build for market sale without hindrance, to a large extent the policy has rested upon the extent to which it can be made legally robust against appeals. As with 'social housing' in the 'anti ghetto act' in France, the definition of 'affordable housing' in the UK also effectively covers a wide range of income groups. 'Affordable housing' does it include social housing but many 'planning for affordable housing' schemes comprise market-price housing or equity sharing tenures, and these do not include the poorest. Bailey and colleagues (2006) also noted instances where allocation of social housing was deliberately skewed away from the most disadvantaged rents in order to appease private developers and home owners.

After 15 years of operation, the more successful planning authorities achieved a high proportion of all 'affordable housing' benefiting from this process. But it is difficult to make such policies work in lower demand markets where the greatest segregation occurs (Kintrea 2008) and since 2008 the slump in housing starts has made it difficult to pursue across the UK (Crook and Monk 2011). There is also evidence that alongside deep public expenditure cuts, local authorities have preferred to take the planning again as a commuted sum of money rather than in the form of affordable housing contributions.

In the mid 2000s the UK government changed it planning guidance to local government in England to include 'mixed communities' as a legitimate object of planning processes (UK Department of Communities and Local Government (2006)). It always seemed doubtful that would add all that much to existing provision. The emphasis was very much on house sizes and types, and not income groups, except that some income differences might be expected to be represented across house types. It was possible for planning authorities, for the first time, to include tenure as a planning concern. However, there was no time for this policy to be tested in practice before the housing market recession in 2008 and the planning guidance was scrapped by the new UK coalition government which came to power in 2010.

## *Mixed Income New Communities*

The final approach to social mix is that, where new settlements are master planned with the involvement of public agencies, they can be constructed as mixed communities from the outset through decisions about housing tenure, housing types, and density. They can also encourage interchange between diverse groups through the provision of public space and an emphasis on walking and cycling as viable means of transport which encourage face to face contact. These 'mixed income new communities' usually start with a site that is more or less open, because former uses have been cleared, or they are greenfield sites. They are then able to overcome the path dependency that has often locked places with greater historical continuity into a particular social status. The idea is that mixed communities can be created successfully because of clear objectives at the point of initiation and, typically, private developers and nonprofit providers have been brought together on that basis. These kinds of developments are closely associated with 'new urbanism', which is essential an urban design-focused movement as a remedy for suburban sprawl and 'unsustainable' car dependent cities (Talen 1999).

In the UK, there is now a body of good practice advice and examples from experience to draw upon best expressed in Bailey and colleagues (2006). In the larger and more ambitious schemes, other local amenities, such as shops, community centres, and schools, are also part of the plan. In the longer term, it is recognised that such communities also need overt management (unlike most residential areas), particularly to consider approaches to letting social housing, tackling crime and antisocial behaviour, and maintaining and managing the physical environment.

The evidence is clear that such schemes can work in the sense that developers and middle income house buyers can be persuaded that their needs can be met in housing areas which also consist of a percentage of social housing. In the US, however, Trudeau and Malloy (2011) who examined in detail the achievements of seven new urbanist developments in one region, report only two of them were affordable to those on less than 30 % of the median family income for the area (and in one of those cases only 8 % of units were affordable).

There are also two other important misgivings. The first is that many developments built on new urbanism principles in inner urban areas, in flatted forms and to relatively high densities present a challenge to attract more affluent households with children. Developers typically build small units which are not suitable for children. In contrast, the social housing element of mixed-tenure developments often has relatively high numbers of children. In the UK this is because, with a general shortage of family-sized homes for rent, housing associations take advantage of new building opportunities to provide larger homes which are then fully occupied (Monk et al. 2008). Similarly in Chicago, Chaskin and Joseph (2011) found that two thirds of relocated public tenants in planned mixed areas had children in the household compared to fewer than 30 % of 'market rate' renters and owners. This implies that socialisation among children and young people in local schools, or in the neighbourhood, may still predominately take place between those who come from relatively poor, social renting backgrounds, as the more affluent still head for the suburbs when they have children. This is important because there is a growing understanding that the impact of neighbourhood effects is greatest during childhood and adolescence (Oberwittler 2007; Galster et al. 2007b). The second misgiving is that, in order to provide confidence to middle income buyers to head off neighbour disputes between owners and tenants, there is a temptation by managers to avoid placing the most disadvantaged households in mixed areas. Tenant screening is practiced both in the US (Goetz 2003) and the UK (Bailey et al. 2006); clearly if the poor are meant to benefit from proximity to the better off, this model is often closed to the most disadvantaged households.

## What Have Social Mix Policies Achieved?

Based on the review here it could be said that social mix policies have been significantly inspired by ideological positioning, political responses to societal tensions or fear of unrest, and their delivery mechanisms are founded more on hope than real expectation of change. To a significant extent, social mix policy in addressing concentration of poverty in social housing -and public housing projects in the US- is in many countries a reaction to the consequences of failure of housing policy, especially within systems which are, or have become, significantly marketised. While to a considerable degree successfully addressing needs for better shelter, social and public housing programmes delivered unacceptable neighbourhood conditions.

While some social mix schemes have been assessed as effective- in that a degree of residential co-location of poor and better off groups has been achieved- others have failed to deliver even at this basic level because their leverage on the housing system have not been strong enough, or they have not included the poorest groups, which cuts across the rationale of social mix. Very few policies have been evaluated in the longer term for their impact on residents, and even those that have been subject to robust evaluation such as the Moving to Opportunity experiment in the USA (Briggs et al. 2010) show only positive impacts on some indicators of life chances, at best, and little impact on the key economic dimensions which were often underpinned the policy thrust. Galster (2013) and Briggs and colleagues (2010) submit that an important reason for this are that relocation only slightly changes the social networks that people have, and that moving by itself (at least in one generation) cannot compensate for other disadvantages that people face, such as poor educational attainment, which might have arisen in part from their original residence in a poor area (see Sampson 2012).

In this collection, two chapters make an evaluation of the impact of social mix approaches in two very different national contexts (Galster on the USA, Bolt and van Kempen on the Netherlands). They both suggest that social mix policies are not working in the sense that they are have managed to do very little to reduce segregation between poor and better off groups. In the USA, the scale of the research effort and the large literature on deconcentration programmes belie the fact that there is nothing that could be called a federal policy on social mix. Galster points to the small scale of the effort compared with the scale of the phenomenon, and the fragmentation of policy delivery across administrative boundaries in cities which are famously balkanised. What is more, Galster's evidence is that the historical concentration of poor people in poor areas, intensified by public housing programmes, has not been significantly changed either by 'vouchering out' or by local projects to improve poor neighbourhoods. Bolt and van Kempen, for their part, recognise that (very costly) selective demolition and rehousing processes aimed to restructure low income neighbourhoods, in a country where national policy has been strong on promoting 'living together', has not led to reduced concentration of low income people and minority ethnic groups. Furthermore, they show that the normal pattern of new suburban development lies outside the net of social mix policy and effectively widens segregation by being marketed at high income households.

In many countries which have social mix policies, it is also clear that polices in related field are often countervailing; they help to segregate rather than integrate. In this volume Bolt and van Kempen argue that emerging reforms to raise rents for better off tenants in Dutch social housing on the of grounds of effective subsidy targeting will act counter to achieving social mix. In the UK, other kinds of housing subsidy reforms are predicted to have the same effect (Fenton 2011) (although in England since the Conservative-Liberal Democrat coalition can to power in 2010 there is no longer a national social mix policy). Meanwhile policies on schools, which are a key driver of middle class families' residential location decisions, have always been generally heedless of neighbourhood impacts.

## The Limits of Social Mix

This all makes for salutary reading but it does not make the problem of poor neighbourhoods go away. Bridge et al., while not denying that concentrations of poverty exacerbate the problems of poverty, maintain that concentrated poverty is more fundamentally a symptom, not a cause, of inequality and (citing van Criekingen) and argue that policy should concentrate on 'the upward social mobility of the incumbent population in working class neighbourhoods' (2012, p.319). However, this does not really provide an answer because it neglects residential dynamics and issues of relative poverty, and begs the question about what is the acceptable level of spatial inequality. For example, targeted employment training programmes aimed at people who live in disadvantaged areas might successfully improve access to jobs and improve their earnings relative to the average. In turn this could serve to thin out poverty concentrations. However, the success of this approach would depend on how many people improved their economic position long term, and then how many of them choose not to change their neighbourhoods, even though they had improved their earnings. In all likelihood, many would probably choose to move to better neighbourhoods if they could afford it as Andersson and colleagues (2009) found even in Sweden, whose poor neighbourhoods offer far better conditions than most of their counterparts in other countries. Similar findings are available from regeneration areas in the UK where there is a typically churn of economically successful households leaving and being replaced by poorer newly formed households and migrant (Robson et al. 2008).

But the fundamental problem with social mix is that, given a choice, people mainly gravitate towards others who are like themselves. Western countries- all of them- are characterised by housing system mechanisms which sort people according to their economic position, and even those which, in principle, are strongly welfare oriented and interventionist such as the Netherlands and Sweden are not exempt. Cheshire (2007) drew some heavy criticism when he appeared to be an apologist for existence of deprived areas as one end of a spectrum of 'specialised neighbourhoods' which exist to cater for the needs of their residents (he now concedes that the most deprived neighbourhoods may not provide many benefits (Cheshire 2012)). But he was always right to point to the pervasive spatial sorting effects of markets. One of the key lessons drawn by Briggs and colleagues in their evaluation of the impacts of MTO as modest is the 'quiet crisis in affordable housing' (2010, p.225). The poor are concentrated in the USA because there is an insufficiency of affordable market housing in economically buoyant areas, and the same conclusion could be drawn from an analysis of spatial patterns of poverty in the UK (Kintrea et al. 2011).

But residential sorting is not all about affordability. Recent UK studies show it is partly cultural, in other words it responds to individual or household level preferences for the kind of neighbourhood they would like to live in. In the UK there has been attention to the extent to which better off people- in a global world- still find the neighbourhood important and settle among people who have the similar tastes and occupations as well as income brackets (Savage et al. 2005; Butler and Robson 2003). But cultural factors also perhaps explain in part why social mix policies may not be effective in changing the residential patterns of the poor as well. In a study in Scotland, it was clear that there was a comfort and familiarly among disadvantaged people in their poor areas (Atkinson and Kintrea 2004). Galster (2013) offers 'individualist' arguments (as well as structuralist and programme design explanations) for the weak achievements of desegregation policies in the US. Attachment to existing social networks, housing search behaviour and preferences appears to be some of the factors that have channeled relocated poor people into areas which are not so different in their social characteristics from the ones they left. Similarly, Sampson's portrait of Chicago neighbourhoods argues that' preferences and constraints thus together sustain the self reinforcing cycle of inequality. Therefore poverty traps are difficult to escape and likely to continue' (2012, p.308).

Where does this leave social mix policy? Since there is still a sense that neighbourhood mix matters to wellbeing, in principle social mix could be made more robust and evaluated better. The academic research that continues apace to better understand neighbourhood effects, in principle, could be used to design better-founded policies, but that is likely to be slow work with an uncertain take up, bearing in mind also the political environments in which neighbourhood policy has been developed. And there is certainly a good case for more systematic cross-national evaluation about the underpinnings, assumptions, methods and outcomes of policy with a view to learning. But the continued contestation of the importance of social mix, and doubts about its basic effectiveness and its connection with beneficial social mixing should cause other agendas to be revisited. Programmes which seek to regenerate poor neighbourhoods are valuable for quality of life improvements and address social justice, even if they do not fully overturn established residential pecking orders. And poor people face many other kinds of disadvantages; it would be as well to reexamine how their opportunities can be extended through actions on education, health, crime, education and employment, as who their neighbors are is not the only thing that matters.

## References

Amin, A. (2002). Ethnicity and the multicultural city: Living with diversity. *Environment and Planning A, 34*(6), 959–980.

Andersson, R. (2006). Breaking segregation: Rhetorical construct or effective policy? The case of the metropolitan development initiative in Sweden. *Urban Studies, 43*(4), 787–799.

Andersson, R., Brama, A., & Holmqvist, E. (2009). Countering segregation: Swedish policies and experiences'. *Housing Studies, 25*(2), 237–256.

Arthurson, K. (2008). Australian public housing and the diverse histories of social mix. *Journal of Urban History, 34*(3), 484–501.

Atkinson, R. (2008). *Housing policies, social mix and community outcomes* (AHURI Final Report 122). Melbourne: Australian Housing and Urban Research Institute.

Atkinson, R., & Kintrea, K. (2000). Owner occupation social mix and neighbourhood impacts. *Policy & Politics, 28*(1), 93–108.

Atkinson, R., & Kintrea, K. (2001). Disentangling area effects: Evidence from deprived and non-deprived neighbourhoods. *Urban Studies, 38*(12), 2277–2298.

Atkinson, R., & Kintrea, K. (2002). Area effects: What do they mean for British housing and regeneration policy? *European Journal of Housing Research, 2*(2), 147–166.

Atkinson, R., & Kintrea, K. (2004). Opportunities and despair, it's all in there: Experiences and explanations of area effects and life chances. *Sociology, 38*(3), 437–455.

Atkinson, R. (2006). Padding the bunker: Strategies of middle-class disaffiliation and colonisation in the city. *Urban Studies, 43*(4), 819–832.

Bacqué, M.-H., & Fijalkow, Y. (2012). Social mix as the aim of a controlled gentrification process: The example of the Goutte d'Or District in Paris. In G. Bridge, T. Butler, & L. Lees (Eds.), *Mixed communities: Gentrification by stealth?* (pp. 115–132). Bristol: Policy Press.

Bailey, N., Haworth, A., Manzi, T., Paranagamage, P., & Roberts, M. (2006). *Creating and sustaining mixed communities: A good practice guide*. Coventry: Chartered Institute of Housing.

Barlow, J., & Duncan, S. (1994). *Success and failure in housing provision: European systems compared*. Oxford: Pergamon Press.

Blanc, M. (2010). The impact of social mix policies in France. *Housing Studies, 25*(2), 257–272.

Blasius, J., Friedrichs, J., & Galster, G. (2007). Introduction: Frontiers of quantifying neighbourhood effects. *Housing Studies, 22*(5), 627–636.

Blokland, T. (2003). *Urban bonds: Social relationships in an inner city neighbourhood*. Cambridge: Polity Press.

Bolster, A., Burgess, S., Johnston, R., Jones, K., Propper, C., & Sarker, R. (2006). Neighbourhoods, households and income dynamics; a semi-parametric investigation of neighbourhood effects. *Journal of Economic Geography, 7*(1), 1–38.

Bolt, G., & van Kempen, R. (2013). Neighbourhood based policies in the Netherlands: Counteracting neighbourhood effects? In D. Manley, M. van Ham, N. Bailey, L. Simpson, & D. Maclennan (Eds.), *Neighbourhood effects or neighbourhood based problems? A policy context* (chap. 10). Dordrecht: Springer.

Bolt, G., Phillips, D., & van Kempen, R. (2010). Housing policy, (De)segregation and social mixing: An international perspective. *Housing Studies, 25*(2), 129–135.

Bond, L., Saukina, E., & Kearns, A. (2010). Mixed message about mixed tenure: Do reviews tell the real storey? *Housing Studies, 26*(1), 69–94.

Bradford, N. (2013). Neighbourhood revitalization in Canada: Towards place-based policy solutions. In D. Manley, M. van Ham, N. Bailey, L. Simpson, & D. Maclennan (Eds.), *Neighbourhood effects or neighbourhood based problems? A policy context* (chap. 11). Dordrecht: Springer.

Bramley, G., et al. (2007). *Transforming places: Housing investment and neighbourhood market change*. York: Joseph Rowntree Foundation.

Bridge, G., Butler, T., & Lees, L. (2012). Afterword. In G. Bridge, T. Butler, & L. Lees (Eds.), *Mixed communities: Gentrification by stealth?* (pp. 319–322). Bristol: Policy Press.

Briggs, X., Popkin, S., & Goering, J. (2010). *Moving to opportunity: The story of an American experiment to fight ghetto poverty*. New York: Oxford University Press.

Buck, N. (2001). Identifying neighbourhood effects on social exclusion. *Urban Studies, 38*(12), 2251–2275.

Busch-Geertsema, V. (2007). Measures to achieve social mix and the impact on access to housing for people who are homeless. *European Journal of Homelessness, 1*, 213–224.

Butler, T., & Robson, G. (2003). *London calling: The middle classes and the remaking of inner London*. London: Berg.

Chamboredon, J.-C., & Lemaire, M. (1970). Proximité spatiale et distance sociale. Les grands ensembles et leur peuplement. *Revue Française de Sociologie, 11*, 3–33.

Chaskin, R., & Joseph, M. (2011). Whose space? Whose rules? Social challenges in mixed income developments. *Research Brief* 4, Mixed Income Development Study. Chicago: University of Chicago/Case Western Reserve University.

Cheshire, P. (2007). *Segregated neighbourhoods and mixed communities: A critical analysis*. York: Joseph Rowntree Foundation.

Cheshire, P. (2012). Why do birds of feather flock together? Social mix and social welfare: A quantitative appraisal. In G. Bridge, T. Butler, & L. Lees (Eds.), *Mixed communities: Gentrification by stealth?* (chap. 2, pp. 17–24). Bristol: Policy Press.

Cole, I., & Goodchild, B. (2000). Social mix and the 'balanced community' in British Housing Policy: A tale of two epochs. *GeoJournal, 51*(4), 351–360.

Cole, I., & Green, S. (2011). Mixed communities: From vision to oblivion'. In I. Anderson & D. Sim (Eds.), *Housing and inequality* (chap. 9). Coventry: Chartered Institute of Housing/Housing Studies Association.

Council of the European Union. (2010). *Joint report of social protection and inclusion*. Brussels: Council of the European Union.

Crook, T., & Monk, S. (2011). Planning gain, providing homes. *Housing Studies, 26*(7–8), 997–1018.

Crook, T., Currie, J., Jackson, A., Monk, S., Rowley, S., Smith, K., & Whitehead, C. (2002). *Planning gain and affordable housing: Making it count*. York: York Publishing Services.

Davidson, M. (2012). The impossibility of gentrification and social mixing. In G. Bridge, T. Butler, & L. Lees (Eds.), *Mixed communities: Gentrification by stealth?* (chap. 15, pp. 233–250). Bristol: Policy Press.

Dorling, D. (2010). *Injustice: Why social inequalities persists*. Bristol: Policy Press.

Durose, C., & Rees, J. (2012). The rise and fall of neighbourhood in the New Labour era. *Policy and Politics, 40*(1), 39–55.

Feinstein, L., Lupton, R., Hammond, C., Mujitiba, T., Salter, E., & Sorhaindo, A. (2008). *The public value of social housing: A longitudinal analysis of the relationship between housing and life chances*. London: Smith Institute.

Fenton, A. (2011). *Housing benefit reform and the spatial segregation of low income households in London*. Cambridge: Cambridge Centre for Housing and Planning Research, Cambridge University.

Forrest, R., & Kearns, A. (2001). Social cohesion, social capital and the neighbourhood. *Urban Studies, 38*(12), 2125–2143.

Fraser, J., De Fillipis, J., & Bazuin, J. (2012). HOPE VI: Calling for modesty in its claims. In G. Bridge, T. Butler, & L. Lees (Eds.), *Mixed communities: Gentrification by stealth?* (chap. 14, pp. 209–232). Bristol: Policy Press.

Galster, G. (2007). Should policy makers strive for neighbourhood social mix? An analysis of the European evidence base. *Housing Studies, 22*(4), 523–545.

Galster, G. (2013). U.S. assisted Housing programs and poverty deconcentration: A critical geographic review. In D. Manley, M. van Ham, N. Bailey, L. Simpson, & D. Maclennan (Eds.), *Neighbourhood effects or neighbourhood based problems? A policy context* (chap. 11). Dordrecht: Springer.

Galster, G., Marcotte, D., Mandell, M., Wolman, H., Augustine, N., et al. (2007). The influence of neighbourhood poverty during childhood on fertility, education and earnings outcomes. *Housing Studies, 22*(5), 723–751.

Goetz, E. (2003). *Clearing the way: Deconcentrating the poor in urban America*. Washington, D.C.: The Urban Institute.

Granovetter, M. (1973). The strength of weak ties. *The American Journal of Sociology, 78*(6), 1360–1380.

Hills, J. (2007). *Ends and means: The future of social housing in England* (Centre for the Analysis of Social Exclusion (CASE) report 34). London: London School of Economics.

Hulse, C., Jacobs, K., Arthurson, K., & Spinney, A. (2010). *Housing public policy and social inclusion* (AHURI Positioning Paper 135). Melbourne: Australian Housing and Urban Research Institute.

Imrie, R., & Raco, M. (2003). *Urban renaissance? New labour, community and urban policy*. Bristol: Policy Press.

Joseph, M., Chaskin, R., & Webber, H. (2007). A theoretical basis for addressing poverty through mixed income development. *Urban Affairs Review, 42*(3), 369–409.

Kintrea, K. (2008). Social housing and spatial segregation. In S. Fitzpatrick & M. Stephens (Eds.), *The future of social housing* (chap. 5, pp. 71–86). London: Shelter.

Kintrea, K., St Clair, R., & Houston, M. (2011). *The influence of parents, places and poverty on educational attitudes and aspirations*. York: Joseph Rowntree Foundation.

Kleinhans, R. (2004). Social implications of housing diversification in urban renewal: A review of recent literature. *Journal of Housing and the Built Environment, 19*(4), 367–390.

Lees, L. (2008). Gentrification and social mixing: Towards an inclusive urban renaissance? *Urban Studies, 45*(12), 2449–2470.

Lees, L., Butler, T., & Bridge, G. (2012). Introduction: Gentrification, social mixing and mixed communities. In G. Bridge, T. Butler, & L. Lees (Eds.), *Mixed communities: Gentrification by stealth?* (chap. 1, pp. 1–16). Bristol: Policy Press.

Levitas, R. (2005). *The inclusive society? Social exclusion and new labour*. Basingstoke: Palgrave.

LévyVroelant, C. (2007). Urban renewal in France: What or who is at stake? *Innovation: The European Journal of Social Science Research, 20*(2), 109–118.

Ley, D. (2012). Social mixing and the historical geography of gentrification. In G. Bridge, T. Butler, & L. Lees (Eds.), *Mixed communities: Gentrification by stealth?* (chap. 6, pp. 53–68). Bristol: Policy Press.

Lupton, R., Tunstall, R., Hayden, C., Gabriel, M., Thompson, R., Fenton, A., Clarke, A., Whitehead, C., Monk, S., Geddes, M., Fuller, C., & Heath, N. (2010). *Evaluation of the mixed communities initiative demonstration projects: Final report*. London: CLG.

Maclennan, D. (2013). Neighbourhoods: Evolving ideas, evidence and changing policies. In D. Manley, M. van Ham, N. Bailey, L. Simpson, & D. Maclennan (Eds.), *Neighbourhood effects or neighbourhood based problems? A policy context* (chap. 8). Dordrecht: Springer.

Manley, D., van Ham, M., & Docherty, J. (2012). Social mixing as a cure for negative neighbourhood effects. In G. Bridge, T. Butler, & L. Lees (Eds.), *Mixed communities: Gentrification by stealth?* (chap. 11, pp. 151–168). Bristol: Policy Press.

McCulloch, A. (2001). Ward-level deprivation and individual social and economic outcomes in the British Household Panel Study. *Environment and Planning A, 33*(4), 667–684.

Monk, S., Clarke, A., & Whitehead, C. (2008). Understanding the demand for social housing. In S. Fitzpatrick & M. Stephens (Eds.), *The future of social housing* (chap. 8, pp. 137–154). London: Shelter.

Murray, C. (1996). *Charles Murray and the underclass: The developing debate*. London: Institute of Economic Affairs.

Musterd, S. (2002). Response: Mixed housing policy: A Dutch perspective. *Housing Studies, 17*(1), 139–143.

Musterd, S., & Ostendorf, W. (2009). Spatial segregation and integration in the Netherlands. *Journal of Ethnic and Migration Studies, 35*(9), 1515–1532.

Oberwittler, D. (2007). The effects of neighbourhood poverty on adolescent problem behaviours: A multi-level analysis differentiated by gender and ethnicity. *Housing Studies, 22*(5), 781–803.

Ponce, J. (2010). Affordable housing and social mix: A comparative approach. *Journal of Legal Affairs and Dispute Resolution in Engineering and Construction, 2*(1), 31–41.

Portes, A. (2000). The two meanings of social capital. *Sociological Forum, 15*, 1–12.

Riots Communities and Victims Panel. (2012). *Five days in August: An interim report on the 2011 English Riots*. http://www.5daysinaugust.co.uk/.

Robson, B., Lymperopoulou, K., & Rae, A. (2008). People on the move: Exploring the functional roles of deprived neighbourhoods. *Environment and Planning A, 40*, 2693–2714.

Room, G. (2005). *Beyond the threshold: The measurement and analysis of social exclusion*. Bristol: Policy Press.

Rowlands, R., Musterd, S., & van Kempen, R. (2009). *Mass housing in Europe. Multiple faces of development, change and response*. London: Palgrave, Macmillan.

Sampson, R. (2012). *Great American City: Chicago and the enduring neighbourhood effect*. Chicago: University of Chicago Press.

Sampson, R., & Raudenbush, S. (1999). Systematic social observation of public spaces: A new look at disorder in urban neighborhoods. *The American Journal of Sociology, 105*, 603–651.

Sarkissian, W. (1976). The idea of social mix in town planning: An historical review. *Urban Studies, 13*(3), 231–246.

Savage, M., Bagnall, G., & Longhurst, B. (2005). *Globalization and belonging*. London: Sage.

Smith, S., & Easterlow, D. (2005). The strange geography of health inequalities. *Transactions of the Institute of British Geographers, 30*(2), 173–190.

St Clair, R., & Benjamin, A. (2011). Performing desires: The dilemma of aspirations and educational attainment. *British Educational Research Journal, 3*(3), 501–517.

Stephens, M. (2008). The role of the social rented sector. In S. Fitzpatrick & M. Stephens (Eds.), *The future of social housing* (chap. 2, pp. 27–40). London: Shelter.

Talen, E. (1999). Sense of community and neighbourhood form: An assessment of the social doctrine of new urbanism. *Urban Studies, 36*, 1361–1379.

Trudeau, D., & Malloy, P. (2011). Suburbs in disguise? Examining the geographies of the new urbanism. *Urban Geography, 32*(3), 424–447.

Tunstall, R. (2013). Neighbourhood effects and evidence in neighbourhood policy: The UK: Have they been connected and should they be? In D. Manley, M. van Ham, N. Bailey, L. Simpson, & D. Maclennan (Eds.), *Neighbourhood effects or neighbourhood based problems? A policy context* (chap. 9). Dordrecht: Springer.

Uitermark, J. (2003). 'Social mixing' and the management of disadvantaged neighbourhoods: the Dutch policy of urban restructuring revisited. *Urban Studies, 40*, 531–549.

UK Department of Communities and Local Government. (2006). *Housing planning policy statement 3*. London: DCLG.

United Kingdom Cabinet Office. (2011). *Opening doors, breaking barriers: A strategy for social mobility*. London: Cabinet Office.

van Gent, W., Musterd, S., & Ostendorf, W. (2009). Bridging the social divide? Contemporary Dutch neighbourhood policy. *Journal of Housing and the Built Environment, 24*(3), 357–368.

van Ham, M., & Manley, D. (2010). The effects of neighbourhood housing tenure mix in labour market outcomes: A longitudinal investigation of neighbourhood effects. *Journal of Economic Geography, 10*(2), 257–282.

van Ham, M., Manley, D., Bailey, N., Simpson, L., & Maclennan, D. (2012). Introduction. In M. van Ham, D. Manley, N. Bailey, L. Simpson, & D. Maclennan (Eds.), *Neighbourhood effects research: New perspectives* (pp. 1–22). Dordrecht: Springer.

van Ham, M., Manley, D., Bailey, N., Simpson, L., & Maclennan, D. (Eds.). (2013). *Understanding neighbourhood dynamics: New insights for neighbourhood effects research*. Dordrecht: Springer.

Venkatesh, S. (2000). *American project: The rise and fall of a modern ghetto*. Boston: Harvard University Press.

Wacquant, L. (2008). *Urban outcasts: A comparative sociology of advanced marginality*. Cambridge: Polity Press.

Wilson, W. (1987). *The truly disadvantaged: The inner city, the underclass and public policy*. Chicago: University of Chicago Press.

Wilson, W. (1997). *When work disappears: The new world of the urban poor*. New York: Random House.

# Chapter 8
# Neighbourhood Revitalization in Canada: Towards Place-Based Policy Solutions

**Neil Bradford**

## Introduction

In recent years there has been growing awareness that today's major public policy challenges play out in local spaces. As Meric Gertler aptly observes, "a central paradox of our age is that, as economic processes move increasingly to a global scale of operation, the centrality of the local is not diminished but is in fact enhanced" (Gertler 2001). Geographers studying innovation in the knowledge-based economy now emphasize the importance of localized knowledge clusters for national economic success. Analysts of social inclusion and community planners encounter the multiple barriers that individuals and families face living in distressed neighbourhoods. Meanwhile, rural areas and smaller centres confront another set of risks altogether, managing industrial change with few assets and declining populations. Common to all of these perspectives is appreciation of how local territorial contexts – the geographic form and social nature of places – shape people's life chances.

For governments these dynamics frame a novel set of challenges. Their policy interventions must work from the ground up to generate solutions rooted in the concerns of local communities, attuned to the specific needs and capacities of residents. But what policy frameworks and institutional arrangements will enable such multi-level collaboration to happen? The conceptual and practical challenges remain daunting for national governments everywhere as they rethink and retool for an era of more intensive global–local interaction.

The purpose of this chapter is to explore Canadian progress in what has come to be known internationally as place-based policy. While place-based approaches have gathered momentum across a range of sectors and scales, our focus is on their application to the particular challenges of neighbourhood revitalization in

N. Bradford (✉)
Huron Universit College, Western University,
1349 Western Road, London, ON N6G 1H3, Canada
e-mail: bradford@huron.uwo.ca

cities. The presentation is organized in three parts. We begin by highlighting the place-based policy movement across the OECD, outlining the rationale and ideas informing the "new localism". Observing that Canada has not been at the forefront of such cross-national experimentation, the second section of the chapter tracks a growing awareness in federal policy communities of the potential of place-based interventions. We identify three key features of an emergent Canadian policy framework for neighbourhood revitalization that aims to balance top-down policy support with bottom-up innovation. The third section of the chapter then offers case studies of the framework in action, detailing the two most significant national neighbourhood policy initiatives in Canada over the last several decades, the Vancouver Urban Development Agreement and the Action for Neighbourhood Change Pilot Learning Initiative. The chapter closes with reflections on prospects for a new era of Canadian place-based policy in neighbourhoods.

## The New Localism and Place-Based Policy: Situating Canada

The new localism is a concept that now resonates across a multi-disciplinary literature analysing how globalization's most important flows – of people, investment, and ideas – intersect in cities around the world (Gertler 2001; Bradford 2005). Three central claims are advanced. First, to deliver on major public policy outcomes such as economic innovation, social inclusion, and ecological sustainability, national governments must engage local actor networks. Joining-up is necessary because 'wicked problems' – entrenched, interconnected, and localized – require holistic interventions addressing multi-faceted causality and capturing spillovers across sectors and governments. Second, features of the local milieux may constitute "neighbourhood effects" that shape individual life chances "over and above non-spatial explanatory social categories such as gender and class, and specific disadvantages such as unemployment or ill health" (Atkinson and Kintrea 2001, p.2277; see also van Ham et al. 2012, 2013). Importantly, such effects often find their origins in national-level policies that disadvantage certain localities and their residents in accessing public services and resources. Here the new localism's third claim comes into focus: effective public policy merges the professional technical knowledge of governments with the experiential know-how of residents living daily with the problems, and street-level service providers organizing opportunities. While problems play-out locally, solutions require multi-level responses leveraging the different policy assets of governments and community-based actors.

Research substantiating these three claims has supplied the analytical foundation for design and delivery of place-based policies across the OECD (Barca 2009). Particular implementation pathways have varied in accordance with national policy traditions and institutional frameworks. In the United Kingdom, for example, the project has been driven by the central government, mandating and orchestrating a complex web of governance networks at neighbourhood and metropolitan scales. The United States exemplifies an alternative strategy. There the push has come from below where a myriad of community organizations and institutional intermediaries have long worked in inner cities (Sirianni and Friedland 2001).

It has been widely observed that Canada lacks a robust tradition of place-based policy making (Bradford 2011; Policy Horizons Canada 2010). Unlike countries such as the United Kingdom or the United States, Canadian public policy has evolved without any powerful 'whole of government' focal point for place-based thinking and action. There has been no Canadian equivalent of the American Housing and Urban Development Department or the recent British combination of the Social Exclusion Unit and the Office of the Deputy Prime Minister. While Canada has an internationally-recognized national statistical agency, it has also lacked the network of spatially-oriented policy research institutions such as the British Joseph Rowntree Foundation or American Anne E. Casey Foundation that provide community-based analysis and program evaluation (Maclennan 2006; Cook 2010). Further, Canadian federalism has long featured an inter-governmental ethos of "rights, order, and control" (Gross Stein 2006). The system runs on competition and conflict between federal and provincial governments (and often among provinces themselves) over resources, responsibilities, visibility and credit. Moreover, the game is only two-level, as municipalities and community organizations have no 'seat at the policy table'. Canadian policy debates often pivot on multiple claims for redistributing money across regions, deflecting attention from a more productive question set: what distinguishes any given place, what are its unique assets, and how might public policy leverage opportunity? The result is a national policy system that is sectorally strong but locally weak – for example, producing housing but not necessarily resilient neighbourhoods, or supporting firms but not necessarily knowledge clusters.

It follows that a recent high-level policy report concluded that "Governments in Canada have lost their sense of place in policy-making" and that "Canada needs to catch up with other countries on the issue of place" (External Advisory Committee 2006, p.15–16). While Canada's various policy legacies have not been conducive to place-based approaches, numerous analysts have started to connect evidence of less than stellar policy performance to the absence of spatially aware governance (Policy Horizons Canada 2010; McMurtry and Curling 2008). The concerns cross key national policy goals, from economic development to ecological sustainability. A major overarching theme has been social cohesion and cultural inclusion, especially in urban contexts. Canadian cities have historically not experienced the kind of spatially concentrated poverty documented in many American and some European cities. However, in recent years a number of studies report a growing population of visible and ethnic minority groups living in the same poor neighbourhoods in Canada's largest cities (Hulchanski 2007; Ross et al. 2004; Walks and Bourne 2006). Such concentrations represent a daunting challenge in Canada as the country's future growth depends entirely on immigration, and newcomers overwhelmingly choose to settle in large cities. Effective integration of newcomers into Canadian urban housing and labour markets is a critical national priority, and equally one that can only be met through collaboration among the three levels of government and front-line settlement service providers with the inter-cultural tools and local knowledge to bridge social, sectoral, and spatial divides.

In fact, Canadian policy making is presently at a moment of transition (Bradford 2011). Opportunities to move along a place-based trajectory are emerging and several high level reports have made the case, drawing on experience from other OECD countries. At the federal level, the Prime Minister's External Advisory Committee on Cities and Communities offered a new vision of the federation that called on all governments to adopt place-based approaches through a "double devolution" of authority and capacity from upper level governments to municipalities and communities. Along the same lines, the Senate of Canada issued two reports on poverty reduction and population health, recommending creative blends of community place-based and individual rights-based approaches. In civil society, the Caledon Institute of Social Policy has supported a pan-Canadian 16 city "Vibrant Communities" network, applying place-based concepts to poverty reduction. Its findings have further empowered local community organizations such as the United Way to partner with municipalities on "comprehensive community initiatives" focused on marginalized neighbourhoods. At the provincial level, similar concepts are resonating. Explicitly drawing on British examples, the Ontario report on the roots of youth violence recommended a "neighbourhood capacity and empowerment focus" implemented through a Neighbourhood Strategic Partnership headed by a Cabinet Committee on Social Inclusion (McMurtry and Curling 2008).

## Public Policy in the Neighbourhood? A Canadian Perspective

New research mapping distressed urban neighbourhoods and grass-roots experimentation with local solutions has driven Canadian policy interest in place-based approaches.

With evidence mounting of worrying forms of socio-spatial segregation in many cities, researchers have begun to investigate the potential operation of neighbourhood effects that amplify the pressures on low-income residents and create complex barriers to progress across the spectrum of well-being outcomes. This work has also been driven by analytical breakthroughs in data sets and multi-level modeling methods that capture the incremental impact of neighbourhood characteristics on outcomes beyond individual- or family-level factors (Dunn et al. 2010).

From a comparative perspective, two aspects of this emerging body of Canadian neighbourhoods research are notable, shaping the particular way in which the place-based social policy movement has evolved in Canada (Beauvais and Jenson 2003). First, the relatively few quantitative studies all conclude that, compared to the United States and the United Kingdom, "neighbourhood effects in Canada are much smaller" (Statistics Canada 2004, p.1; Tremblay et al. 2001; Willms 2002). Second, despite growing residential segregation by income in Canadian cities (Ross et al. 2004), it remains true that only a small proportion of Canadians living in poverty reside in areas that would be targeted in place-based initiatives to reduce inequities. The overarching policy message from this research highlights the centrality of family and individual factors in shaping life chances for those at risk of social exclusion regardless of where they live (Oreopoulos 2002; Seguin and Divay 2003). While not denying negative neighbourhood effects, the Canadian researchers underscore the ongoing importance of generally available or 'aspatial' policies providing income support and access to health, education, and employment. As one study of labour force attachment for public housing residents summarized: "the results strongly suggest that policies aimed at improving outcomes among children from low-income backgrounds are more likely to benefit by addressing causes of household distress and family circumstance than by improving residential environmental conditions" (Oreopoulos 2002, p.21).

Taking stock of the Canadian neighbourhood effects research, Christa Frelier offered a balanced assessment of the findings and their policy implications:

> Knowing whether there are neighbourhood effects and how they operate may not be as important as we think since there are, arguably, other reasons for focusing on neighbourhoods or area-based initiatives more generally ... these include : ensuring a fairer distribution of resources; piloting new approaches to service delivery or community development; having a greater impact by focusing activity; increasing people's confidence and capacity to participate in the community; and promoting social cohesion and 'bottom up' approaches to neighbourhood revitalization. Some of these are the rationales behind current neighbourhood initiatives in Canada and other countries (Frelier 2004).

Given the ambiguity of the research findings, it is not surprising that the Canadian place-based policy 'turn' has been distinguished by three main features.

- *Incrementalism.* The place-based policy roll-out has been step-by-step, relying on small-scale, discrete pilot projects in selected neighbourhoods. Unlike some European countries where ambitious national renewal programs have targeted dozens of localities, Canadian efforts have remained modest in scope and scale.
- *Interscalar Links.* The discourse has consistently emphasized that place-based policy is *not a panacea* for urban poverty; rather systematic connections must be made between targeted neighbourhood initiatives and macro-level, universalistic social measures for income support and well-being.
- *Learning from the Local.* In the absence of strong and widespread evidence of neighbourhood effects, policy interventions must be designed carefully on the basis of *fine-grained qualitative knowledge* of neighbourhoods, their community dynamics and individual and family pathways of connection.

On each dimension, Canadian place-based researchers have recently contributed useful policy knowledge. Canadian Policy Research Networks (CPRN), a national social policy think tank, commissioned a series of research papers investigating what it termed the "the right policy mix" in tackling social exclusion, illustrating the ways in which targeted interventions can reinforce the positive effects of general social programs while emphasizing that the mix will vary from community to community (Seguin and Divay 2003). Scholars of inter-governmental relations have explored various mechanisms for "learning from the local" such that federal and provincial policy frameworks incorporate local knowledge and municipal or community priorities (Sandercock 2004; Bradford 2005; Torjman 2007).

Mechanisms include formal policy dialogues convening local actors and policy makers, and framework agreements committing different levels of government to common planning, aligned programming, and joint evaluation. In some cases, the framework agreement specifies a community agency to act as an intermediary coordinating local efforts while in other instances an inter-governmental secretariat is established to steer the process (Public Policy Forum 2008). Our two case studies below illustrate examples of each of these multi-level neighbourhood-based governance structures.

Finally, Canadian researchers have made progress in shifting the analytical focus away from aggregate, quantitative neighbourhood profiles toward more qualitative and institutionally-grounded portraits (Bernard et al. 2007; Dunn et al. 2010) The concept of the "local opportunity structure" has been used for interpreting individual and family capacities to access key resources in their neighbourhoods (Bernard et al. 2007). Grouping resources into different institutional domains such as the market, community, and physical, different types of "access rules" to particular domain resources are identified, supplying policy guidance in tackling the place-specific mix of barriers ranging from inadequate transit to social service gaps or limited entry-level employment. Most importantly, the framework captures institutional and sectoral connections at the neighbourhood scale, making clear the interaction among the domains and the need to improve individual and family access to resources across several domains simultaneously. Applying this place-based model reveals the specific pathways that are most conducive to opportunity in differently structured neighbourhoods. A less formalized but equally focused qualitative approach to understanding neighbourhood assets and policy interventions has come through the 16 city Vibrant Communities network (Born 2008). With each local community developing its own multi-sectoral projects for poverty reduction, a pan-Canadian policy learning community was regularly convened to share lessons and consider scaling-up promising innovations

These three features of the Canadian place-based policy dialogue – incrementalism, interscalar links, and learning from the local – are evident in the two most prominent national neighbourhood revitalization initiatives over the last several decades. The Vancouver Agreement (VA) and Action for Neighbourhood Change (ANC) aimed to help revitalize distressed urban neighbourhoods, with the federal government playing a leadership role in bringing together the key players and facilitating joint work. For time-limited pilot projects, both the VA and the ANC generated considerable national and international policy recognition as governance innovations tailored to the complexities of a federal state (Dunn et al. 2010). At the same time, they followed different institutional designs: the VA focused on one neighbourhood and relied on an inter-governmental secretariat for *joined-up government*. The ANC worked in five cities across the country and engaged Canada's leading third sector organization for *federal-community partnership*. Presenting an interesting mix of similarities and differences, the VA and the ANC offer valuable lessons about the design and delivery of place-based strategies in federations where the challenges of collaboration are as daunting as they are necessary.

## Joined-Up Government: Vancouver Agreement

### *Background*

The VA was a 5 year agreement signed by the federal and provincial governments and the City of Vancouver in March 2000 (Bradford 2008). In 2005, the three governments renewed the VA for another 5 years, signalling mutual recognition of the longer term nature of the change processes and relationship building. The VA committed the three governments to work together and with community organizations "to develop and implement a coordinated strategy to promote and support sustainable economic, social, and community development". While conceived with a city-wide mandate, the VA's priority quickly became Vancouver's Downtown Eastside (DTES), an inner-city neighbourhood of about 16,000 residents that was Canada's poorest postal code and experiencing severe social strain. In the 1990s an intersection of public health and economic crises had left the DTES with what were described as disease rates rivalling third world countries and a local economy worse than the Great Depression. By the late 1990s it was widely observed that these conditions existed despite years of policy activity, with some 25 departments from the three levels of government all 'present' in the neighbourhood. In addition, it was estimated that nearly 300 organizations were engaged in various forms of community development, ranging from service delivery to crisis supports and economic development. Not surprisingly, a consensus began to emerge especially in government circles that residents in the DTES were not well-served by the existing pattern of multiple, diffuse interventions – marginalized and multi-barriered residents fell through the cracks as "service offerings and impact were fractionated" (Macleod Institute 2004).

City officials moved first to find better ways to deliver municipal programs and services. They introduced two coordination strategies. Neighbourhood Integrated Services Teams utilized an interdisciplinary and collaborative approach to problems, designating representatives of various City departments to find solutions across traditional lines of authority while providing a single point of contact for the community. Complementing this integrated structure, the City embraced a grass-roots "Four Pillars" healthy community policy vision that emphasized a continuum of supports for at-risk DTES residents. While these local innovations were a good start, it was apparent that the breadth and depth of DTES issues reached well beyond municipal authority and resources.

### *Strategy: Incrementalism*

To tackle the problems in the DTES, a new inter-governmental partnership was required, and two federal departments came forward. At Health Canada, policy work on the social determinants of health was proceeding and the DTES offered a prime setting for investigating how individual health outcomes are influenced by employment opportunities, education levels, social networks and the like. To coordinate the multiple actors implicated in such an expansive health policy table, another federal agency, Western Economic Diversification Canada, offered its experience with urban development agreements in Winnipeg and Edmonton as a relevant model for tri-level government collaboration in Vancouver. At the same time, the British Columbia provincial government had just released a high profile report on heroin addiction and spread of HIV/AIDs, recommending that the problem be addressed from a public health rather than criminal perspective. Thus, all three governments rallied around the principle of inter-sectoral and cross-jurisdictional coordination.

A draft tri-partite VA was negotiated in July 1999 announcing three priorities: community health and safety; economic and social development; and community engagement and capacity building. Speaking to the third priority, the draft agreement was translated into four languages and discussed at 11 public meetings in the fall of 1999 with more than 200 people attending. A report titled "Community Review of the Draft Vancouver Agreement" was published capturing the community's views and priorities. In general, community members and representatives voiced support for the VA, its inter-governmental process and policy priorities. However, the community made several challenging observations: problems in the DTES reflected not simply a lack of policy coordination or service integration but also government cutbacks and program barriers in social assistance, affordable housing and other broader policy areas; the VA should aim to link its targeted DTES interventions to an enhancing of these more general social and economic policies that profoundly influenced opportunities for marginalized people; DTES residents possessed the local knowledge, community experience and skills to be partners in the VA and "outside experts" must tap these resources and assist marginalized individuals and groups to participate; finally, there was skepticism that an unfunded agreement could produce meaningful change and concern that the VA might become more talk and study than action (Final Report 1999).

The VA's model of joined up government distributed authority across a wide governance network. At the executive level, decision-making power rested with a Policy Committee of the Mayor and the relevant federal and provincial Ministers. Policy decisions required unanimous consent. At the advisory level, the political body was supported by a Management Committee comprised of nine senior officials, three from each of the governments. At the operational level, a Coordination Unit with an Executive Director and small secretariat facilitated problem-focused multi-sectoral task teams working on issues such as harm reduction, housing, women's safety and opportunity, community economic development, and Aboriginal well-being. The flow of ideas through the VA was designed to be bottom-up with the community-driven, street-level task teams identifying specific opportunities and preparing project proposals for consideration by the Management and Policy Committees. It was through the task teams that the VA achieved a form of community representation and engagement, although questions about a more substantive decision-making role for the community remained (Macleod Institute 2004).

The VA began as an unfunded collaboration, without the capacity to implement its own services or programs. Each government would continue to work within its own jurisdiction, mandates, and accountabilities. Through institutionalized dialogue existing resources would be redirected around common priorities. The animating vision was a policy division of labour based on problem solving strengths of each government. Federal, provincial, or municipal levels could take the lead on a shared priority based on its specific jurisdiction and experience. The other two governments would then contribute their own resources as appropriate. On the problem of substance misuse, for example, a senior City official explained "while the Federal government was dealing with drug issues from the justice perspective, Vancouver from enforcement (policing), and the Province from health (treatment), a coordinated effort on drug efforts would be more effective" (Rogers 2001). More generally across different VA priorities, an appropriate division of labour respecting comparative advantages would find federal government leading on the economic development aspects, the provincial government on housing or health dimensions, and the City on safety and public realm improvements.

## *Projects: Interscalar Links*

The VA's unfunded status was initially viewed by government players as beneficial since it 'took off the table' resource competition and avoided the accountability challenges of integrated policy (Bradford 2008). Officials gained better awareness of different organizational cultures, forged new policy relationships, and learned about the specific strengths of other governments. However, the lack of dedicated funding also came to be seen as an obstacle to substantive progress. Better coordination of existing measures could not change policy frameworks or program criteria ill-suited to DTES challenges. Moreover, the community had long argued that the problems of the DTES demanded both better service delivery and more public investment. In 2003, the provincial and federal governments allocated a combined $20 million to fund VA projects with in-kind support from the municipal government. An Integrated Strategic Plan was prepared by the three governments in consultation with the community. With this plan in place, the VA's focus shifted toward implementation of specific revitalization projects. Across each of the VA's social, economic, and health priorities notable innovations took shape, each involving interscalar policy links (Donovan and Au 2004).

With public health, the VA oversaw implementation of North America's first Safe/Supervised Injection Site. The federal department, Health Canada, granted the responsible Vancouver Coastal Health Authority a 3-year operating exemption under federal drug legislation. The provincial Health Ministry provided funding to renovate the space and operate the service. City police were redeployed to ensure safety and order in the immediate area. Together, the three governments also delivered a prevention and enforcement strategy that drew together criminal justice and business regulation resources for a concerted attack on the infrastructure of the DTES drug trade.

For economic development, the VA created a non-profit community economic development organization – Building Opportunities with Business (BOB) – to champion inclusive revitalization. It facilitated DTES business clusters, negotiated community benefits agreements with the 2010 Olympic Games developers, and supported a social purchasing portal for DTES goods and services suppliers. Working with all levels of government it sponsored training programs to support DTES resident employment in clusters and construction projects, and oversaw customized training and accessible integrated employment strategies for multi-barriered people.

Concerning safety for the DTES's most vulnerable, the VA's Women's Task Team put together a Mobile Access Project for Sex Trade Workers. A converted ambulance vehicle made available first aid, peer counseling, and service referrals to sex trade workers, many of whom were the most vulnerable DTES residents such as Aboriginal youth. The project was led by the provincial government working closely with two community organizations and current and former sex trade workers who received training in front-line services. The federal Justice Department provided support in dealing with the sexual exploitation of aboriginal youth and the Vancouver Police contributed to harm reduction – offering self defense for sex trade workers and enhanced intelligence gathering on sex trade consumers.

With housing and revitalization, the VA coordinated the $183 million redevelopment of a former department store on principles of mixed used revitalization and community participation. Leveraging market housing as a catalyst for inclusive revitalization, the project combined market priced and subsidized units, and featured a variety of other uses – commercial/retail, arts and theatre, educational, childcare, and rooftop greenspaces. Consistent with the VA's social purchasing strategy, procurement of materials and services through DTES suppliers was a guiding principle, as was local employment in retail, security, gardening and cleaning.

On the strength of such multi-faceted interscalar policy, the VA's influence reached well beyond the neighbourhood (Bradford 2008). When the provincial government was looking to implement a province wide accessible communities project for people with disabilities it worked with and through the VA's networks and expertise. When the federal government in the mid 2000s began to extend urban development agreements to other cities it drew on the VA experience as a promising model. Finally, at an international scale, VA leaders contributed to knowledge transfer projects with cities in South America, partnering with the Institute of Public Administration of Canada.

## *Outcomes: Learning from the Local*

The VA brought together the three levels of government to work with communities for better delivery of existing services and programs, and implementation of innovative projects. The reach of the inter-government collaboration was remarkable. The federal government involved 12 departments, the provincial government involved 19 Ministries and agencies, and the municipality involved 13 organizations including the Vancouver School Board and the Vancouver Coastal Health Authority. As such, there are important lessons to learn about neighbourhood revitalization policy in a federal state. A comprehensive 2004 review concluded that: "The VA's successes provide powerful evidence that coordination and strategic decision making can indeed produce positive results in a large Canadian urban setting" (Macleod Institute 2004). A survey of participants from government found that 85 % of respondents very often supported one another's goals across levels and departments, and 72 % of respondents said that they were very often changed their work based on lessons learned through collaborating.

In addition to these 'process' benefits, a review of VA official documents and related commentary highlights a number of substantive neighbourhood improvements (Dunn et al. 2010). These include: connecting more than 200 DTES businesses with social purchasing opportunities; community benefit agreements with private sector developers for new jobs and training for DTES residents; heritage restoration through leveraged investments of over $400 million for 23 mixed used developments, including housing units for multi-barriered people, and other vulnerable DTES residents; and a significant decline in death rates due to drugs, alcohol, suicides, and HIV/AIDs as substance abusers accessed the supervised injection site and VA facilitated primary health care services; and with community safety, a survey of sex trade workers found that 90 % of respondents felt that the Mobile Access Project had prevented physical assaults and limited sexually transmitted diseases.

Not surprisingly, the VA has been the subject of much policy commentary, analysis, and recognition. It received three public management awards for governance innovation from national and international bodies (Western Economic Diversification Canada 2004). The Auditor General of Canada, in a review of federal joined-up initiatives, described the VA as the most "promising governance model of collaboration to meet community needs" and identified it as the "benchmark" for Canadian urban development agreements (Auditor General 2005, Chapter 4). Academic analysis has also been positive. A Danish-Canadian comparative study of collaborative governance concluded about the DTES that the "undoubted stabilization of the area can be seen to be in part due to the Agreement" (Walker and Sankey 2008). Michael Mason found that the VA met the key criteria established by urban governance scholars for successful revitalization partnerships (Mason 2006). Neil Bradford used a social learning perspective to highlight how the VA's multiple knowledge flows contributed to public policy innovation (Bradford 2008). Herman Bakvis and Luc Juillet (2004) acknowledged the innovative features of the VA while also underscoring the importance of support from government central agencies. Along similar lines, the Auditor General of Canada warned that such governance experiments in Canada too often proceed without the "specialized guidance" required from senior government officials (Auditor General of Canada 2005, Chapter 4).

## Federal-Community Partnership: Action for Neighbourhood Change

### Background

The ANC was a 2-year action learning project to explore, test and articulate a resident-led approach to neighbourhood revitalization. According to federal officials, the ANC represented "a new commitment by federal policy makers, not only to *listen* to what Canadian communities need to make their neighbourhoods healthier, but to get right down in the trenches *with the people who actually live there* as they work together to solve the problems they face" (Minister of Labour and Housing 2005, emphasis in original). Working through community intermediaries, the federal government partnered with the United Way Canada/Centraide to support neighbourhood revitalization in five cities across the country. The ANC was launched in 2005, surviving a difficult 2 month funding hiatus amid shifting government priorities in 2006, to complete its work in 2007.

The ANC had its origins in a growing recognition among federal civil servants associated with anti-poverty policy in the late 1990s that "distressed neighbourhoods" were both growing in number in Canadian cities and complex in their causes (Bulthius and Leviten-Reid 2005). Two initiatives brought the issues into focus. First, the National Homelessness Initiative (NHI) was important in demonstrating the value of community-driven planning and also the importance of linking shelter strategies to 'upstream' preventive policies addressing poverty, crime, health, and education. Second, the Vibrant Communities network had underscored the role of intermediary organizations in leading community development and facilitating local participation in public policy. Both of these initiatives pointed to the potential synergies between local activities and place-sensitive federal programming.

In 2004, the National Homelessness Secretariat took the lead in convening relevant departments and agencies to consider how the federal government might become a catalyst and partner in neighbourhood revitalization. The ANC emerged as 2-year action learning project targeting high poverty neighbourhoods in Halifax, Toronto, Thunder Bay, Regina, and Vancouver. Officials from four federal programs, representing three federal departments came forward with funding commitments. Rather than the usual four or five contribution agreements, a single reporting relationship was hammered out, pooling cross-departmental funds and streamlining burdens on local organizations. While not formally involving provincial and municipal governments, it was hoped that ANC 'start-up' projects would attract inter-governmental support when the 2-year federal pilot wound down (Bradford and Chouinard 2009).

## *Strategy: Incrementalism*

The federal government engaged three national community organizations as project partners, with each playing leadership roles in different areas based on comparative advantage in task expertise and experience. United Way Canada/Centraide was responsible for overall project management and accountability. Through its local affiliates in each of the five cities, neighbourhoods were selected and local residents and organizations mobilized. The Tamarack Institute for Community Engagement was the second national intermediary engaged, bringing practical knowledge in community capacity building. Tamarack produced a guide for residents and organizations on comprehensive community initiatives and also organized tele-learning conferences across the five sites. The third national partner was the Caledon Institute of Social Policy a leading think tank. It directed the learning process, generating a steady stream of policy research, practical tools, and progress reports. In addition, Caledon organized regular policy dialogues between government officials and community representatives, and a national policy forum drawing on the expertise of practitioners, policy makers, and scholars from Europe and United States (Maclennan 2006).

The ANC inspired "deep interest and excitement" among the different participants (Jamieson and Kinnon 2007). Government officials were motivated to learn how their respective departmental mandates for health, crime prevention, literacy and so forth could be advanced through collaborative, place-based work. For the United Way Canada/Centraide, the ANC was well-timed as the organization broadened its service mission to include community building beyond traditional grant making. Both Caledon and Tamarack were already engaged in research and learning about the 'right policy mix' for place-based approaches. Their projects consistently focused on how community driven innovations could address service gaps or access barriers in macro-level aspatial social policies (Torjman 2007). Building on existing initiatives, the ANC engaged national partners with a track record in place-based approaches and organizational capacity on the ground in neighbourhoods. To enable systematic comparison across neighbourhoods, each of the sites followed the same steps and had access to similar resources and support. Initial neighbourhood profiles, drawing on census data and other indicators of well-being, established baselines for local priority setting. Created through participatory processes, the profiles involved residents in learning about community organization, partnership formation, grant writing, and evaluation frameworks. For action on the profiles, the ANC made available seed money for neighbourhood mobilization around identified priorities.

As its work rolled out, the ANC brought into focus several key tensions when residents become co-producers of strategies and solutions rather than passive recipients of predetermined services. Under such conditions, program outcomes could not be firmly specified in advance and it would be challenging to link individual departmental mandates for reductions in crime or substance misuse to local activities such as community capacity building or resident networking. Accountability and evaluation frameworks required the flexibility to capture such processes, specifically the incremental nature of community-driven progress, and the indirect causal connections in preventive strategies. Meeting these challenges became a priority in the ANC's second year as projects were implemented across the neighbourhoods.

## *Projects: Interscalar Links*

All of the ANC activities were guided by the principle of resident-led change. As such the specifics of the initial engagement, neighbourhood mobilization, and priority projects varied across the five sites. Site reports from the ANC's first year convey the diversity (Makhoul 2007a). In Halifax, an initial challenge came from a municipal councillor concerned about stigmatization of the ANC neighbourhood. Overcoming this tension, the Halifax team focused on advancing a multi-service children's centre and seeking more supportive provincial families policy. In Toronto, the ANC usefully linked with existing work done by the municipal Strong Neighbourhoods Task Force, and worked to engage racial minorities and immigrant women in service networks, while training residents for leadership through "community animation". In Thunder Bay, the ANC's work was distinguished by a concern with youth engagement and the establishment of a "policy validation group" to support resident voice on public policy issues. In Regina, the ANC had to coordinate with an existing federal neighbourhood initiative, advancing understanding of how to incorporate a resident engagement model of change into a more traditional government programming. In Surrey, the ANC helped integrate services for families, with the neighbourhood being selected as a pilot site for a provincial literacy program. Another project involved a neighbourhood cleanup that leveraged a wider Adopt a Street campaign.

Across its second year, the ANC brought to these local sites a "neighbourhood theory of change" (Gorman 2006). Designed to capture the iterative nature of revitalization work and guide government investment decisions, the theory of change was built around five key assumptions: first, the well-being of residents and neighbourhoods depends on local control over social, cultural, physical, environmental, and economic assets; second, this control requires collaborative neighbourhood governance joining residents with extra-local "system-wide supports"; third for such governance relations to flourish in distressed neighbourhoods "transformational change" is necessary through "financial investment, technical assistance, research data, and policy changes"; fourth, there is no single neighbourhood change starting point, rather it will vary by context; fifth, the change process must allow for course corrections over time as new learning from participants enriches understanding of complexity. In guiding ANC projects, the theory emphasized interventions to support neighbourhood governance networks capable of building-up community assets in four "outcome domains": inclusion and engagement; housing; health and safety; and the economy (Gorman 2006, p.16).

The theory was field-tested across the different ANC sites as the action grants rolled-out (Makhoul 2007a, b). The Toronto experience illustrates the dynamic. Working in a Scarborough neighbourhood where more than half the population was born outside Canada, and 26 languages were spoken, the ANC hired nine community animators, training them in neighbourhood revitalization and directing them to engage with their communities on priorities. Identifying the neighbourhood's key challenge as a failure to engage with and leverage its own ethno-cultural diversity, the community animators created network-based projects for language training, youth, and immigrant women. These networks came together through a new community hub to integrate services and governance resources. According to participants, the ANC community animators delivered across several theories of change priorities related to inclusion and engagement outcomes. These included the training a cohort of community builders, providing civic engagement for marginalized residents, and strengthening neighbourhood capacity to develop its socio-cultural assets. Further, the process included adjustments along the way to help ensure the project's sustainability as the ANC pilot wound down.

A notable example of the ANC facilitating interscalar links was the Policy Dialogue. In its study of local governance and partnership, the OECD has emphasized the need for "a mechanism through which local and regional experience is fed back to the top to highlight deficiencies in the national policy framework" (OECD n.d., p.14). In Canada there was no such federal-local policy machinery that permitted "this kind of sustained discussion and that adds to the body of knowledge regarding effective collaboration between government and communities" (Torjman 2005, p.19). Here, the ANC's Policy Dialogues were a significant addition to the Canadian repertoire. The Caledon Institute sought to move the various players along a continuum of joint work, from the relatively routine aspects of horizontal and vertical collaboration such as information sharing on mandates and expectations, to tackling the challenges of actually changing government procedures and policies based on community feedback for solving problems. In focus were interscalar links: the macro-level social "policy domains that relate to the revitalization of neighbourhoods" and the related "administrative barriers that make it difficult for communities to do their work" (Torjman 2005, p.17–18).

## *Outcomes: Learning from the Local*

For a 2-year pilot project with modest funding, the ANC partners set an ambitious goal to forge a national policy network for strengthening neighbourhoods through resident-led, federal-local partnerships. Through its website, documents, tools, workshops, teleconferences, site visits, and local projects, the ANC forged relationships at three levels: between local residents and community organizations; between community organizations and national partners; and between national partners and federal departments. Further external connections were made as the ANC pursued its knowledge dissemination mandate, with the partners presenting at numerous venues: the academic gathering of the American Urban Affairs Association; United Way Canada/ Centraide conferences; the Inclusive Cities Canada network; and the United Nations World Urban Forum.

The ANC's main contribution was to identify challenges in place-based policy and then explore solutions through neighbourhood action-research. A key issue concerned accountability and evaluation. The "abstract" nature of the ANC early outputs and their focus on "upstream prevention" of problems through various kinds of community building supports proved challenging for project champions inside government to defend (Levitan-Reid 2006). Or as one civil servant put it: "It is hard for politicians to connect collecting garbage on Saturday to crime prevention and defend it in the House of Commons" (Jamieson and Kinnon 2007, p.E-6).

Here the ANC partners pursued several creative responses. They extended the timelines on reporting (from monthly to quarterly updates) and developed novel ways to capture progress, using qualitative, experiential data in the 'resident voice'. Published as ANC *Community Stories*, these neighbourhood narratives conveyed important contextual insights into local change processes and resident interpretations of the ANC's impact on their daily lives. Accessible to wide readership, they were published at the beginning, mid-point, and end of each site project. Complementing the resident-based narratives, the ANC developed its theory of change to guide government investments in neighbourhood assets and impact evaluation over the longer term. Taken together, the narrative stories and theory of change were establishing "a crisper evidence base for neighbourhood revitalization" that could be "linked to investment decisions" (Gorman 2007).

The second broad issue where the ANC generated better understanding was about the role of the federal government in community governance. While the ANC community partners praised the "exemplary" efforts of individual civil servants involved in the project (Levitan-Reid 2006, p.17), they also emphasized that these officials worked within a wider public sector environment ill-suited to the collaborative and long-term nature of neighbourhood revitalization. On these challenges, the ANC produced useful feedback. Peer-to-peer "reflection sessions" were convened on key lessons. There were calls for civil service training and mentorship in community change processes, and creation of dedicated project secretariats to generate progress indicators for line departments and arrange neighbourhood site visits for senior officials. Echoing lessons from the VA experience, the reflection session also emphasized the need to connect place-based policy making with government central agencies such as the Privy Council Office and Treasury Board.

Overall, the ANC left three policy legacies. First, it built local capacity, leaving in place organizational structures and networks for further work. Commitments have been made by each of the local United Ways to continue in the five neighbourhoods, and in some cases municipal and provincial governments have followed through. In the case of Toronto, the Greater Toronto United Way and the City of Toronto extended the ANC model to a 13 neighbourhood revitalization strategy that features place-based provincial investments in community hubs, health centres, and youth empowerment. Second, the ANC's intellectual capital is a resource for future policy makers and community activists. In addition to the concepts and tools, the

ANC experience contains valuable insights about how joint work and collaboration actually occurs in terms of operational details like clear communication, transparent decision making, regular information-sharing and knowledge dissemination, continual process monitoring and course corrections. Finally, the ANC contributed to formation of a federal policy 'community of practice' on neighbourhood revitalization in Canada. Its action research fed into emerging policy networks such as the Federal Family on Community Collaboration and the federal Policy Research Initiative exploring community roles in social innovation and place-based approaches (Policy Horizons Canada 2010).

## Conclusion

This chapter has explored Canada's recent engagement with neighbourhood revitalization policy through case studies of two prominent federal initiatives. Arguing that the themes of incrementalism, interscalar links, and 'learning from the local' have distinguished the nascent Canadian approach to place-based initiatives, the chapter tracked the federal government's efforts to find a policy middle ground between what has been viewed by most Canadian observers as overly-decentralized American localism and perhaps a too top-down European, particularly British, approach to neighbourhood change (Seguin and Divay 2003; Saint-Martin 2004). Canada has been a latecomer to the OECD policy interest in the new localism and place-based policy. However, as the VA and ANC demonstrate, Canadian governments and community networks are now reflecting on international practices and using pilot projects to explore the 'right policy mix' in place-based approaches.

It remains unclear whether these promising practices will embed a new national-local governance partnership. In a complex federation, almost all individual government initiatives depend for their success on alignment with policies and programs at other levels. Much rests with the quality of political leadership in translating creativity in the realm of ideas and practice into a coherent policy framework. Along these lines, it is noteworthy that the federal Conservative government chose neither to renew the VA in 2010 nor to extend the model to other large cities beyond Western Canada. Similar neglect befell the ANC as there has been no federal support in taking further its lessons for community partnerships and resident mobilization.

At the same time, the place-based framework has not been abandoned. In its 2010 Speech from the Throne, the Conservative government acknowledged that "the best solutions to the diverse challenges confronting Canada's communities are often found locally [where] the power of innovation is seen at work in communities across this country, as citizens, businesses and charitable groups join forces to tackle local problems" (Government of Canada 2010). Place-based policy development work continues across the federal departments (Bellefontaine 2011). A main finding from this chapter is that meaningful progress in Canadian neighbourhood revitalization policy now requires a creative mix of joined-up government as exemplified in the VA, and the broad-based community mobilization tapped through the ANC. At present, Canadian public policy demonstrates strength in pilot projects but much less capacity for institutionalizing innovation. As the Auditor General of Canada concluded, such a "case-by-case approach" generates 'one offs' rather than a policy framework. There is now a need for a "coherent, integrated body of policies and guidance" that would see local and national partners "learning, taking corrective action, and following up on weaknesses" (Auditor General of Canada 2005, Chapter 4).

This chapter's review of Canadian knowledge and practice informing neighbourhood revitalization reveals an analytical and practical foundation on which to construct a distinctive national approach to poverty reduction and social inclusion in an increasingly urbanized society and economy. In a vast and diverse country such as Canada, place-based policy engaging neighbourhood residents touches deep-seated values of autonomy, inclusion, and community. There is a pan-Canadian political coalition yet to be mobilized in support of this vision and these policies. Governments at all levels should take heed.

# References

Atkinson, R., & Kintrea, K. (2001). Disentangling area effects: Evidence from deprived and non-deprived neighbourhoods. *Urban Studies, 38*(11), 2277–2298.

Auditor General of Canada (2005). *Chapter 4 – Managing horizontal initiatives*. Office of Auditor General of Canada Report.

Bakvis, H., & Juillet, L. (2004). *The horizontal challenge: Line departments, central agencies and leadership*. Ottawa: Canada School of Public Service.

Barca, F. (2009). *An Agenda for a Reformed Cohesion Policy: A place-based approach to meeting European Union challenges and expectations*. Independent Report prepared at the request of Danuta Hubner, Commissioner for Regional Policy. Rome: Ministry of Economics and Finance.

Beauvais, C., & Jenson, J. (2003). *The well-being of children: Are there "neighbourhood effects"?* (Discussion Paper F/31). Ottawa: CPRN.

Bellefontaine, T. (2011). *Bringing place-based tools to policy: connecting information, knowledge, and decision-making in the Federal Government*. Policy Horizons Canada, Workshop Report.

Bernard, P., Charafeddine, R., Frohlich, K. L., Daniel, M., Kestens, Y., & Potvin, L. (2007). Health inequalities and place: A theoretical conception of neighbourhood. *Social Science & Medicine, 65*(9), 1839–1852.

Born, P. (Ed.). (2008). *Creating vibrant communities*. Toronto: BPS Books.

Bradford, N. (2005). *Place-based public policy: Towards a new urban and community agenda for Canada* (Research Report F/51). Ottawa: CPRN.

Bradford, N. (2008). *Rescaling for regeneration? Canada's Urban Development Agreements*. Paper prepared for Canadian Political Science Association Annual Meetings, Vancouver B.C.

Bradford, N. (2011). Public policy in Canada: Bringing place in? In S. Chisholm (Ed.), *Investing in better places: International perspectives* (pp. 22–36). London: The Smith Institute.

Bradford, N., & Chouinard, J. (2009). Learning through evaluation? Reflections on two federal community-based learning initiatives. *Canadian Journal of Program Evaluation, 24*(1), 51–77.

Bulthius, M., & Leviten-Reid, E. (2005). *Partnering for strong neighbourhoods*. Paper presented to International Conference for Integrating Urban Knowledge & Practice, Gothenburg, May 29–June 3.

Cook, D. (2010). *Donec prohibiti, procidite: Building a knowledge infrastructure to support place-based policy*. Ottawa: Policy Research Initiative.

Donovan, E., & Au, W. (2004). *The Vancouver agreement: Moving ideas forward together* (Vancouver).
Dunn, J., Bradford, N., & Evans, J. (2010). *Place-based policy approaches: Practical lessons and applications*. Report submitted to Human Resources and Skills Development Canada, Government of Canada.
External Advisory Committee on Cities and Communities (2006). *From restless communities to resilient places*. Final Report. Ottawa: Infrastructure Canada.
Final Report (1999). *Community review of the draft Vancouver Agreement*. City of Vancouver, Province of British Columbia, and Government of Canada.
Frelier, C. (2004). *Why strong neighbourhoods matter: Implications for policy practice*. Toronto: Strong Neighbourhoods Task Force.
Gertler, M. (2001). Urban economy and society in Canada: Flows of people, capital & ideas. *Isuma, 2*(3), 119–130.
Gorman, C. (2006). *Orienteering over new ground: A neighbourhood theory of change*. Ottawa: Caledon Institute of Social Policy.
Gorman, C. (2007). *The transition: Final reflections from the action for neighbourhood change research project*. Ottawa: Caledon Institute of Social Policy.
Government of Canada, Govenor General (2010, March 3). *A stronger Canada. A stronger economy. Now and for the future*. Speech from the Throne.
Gross Stein, J. (2006). *Canada by Mondrian: Networked federalism in an era of globalization*. Ottawa: Conference Board of Canada.
Hulchanski, D. (2007). *The three cities within Toronto: Income polarization among Toronto's neighbourhoods, 1970–2000*. Toronto: University of Toronto Centre for Urban and Community Studies.
Jamieson, W., & Kinnon, D. (2007). *Action for Neighbourhood Change (ANC) Project: Summative evaluation*. Ottawa: JHG Consultants.
Levitan-Reid, E. (2006). *Asset-based, resident-led neighbourhood development*. Ottawa: Caledon Institute of Social Policy.
Maclennan, D. (2006). *Remaking neighbourhood renewal: Towards creative neighbourhood renewal policies for Britain*. Ottawa: Caledon Institute of Social Policy.
Macleod Institute (2004). *In the spirit of the Vancouver Agreement: A case study in governance*. Evaluation report prepared for Western Economic Diversification Canada.
Makhoul, A. (2007a). *ANC in Scarborough village sets a good example for Toronto neighbourhoods*. Ottawa: Caledon Institute of Social Policy.
Makhoul, A. (2007b). *ANC sketches: Building a neighbourhood renewal process*. Action for Neighbourhood Change. Ottawa: Caledon Institute of Social Policy.
Mason, M. (2006). *Collaborative partnerships for urban development: A study of the Vancouver Agreement* (Research papers in environmental & spatial analysis, No. 108). London: London School of Economics and Political Science.
McMurtry, R., & Curling, A. (2008). *The review of the roots of youth violence*. Toronto: Queen's Printer for Ontario.
Minister of Labour and Housing (2005, May 9). Government of Canada, United Way and partners establish Action for Neighbourhood Change. *News Release*.
OECD. (n.d.). *Local governance and partnerships: A summary of findings of the OECD study on local partnerships*. Paris: LEED Programme.
Oreopoulos, P. (2002). *Do neighbourhoods influence long-term labour market success? A comparison of adults who grew in different public housing projects* (Research Paper, Catalogue No. 11F0019MIE, No. 185). Ottawa: Statistics Canada.
Policy Horizons Canada (2010). *Sustainable places: Compendium of place-based initiatives in Canada*. Ottawa: Government of Canada.
Public Policy Forum. (2008). *Collaborative governance and changing federal roles*. Ottawa: PPF and PRI Outcomes Report.
Rogers, J. (2001). *Governments in progress*. Panel presentation to FCM Symposium, "Cities in an Urban Century."
Ross, N. A., Houle, C., Dunn, J. R., & Aye, M. (2004). Dimensions and dynamics of residential segregation by income in urban Canada, 1991–1996. *The Canadian Geographer, 48*(4), 433–445.

Saint-Martin, D. (2004). *Coordinating interdependence: Governance and social policy redesign in Britain, the European Union and Canada* (Social Architecture Papers F/41). Ottawa: CPRN.

Sandercock, L. (2004). *Sustaining Canada's multicultural cities: Learning from the local*. Ottawa: Canadian Federation for the Humanities and Social Sciences.

Seguin, A., & Divay, G. (2003). *Urban poverty: Fostering sustainable and supportive communities*. Ottawa: CPRN.

Sirianni, C., & Friedland, L. (2001). *Civic innovation in America: Community empowerment, public policy, and the movement for civic renewal*. Berkeley/Los Angeles: University of California Press.

Statistics Canada (2004, October 18). Study: neighbourhood influence on health in Montreal. *The Daily*.

Torjman, S. (2005). *Policy dialogue*. Ottawa: Caledon Institute of Social Policy.

Torjman, S. (2007). *Shared space: The communities agenda*. Ottawa: Renouf.

Tremblay, R., et al. (2001). *Multi-level effects on behaviour outcomes in Canadian children*. Ottawa: Applied Research Branch, Human Resources Development Canada.

van Ham, M., Manley, D., Bailey, N., Simpson, L., & Maclennan, D. (2012). Introduction. In M. van Ham, D. Manley, N. Bailey, L. Simpson, & D. Maclennan (Eds.), *Neighbourhood effects research: new perspectives* (pp. 1–22). Dordrecht: Springer.

van Ham, M., Manley, D., Bailey, N., Simpson, L., & Maclennan, D. (2013). Understanding neighbourhood dynamics: new insights for neighbourhood effects research. In M. van Ham, D. Manley, N. Bailey, L. Simpson, & D. Maclennan (Eds.), *Understanding neighbourhood dynamics: new insights for neighbourhood effects research* (pp. 1–22). Dordrecht: Springer.

Walker, B., & Sankey, S. (2008). *International review of effective governance arrangements for employment-related initiatives* (Centre for Urban and Regional Studies, Research Report No. 543). London: Department of Work and Pensions.

Walks, R. A., & Bourne, L. S. (2006). Ghettos in Canada's cities? Racial segregation, ethnic enclaves and poverty concentration in Canadian urban areas. *The Canadian Geographer, 50*(3), 273–297.

Western Economic Diversification Canada (2004, September 1). Vancouver Agreement wins highest award. *Government of Canada News Release*.

Willms, J. D. (2002). *Vulnerable children: Finding from Canada's National Longitudinal Survey of Children and Youth*. Edmonton: University of Alberta Press.

# Chapter 9
# Neighbourhood Effects and Evidence in Neighbourhood Policy in the UK: Have They Been Connected and Should They Be?

**Rebecca Tunstall**

## Introduction

This book is based on the idea that neighbourhood policies are at least partly linked to ideas and information on the existence and nature of 'neighbourhood effects'. The seminar series on which the book is based had two premises:

> There is a strong belief that living in deprived neighbourhoods has a negative effect over and above the effect of individual characteristics on residents health, labour market outcomes, and social values; so called neighbourhood effects.
>
> This belief has a major impact on urban, neighbourhood and housing policies designed to tackle poverty and to improve the lives of residents in deprived neighbourhoods (www.neighbourhood.org.uk).

The first premise is stated in a noticeably passive voice. Who are the 'believers'? Are they researchers, policy makers, policy entrepreneurs or others? (Are we included?) What is the relationship between this 'belief' and evidence? The second premise is about the policy process, and asserts that the belief has influenced policy. Is this so? And if so, should it have done so, and what are the consequences?

This chapter argues that 'evidence based policy' is attractive to policymakers and researchers alike, and that research and evidence on neighbourhood effects are a particularly attractive form of research for both groups. It also argues, however, that evidence on neighbourhood effects and neighbourhood policy have not, hitherto, been closely connected, at least in the UK. This is partly due to the absence of sufficient research on neighbourhood effects based on UK neighbourhoods, although there is a considerable amount of research on neighbourhood effects from outside the UK. It is also partly because the evidence which has been developed has been in policy areas such as health, criminology and other fields more remote from neighbourhood policy.

---

R. Tunstall (✉)
Centre for Housing Policy, University of York, Heslington, York YO10 5DD, UK
e-mail: becky.tunstall@york.ac.uk

The chapter also argues that policymakers and researchers interested in neighbourhood effects and neighbourhood policy risk misunderstanding each other, the evidence there is, and its policy implications. Over the past few years, as evidence on neighbourhood effects has gradually accumulated, it appears that a broad, if latent, consensus has developed among researchers on the extent and nature of policies which can be supported by this evidence. In brief, through a combination of absence of good evidence and, increasingly good evidence of weak effects, most researchers appear to be generally sceptical that neighbourhood effects alone provide a justification for costly or disruptive neighbourhood policies. Meanwhile, amongst policymakers, beliefs about neighbourhood effects, and the potential benefits of neighbourhood policies based on them, have become established as a promising or even central element of housing and urban policy. The result is a disconnect between these two groups, and between beliefs about evidence and beliefs about the policy implications. From a researchers' point of view, it suggests that evidence on neighbourhood effects and neighbourhood policy should perhaps be less connected than they are.

## Methodological Note

This chapter is based on studies I have been personally involved with, published reviews (see for instance Galster 2007; Bond et al. 2010), papers presented at seminars in the series this book is based on, discussions at these seminars, and discussions with policymakers and research colleagues in the course of projects over several years (e.g. Feinstein et al. 2008; Lupton et al. 2009; Tunstall et al. 2011a, b). It is not a systematic review of research and evidence on neighbourhood effects. Some of the points it contains are conjectural. It is not based on evidence about the nature of the policymaking process, and would benefit from a programme of interviews with policymakers. It also contains the results of a small scale participatory action research project (see Box 9.7).

## Evidence-Based Policy

'Evidence based policy' is attractive to policymakers and researchers alike (Dabinett 2001; Pawson 2006). In urban policy in the UK, the idea reached a recent high water mark in the late 1990s and 2000s (Dabinett 2001). Research and evidence on neighbourhood effects may be or may appear particularly attractive to both groups too. Some policymakers appear to assume that neighbourhood effects research is exclusively quantitative, and that quantitative neighbourhood effects research can produce unambiguous evidence of causation, with clear policy implications. Researchers, too, may be attracted by the prospect of 'hard' evidence for housing and neighbourhood policy, and the influence on policy it might offer, although in practice neighbourhood effects could be researched though a wide range of methods.

> **Box 9.1** Possible outcomes from a research review on neighbourhood effects
>
> 1. An absence of evidence (or an absence of high quality evidence);
> 2. Evidence of no effect;
> 3. Mixed evidence (with variation between studies, neighbourhood sizes or other characteristics, or outcome variables);
> 4. Evidence of a 'beneficial' effect;
> 5. Evidence of a 'harmful' effect.
>
> Source: Adapted from Bond et al. (2010)

Bond and colleagues suggested five possible outcomes from a research review on tenure mix, which could be applied to reviews of any area of neighbourhood effects or other research (See Box 9.1).

## The Connection Between Evidence on Neighbourhood Effects and Neighbourhood Policy to Date

Standard models of the public policy process assume that governments will set the agenda by identifying a policy problem and then carrying out an appraisal of possible options. Researchers and evidence play a role at this stage. The ideal type of evidence-based policymaking depends on systematic evidence review by researchers, followed by systemic policy options appraisal by policymakers (Pawson 2006). Ideally, the government and other makers of public policy should follow principles as outlined, for example, in the UK government's Treasury Green Book. This aims to ensure that: *"public funds are spent on activities that provide the greatest benefits to society, and that they are spent in the most efficient way"* (HMT 2003, pv). The Green Book states that:

> all new policies, programmes and projects, whether revenue, capital or regulatory, should be subject to comprehensive but proportionate assessment, wherever it is practicable (HMT 2003, p1).

It provides methods for appraising proposed policies and constitutes binding guidance for all central government departments and agencies in the UK:

> The essential technique is option appraisal, whereby government intervention is validated, objectives are set, and options are created and reviewed, by analysing their costs and benefits (HMT 2003, p2-3).

Assuming a single, agreed social problem, policymaking may look like Box 9.2. Evidence and researchers play a fairly straightforward role. More research and evidence is needed on how housing and neighbourhood policy are made, the extent to which the process differs from that seen in Box 9.1, and the roles that any body of evidence or other factors play in practice.

> **Box 9.2** The role of evidence and researchers in policymaking (policymaking in series and in a specialist field)
>
> - P is a social/economic problem
> - Is P associated with/possibly caused by factor X? (answering this question is the role of specialist X factor researchers)
> - If the answer to (2) is 'yes', specialist researchers are likely to suggest that the policy implication is to change or remove X
> - Policy makers and implementers need to go through an options appraisal process: Is changing or removing X practical? Does it have any costs? Do benefits justify costs? This may result in a policy to change or remove X, a different policy, or no action

However, it appears that the evidence on neighbourhood effects and 'neighbourhood' policy, at least in the UK, have not hitherto been closely connected. The UK's history of research and evidence on neighbourhood effects is much more recent and more limited than its neighbourhood policy, which became established as a specialist area in the late 1960s. At least until recently, a systematic review of UK literature would have concluded, in Bond and colleagues' terms above, that there was an absence of evidence.

Thus for much of its history, neighbourhood policy cannot have been based on this evidence and even in recent years the evidence on neighbourhood effects has been limited and partial. There may be other, and additional, reasons for the lack of connection between evidence and policy in this field.

## Developments in the Evidence on Neighbourhood Effects

In 2007 Galster completed a comprehensive review of European evidence on neighbourhood effects, concluding that if all European material was taken together, there was an emerging evidence base which provided complex but useable implications for policy (2007). Since Galster's review, the UK and international evidence base has expanded, as this book demonstrates (see also the previous volumes van Ham et al. 2012, 2013), and in Bond and colleagues' terms, there is no longer an 'absence of evidence'. Three studies I have been involved in are summarised here. Like other examples in this book they exemplify the additional evidence developed since 2007. However, they are introduced primarily to illustrate the pitfalls of attempts to link evidence to policy

The first pair of studies develop the findings of two previous reports (Feinstein et al. 2008; Lupton et al. 2009) on housing in childhood for those born in 1946, 1958, 1970 and 2000, and the relationship with adult outcomes across a range of measures (except for those born in 2000). They found as yet unexplained connections between being 'ever' in social housing in childhood and worse adult outcomes on an overall measure of deprivation and on a range of individual measures for those born in 1958 and in 1970 (but not for those born in 1946) (Feinstein et al. 2008; Lupton et al. 2009). These results appear to suggest a "tenure effect". The new studies aimed to explore whether the characteristics of the neighbourhoods the different housing tenures were found in might play a role in explaining the relationships found between tenure and outcomes.

## *Teenage Housing Tenure and Neighbourhoods and the Links with Adult Outcomes*

This was a study of the relationship between neighbourhood deprivation and tenure mix at age 16 in 1986, and a wide range of outcomes at age 34 in 2004, for members of the British Cohort Study (Tunstall et al. 2011a).

The study aimed to explore the extent to which neighbourhood characteristics modified the relationship found between individual housing tenure at age 16 and outcomes at age 34 in this study, and in other previous studies exploring the same data (Feinstein et al. 2008; Lupton et al. 2009). It considered neighbourhood deprivation, using a measure based on the Carstairs index, with 1991 data, and also neighbourhood tenure mix, distinguishing neighbourhoods with 50 % or more social housing from other neighbourhoods. It defined neighbourhoods at both small and larger neighbourhood scales, using enumeration districts and wards. It examined the following outcomes at age 34: life satisfaction, malaise, depression, self-rated health, low self-efficacy, literacy and numeracy problems, receipt of means tested benefits, employment, and smoking behaviour. As mentioned above, most of these are not conventional outcome measures for housing or urban policy.

The members of the British Cohort Study who were in both social renting and private renting at age 16 in 1986 were more likely to have less desirable outcomes at age 34 in 2004 than those in home ownership at 16, in terms of all the 11 measures examined. This complements the results found for this cohort in Feinstein and colleagues (2008) and Lupton and colleagues (2009). Those in social renting at 16 had worse raw outcomes than those in private renting at 16 in terms of chances of claiming means-tested benefits, having literacy and numeracy problems, smoking, qualifications, self-rated health, and self efficacy than those in private renting at 16. However, those in social renting at 16 actually had slightly better raw outcomes in terms of chances of depression, malaise and life satisfaction at 34 than those in private renting.

There were numerous differences between those who were in different tenures as teenagers, including in the characteristics of their parents and homes. The kinds of neighbourhoods cohort members lived in as teenagers also varied sharply by tenure. Teenagers in social housing were highly concentrated in the most deprived neighbourhoods and those dominated by social housing. 62 % of teenagers in social housing were in the three most deprived deciles of small neighbourhoods, compared to 18 % of those in home ownership. Teenagers in social housing were largely excluded from less deprived neighbourhoods. 33 % of teenagers in social housing were in social housing dominated small neighbourhoods, compared to just 4 % of those in owner occupation.

After both a small and large set of controls for individual and family circumstances in childhood had already been applied, taking neighbourhood deprivation and tenure mix into account modified somewhat the relationship between housing tenure at age 16 and outcomes at age 34. In the case of life satisfaction and malaise, the combination of individual, family and neighbourhood characteristics removed the relationship seen between tenure at 16 and outcomes.

One interpretation is that there were some signs of a 'neighbourhood effect' mediating associations between teenage housing tenure and adult outcomes. Thus the links found can be described as partly due to teenage neighbourhood characteristics in terms of deprivation and tenure mix (Tunstall et al. 2011a). However, individual and family factors were more important than tenure or neighbourhood characteristics. Family and individual characteristics have been found to be more important than any neighbourhood effects in all studies of neighbourhood effects which also reported on family and individual characteristics. Not all studies do this, however. It is also possible that not all readers grasp the significance and implications of the findings (see also van Ham and Manley (2010)).

## *Growing up in Social Housing in the New Millennium*

This was a study of the relationship between neighbourhood deprivation at age 5 in 2005 for members of the Millenium Cohort Study, and the scores children achieved at age 5 on tests used to indicate school-readiness (Tunstall et al. 2011b). Like the above study, this one aimed to explore the extent to which neighbourhood characteristics, amongst other factors, modified a relationship seen between individual housing tenure and outcomes.

Children in social renting at age 5 scored lower on vocabulary and pattern tests administered at age 5 than those in private renting and lower still than those in owner occupation. There were numerous other differences between children in different tenures, including in the characteristics of their parents and homes. Compared to those in other tenures, 5 year olds in social renting were markedly disadvantaged in terms of their neighbourhoods. Forty-seven percent lived in the most deprived 20 % of neighbourhoods, and only 20 % were in the least deprived 50 % of neighbourhoods. Only 46 % of social renting parents thought their area was 'excellent' or 'good' for raising children, compared to 72 % for parents overall.

Analysis found that more than half the gap in test scores between 5 year old children in both social and private rented tenures and those in home ownership was removed by controlling for a small number of family and individual factors, and neighbourhood deprivation. The factors involved were: whether or not the child lived in the most deprived tenth of neighbourhoods, their parents' education and occupational status, the mother's age at first birth, family structure, and the number of siblings children had.

Again this can be interpreted as evidence of a 'neighbourhood effect'; thus the typically high deprivation levels of social housing neighbourhoods appeared to account for part of the link seen between tenure and early childhood outcomes. Again, however, individual and family factors were more important than tenure or neighbourhood characteristics.

These two studies (1, 2) examined the extent to which neighbourhood characteristics modified a relationship seen between tenure and outcomes. They were not direct tests for the existence and size of neighbourhood effects. Policies including demolishing homes and moving families have high monetary costs and social costs, which could include, for example, minor adverse effects on child test scores, and substantial opportunity costs which could include, for example, extra educational programmes to boost test scores.

## *The Task for Disadvantaged Young Job Seekers*

This was a study to investigate evidence for 'postcode discrimination' against residents of neighbourhoods with poor reputations in employment (Tunstall et al. 2012).

There are geographical variations in employment rates, which some studies have been unable to explain through the characteristics of residents in the areas and which have been attributed to residual 'neighbourhood effects'. It has been suggested that 'postcode discrimination' by employers against those from areas with poor reputations could be part of the explanation for geographical variations in employment and neighbourhood effects on employment. The third study used an experimental method to test for postcode discrimination, making job applications by matched pairs of fictional candidates coming from different neighbourhoods.

Over the course of 2010–2011, 3 matched applications were made to 667 jobs advertised via the UK government's Jobcentre Plus employment service website www.direct.gov.uk. The jobs applied to required limited education and experience and generally paid close to the minimum wage. They included cleaner, chef, kitchen hand, security, office admin and accounts roles. They were based in three labour markets, with unemployment levels ranging from high to average. For each job on application came from a fictional candidate with an address in a neighbourhood with a 'bland' reputation, and two came from candidates addresses in two neighbourhoods with 'poor' reputations. Neighbourhood reputations were identified through key informants, street interviews and media searches. All three candidates had CVs demonstrating equitable education and experience and were generally well qualified for the jobs concerned. Only jobs in which a key decision maker appeared to be locally based and could be applied to electronically were selected. Fictional candidates' email addresses and voicemails were monitored for employer responses.

This study was a direct test of one sort of exogenous neighbourhood effect. In 192 of the 667 jobs at least one of the candidates received a positive response (usually a call to interview rather than a job offer). These are the cases that can be used to test for employer preference. One hundred and twenty or 62.5 % of the candidates from the 'bland' reputation neighbourhoods received positive responses compared to 230 or 59.9 % of the candidates from the 'poor' reputation neighbourhoods. The difference was not statistically significant.

## Creating Policy Implications from Evidence

Bond and colleagues criticised researchers for summarising their own evidence inaccurately, and for potentially distorting the policy process:

> It seems likely… that the reviews used by policymakers could be less thorough or rigorous than they might be (Bond et al. 2010, p71).

Bond and colleagues were right to identify problems in connecting evidence on neighbourhood effects and policy, but the problem appears broader and more difficult than they imply. It appears that problems in communicating evidence, and neglect of the process of creating policy implications from evidence, has allowed a divergence to develop between researchers' and policymakers' understanding of the body of evidence and its implications in the UK. In this process, researchers have committed sins of omission (and failure to go beyond the call of role and duty), rather than of commission, or role expansion.

The two complete studies described above can be used to provide an example of the difficulties in communicating research on neighbourhood effects (Tunstall et al. 2011a, b). In Bond and colleagues' terms (see Box 9.1), for some outcomes, there was evidence of no neighbourhood effects (2). For others, there was evidence of 'positive' neighbourhood effects (moving towards outcomes widely seen as beneficial) (4) and for some others, of 'negative' ones (5). Overall, the results could perhaps be summarised as 'mixed' (3).

In medical research, the problem of 'positivity bias' in reporting, publishing and the commissioning of further research is well-established. Findings in categories (3) and (4), in Bond and colleagues' terms, may be easier to explain, and are also more widely reported. Similarly, in the neighbourhood effects field, there may be a tendency amongst researchers to report *any* evidence of neighbourhood effects found in study for any variable at any stages it may be reported, even if trivial in size, and even if the evidence relates to an obscure outcome, or is an exception amongst numerous variables for which no effect was found. There may then be a tendency for policymakers to focus on evidence of positive neighbourhood effects (Finding 4), rather than evidence of no effects (Finding 2), There may also be a tendency to feel that any effect demands or justifies a policy response.

One of the attractive and difficult things about 'neighbourhood effects' are that they can be effects on anything. The premise for the seminar series on which this book is based mentions a broad range of outcomes: 'health, labour market outcomes and social values'. Some other recent studies of neighbourhood effects in the UK, mostly using quantitative multivariate analysis, consider even more diverse topics, topics those interested in neighbourhood policy may be less aware of:

- Fear of crime – Brunton-Smith and Jackson – (2013 see this volume)
- Frequency of walking in the neighbourhood – Mason and colleagues (2011)
- Social cohesion – Robinson (2010)
- Alcohol-related mortality – Connolly and colleagues (2011)
- Muslim electoral registration – Fieldhouse and Cutts (2008)

The range of outcomes researched outside the UK is even more extensive. Outcomes such as frequency of walking or electoral registration have not usually been seen as part of the core goals of housing and neighbourhood policy, and the extent to which it is, or should be, the role of housing and neighbourhood policy to influence these outcomes is at best, debateable. However, it is possible that the open-ended nature of neighbourhood effects might actually be another attraction to housing and neighbourhood policymakers, appearing to offer wide-ranging outcomes from their policy. It should be noted that the more numerous are the outcome variables used, the more likely it is that at least some will show evidence of positive or negative effects.

Even if positivity bias is successfully avoided, it can be difficult to summarise research, and it may not be possible to summarise results in a fully 'neutral' way, and without incorporating or creating policy implications (Pawson 2006). A seemingly-trivial issue is that policymakers may interpret researchers' description of results as 'statistically significant' to mean 'big effects', presumably with big policy implications, rather than 'patterns unlikely to have occurred by chance'. In this case, to avoid misinterpretation, '(statistically significant) evidence of a very small effect' (in Bond et al.'s terms, 4 or 5) might best be presented to policymakers as '(statistically significant) evidence of no effect (or none large enough to have policy implications)' (in Bond et al.'s terms, 2). Overall, the results of Tunstall and colleagues (2011a, b) could perhaps be summarised as 'mixed', but the authors chose to emphasise the scale of effect, including its scale in comparative terms. They pointed out that in their opinions, evidence of neighbourhood effects did not appear sufficient to justify disruptive or costly neighbourhood policies. But what counts as a 'significant' size of effect for policy purposes, and whose responsibility is it to discuss this?

Another issue is that, in practice, policy making does not take place as an isolated linear process (as in Box 9.2), but in the context of multiple policy goals, multiple evidence bases, and multiple possible policy options in parallel. This presents challenges for the relationship between evidence and policy, and the roles of policymakers and researchers (Box 9.2).

Next, as the Treasury Green Book states (2003), and as Box 9.3 shows, even once it has been established that there is an 'effect' of some kind, policy appraisal requires further assessment. Are these effects for outcomes that are considered to be important policy problems, are they sizable, and are there any policies that might be able to change these effects, is it worth trying to change them, and do these policies offer overall cost benefit when compared to alternative policies (Box 9.3)?

**Box 9.3** Evidence on factor X – neighbourhood effects – and the policy process in more complex and realistic contexts

1. Inequality in current or longitudinal outcomes between people in different neighbourhoods is a social or economic problem (P)
2. Inequality in opportunities (P1), incomes (P2), other factors (P3, P4 etc) are also social/economic problems
3. Is P possibly caused by neighbourhood effects? (answering this question creates a role for neighbourhood effects researchers)
4. If the answer is 'yes', the specialist researchers will suggest that the policy implication is to change or remove neighbourhood effects
5. Is P also possibly caused by selection operating via housing markets, family disadvantage, other factors A, B etc (identifiable by separate regressions or as interaction effects) (this provides a role for other types of researchers)
6. If the answer is 'yes', the researchers will suggest that the policy implications are to change or remove neighbourhood effects and/or housing markets and/or family disadvantage and/or A, B etc
7. How do X, Y, Z, A, B etc compare to each other as potential causes of P?
8. How do P, P1, P2, P3 etc compare to each other as problems?
9. How do X, Y, Z, A, B etc compare to each other as associated/potential causes of P?
10. Similarly, the causes of P1, P2, P3 etc.
11. Also, how do X, Y, Z, A, B etc interact with each other?

Policy makers and implementers need to go through an options appraisal process, taking into account X, Y, Z, A, B, P1, P2, P3 etc.

To use the example of the study *Growing up in social housing in the new Millenium*, researchers and policymakers need to think about how much social resource should be devoted to trying to improve the test scores of less advantaged children at age 5, and what role neighbourhood policies should play compared to other policies (Tunstall et al. 2011b).

The results at age 5 on similar tests to those covered in Tunstall and colleagues (2011b) were associated with achievement in reading and maths at age 10 and qualifications and wages at age 30 for the previous generation (Feinstein and Duckworth 2006). How much of the differences in scores did they explain? Housing and neighbourhood characteristics are not the only factors associated with difference in child test scores seen in Tunstall and colleagues (2011b). For example, child test scores are also associated with gender (Schoon et al. 2010), ethnicity (Dearden and Sibieta 2010), parents' employment status and financial situation, parenting practices, parents' relationship quality, mother-child relationships and mother's wellbeing and self-esteem (Jones 2010). The study of children in the most deprived decile of neighbourhoods found that they scored about 1.5 points lower on the test (which had a mean score of around 50 points) than other children after controls. How important is this difference and how does it relate to the gaps found by Feinstein and Duckworth and other authors? In addition, housing and neighbourhood policies – especially the highest economic/social cost ones – are not the only options. The other research suggests, for example, that projects on parent's employment status and relationship quality may offer potential for altering child test scores too.

There is a role to play in drawing out the policy implications of research. The passive tense in the previous sentence is intentional: both in models of policymaking and in practice, it is often unclear whose responsibility this is. In practice, it may be carried out by researchers, policymakers, no-one at all, or by third parties such as think tanks, lobbyists or 'policy entrepreneurs' (Evans 2004). There is also a role for ambiguity amongst and between researchers. Bond and colleagues (2010) also criticised researchers for being insufficiently systematic in finding and assessing literature. Specialist researchers working in one narrow area may not fully grasp the relative importance of policy problems in their area compared to others, and may not be able to or actually consider the implications of their particularly findings in the contexts of numerous other policy problems and potential solutions, in order to carry out steps 4 and beyond (Box 9.2).

## A Sceptical Consensus?

In 2007 Galster asked the policy question, 'Should policymakers strive for neighbourhood social mix?' (2007). At past seminars in this series, as noted, numerous points were made about gaps and weakness in the evidence base and ways to fill them, but few points about the implications of these gaps and weaknesses for policymaking. In fact there has been relatively limited explicit consideration of policy implications of neighbourhood effects research to date. Nevertheless, there appears to be a sceptical consensus about the extent to which the evidence on neighbourhood effects can be connected to policy, and in particular whether it can be said to support the more costly and disruptive policies.

In 2007 Galster said that the 'policy thrust' towards the most active neighbourhood policies, creating more socially mixed neighbourhoods for poorer residents "has not been without its skeptics" (2007, p524). He went on to list large numbers of those researching both neighbourhood policy and neighbourhood effects in the UK as well as across Europe (including Kearns, Atkinson, Meen, Musterd, Ostendorf, Freidrichs, Cole, Goodchild, and Delorenzi). In 2010, after further additions to the evidence base, there remained signs of a latent sceptical consensus amongst researchers, on the strength of evidence of neighbourhood effects, and what types of policy they may justify. There was also a consensus that policymakers and researchers run the risk of misunderstanding each other.

Using the documentation from the previous seminars (see www.neighbourhoodeffects.org.uk and van Ham et al. 2012; 2013) it is clear that, at seminars in the

> **Box 9.4** Neighbourhood effects researchers' own assessments of gaps in and weaknesses of current research (some summarised)
>
> - Need more ethnographic research – Small and Feldman (2012)
> - Need more sophisticated hypotheses – Small and Feldman (2012) (and need to avoid 'descriptive' hypotheses in favour of 'relational' hypotheses)
> - Need to understand and explore potential causal mechanisms more – Galster (2012), Small and Feldman (2012), Lupton and Kneale (2012)
> - Some indicators used by people not familiar with field (e.g. teen parenthood); absence of data on e.g. neighbourhood size – Lupton and Kneale (2012)
> - Need to understand exposure to neighbourhood over time ('dosage') (Galster) and thus need to understand population mobility between neighbourhoods – Hedman and van Ham (2012)
> - Need more longitudinal data (to understand dosage but also effects which emerge or increase over time) – Bailey et al. (2013)
> - Need to be aware of potential variation in neighbourhood effects between neighbourhoods – Manley and van Ham (2012)
> - Need to integrate study of neighbourhood effects with other areas, such as the study of social capital – Volker et al. (2013) or the study of mobility – Hedman and van Ham (2012)
> - Need to be more specific about differences in neighbourhood effects between outcomes, which may vary – for example, there may be a difference between effects on quality of life and on life chances – Cheshire (2012)

series on which this book is based, and in related publications, participants' views on the ability of the evidence to support policy and its policy implications of evidence have been cautious or sceptical (see Boxes 9.4, 9.5, 9.6, and 9.7 below).

The next section presents the results of a small, informal consultation amongst 17 researchers interested in neighbourhood effects in 2011, attendees at the third seminar in the series on which this book is based (Box 9.7). About half of those responding to the questionnaire have carried out research on neighbourhood effects. The responses showed consensus on some points, and disagreement on others. For example, a similar proportion of people thought that there was evidence of neighbourhood effects on outcome variables and of sizes relevant to policy in the UK and thought that the evidence on neighbourhood effects in the UK was too contradictory or varied to form an overall opinion. On the other hand, there was a consensus that the evidence on neighbourhood effects was strong enough to provide grounds for at least some types of neighbourhood polices, and that there were also other grounds, independent of neighbourhood effects, for these policies (Box 9.7). There was also a consensus that there were grounds for 'mixed communities policies' (involving

**Box 9.5** Neighbourhood effects researchers' own assessments of the policy implications of research, 2010

- Not substantial evidence of statistically significant, sizeable neighbourhood effects on policy-relevant outcome variables that remain after controls to date in UK – Cheshire (2012)
- May be neighbourhood effects on other outcomes e.g. quality of life – Cheshire (2012)
- Anti-segregation or anti-neighbourhood effects policies have potential costs – these include loss of positive neighbourhood effects (benefits of homogenous neighbourhoods to rich and poor) – Cheshire (2012); and the opportunity costs (of not pursuing other policies) – Cheshire (2012); Meen et al. (2013)
- Positive neighbourhood effects of homogenous neighbourhoods may be overstated – homogenous neighbourhoods may be sub-optimal in economics terms – Meen et al. (2013)
- Mixed community policy misguided – Cheshire (2012)
- Non-spatial anti-poverty policy is a better way to reduce poverty than neighbourhood-based 'mixed community policy' – Cheshire (2012)
- There may be precautionary grounds for some types of anti-neighbourhood effects/mixed community policies, and it should be noted that the evidence base on individual-based anti-poverty policies is also limited, and not very encouraging – Meen et al. (2013)

**Box 9.6** Explicit comments on the absolute and relative size of neighbourhood effects found, and thus potential cost-benefit of any policy to alter them, drawn from selected recent research, 2007–2011. (Details in square brackets added.)

evidence.... does not support the conclusion that neighbourhood effects [on incomes] are quantitatively all that important... the conclusion for policy is to reduce overall income inequality rather than to promote mixed communities (Cheshire 2007, p34)

the main sources of low incomes are to be found in earnings, employment, and demographics, not in neighbourhood characteristics (Bolster et al. 2007, p234)

For most variables, controlling for the characteristics of the neighbourhood in which the teenager lived... had a moderate additional effect on the size of the association between tenure and outcomes [of various types, including employment and benefit claiming]... much smaller than the effect of the Index of Family Advantage alone (Tunstall et al. 2011a, p30)

**Box 9.7** Views on the policy implications of neighbourhood effects research to date of attendees at the third ESRC seminar on neighbourhood effects (nbh eff), 2011.

**(Negative) unanimity**

0/17 There is evidence that neighbourhood effects on policy-relevant outcome variables and of sizes relevant to policy do not exist in the UK

0/17 The evidence on neighbourhood effects provides grounds for stopping or reversing mixed communities policies in the UK

0/17 The evidence on neighbourhood effects provides grounds for stopping or reversing other neighbourhood policies in the UK

**Majority**

12/17 Other social, housing or neighbourhood policies should have a higher priority than mixed communities policies (e.g. involving demolishing homes or moving residents)

10/17 The evidence on neighbourhood effects does provides grounds for at least some types of other neighbourhood polices (e.g. redistributing resources or altering services) in the UK

10/17 Other rationales do provide grounds for at least some types of mixed communities policies in the UK

10/17 Other rationales do provide grounds for at least some types of other neighbourhood policies in the UK

**Minority**

8/17 There is evidence of neighbourhood effects on outcome variables and of sizes relevant to policy in the UK

8/17 The evidence on neighbourhood effects in the UK is too contradictory or varied to form an overall opinion

7/17 There is evidence of neighbourhood effects on policy-relevant outcome variables and of sizes relevant to policy for countries other than the UK

6/17 The evidence on neighbourhood effects does provides grounds for at least some types of other neighbourhood polices in other countries

5/17 The evidence on neighbourhood effects does provide grounds for at least some types of mixed communities polices in the UK

5/17 The evidence on neighbourhood effects does provide grounds for at least some types of mixed communities polices in other countries

5/17 Other social or housing policies should have a higher priority than neighbourhood policies

**Near unanimous disagreement**

2/17 The evidence on neighbourhood effects does not provide grounds for at least some types of mixed communities polices (e.g. involving demolishing homes or moving residents) in the UK

1/17 There is no evidence of neighbourhood effects on policy-relevant outcome variables and of sizes relevant to policy in the UK

1/17 The evidence on neighbourhood effects does not provides grounds for at least some types of other neighbourhood polices (e.g. redistributing resources or altering services) in the UK

demolishing homes or moving residents) in the UK, but independent of neighbourhood effects. Only a minority felt that the evidence on neighbourhood effects does provided grounds for at least some types of mixed communities polices in the UK. However, no-one thought that the evidence on neighbourhood effects provides grounds for stopping or reversing mixed communities or other neighbourhood policies in the UK. There was a consensus that other social, housing or neighbourhood policies should have a higher priority than mixed communities policies (Box 9.7).

Typical academic conferences and seminars do not generally aim at a summative consensual position on a single question. Instead they consist of exploratory and developmental work on a number of questions, albeit within a small field. However, in theory, the existence and degree of consensus suggested above could be explored formally, and made explicit through a systematic evidence re-review by researchers (Pawson 2006), or through an alternative process designed to solicit views of experts on this topic, such as a Delphi review (Linstone and Turoff 1975; Adler and Ziglio 1996; Mullins 2006). A systemic policy options re-appraisal by policymakers, on the lines advocated by the Green Book, would provide an alternative or complement. However, it is unclear when, how or by whom any such review should be triggered, and how emerging latent consensus or near-consensus could be made overt, and introduced to the policy process.

In the absence of any other trigger, the task of connecting neighbourhood effects evidence and any consensus on it to policy may fall to researchers. Specialist researchers may commit sins of omission of not pointing out implementation and opportunity cost issues and their feelings about them, from more generalist knowledge of related areas. Any vacuum, or other gap, may be filled by 'policy entrepreneurs' who have less interest or awareness of evidence or interest in it. At the very least, the Social Research Association implies that researchers have a responsibility to comment when they feel policymakers have misunderstood the implications of their own research:

> Social researchers may not be in a position to prevent action based on their findings. They should, however, attempt to pre-empt likely misinterpretations and should counteract them when they occur (SRA 2003, p18).

## Conclusion

UK neighbourhood policies over the past 40 years have not largely been based on evidence about neighborhood effects to date. Even over the most recent past government, 1997–2010, there was little timely neighbourhood effects evidence, and neighbourhood policy was justified in the absence of positive evidence.

In recent years, policymakers appear to have increased their reliance on putative neighbourhood effects as justification for neighbourhood policies, and this appears to have been in contrast to researchers in the areas. Many researchers are at least sceptical that evidence of neighbourhood effects provides support for neighbourhood policy, particularly the most costly or disruptive policies. There appears to be a consensus amongst researchers that other rationales continue to justify at least some elements of neighbourhood policy.

This does not mean there is no connection, however, between neighbourhood policy and neighbourhood effects. There may be a link between neighbourhood policy and neighbourhood effects in the other direction. It seems likely that past UK neighbourhood policies have had some effect on UK neighbourhood effects by assessing neighbourhood characteristics which might generate then. For example, between 1997 and 2010, there were some, if limited, effects on deprived neighbourhood outcomes such as education and employment. There were some, limited, changes in characteristics of neighbourhoods we attribute neighbourhood effects to, including population composition, neighbourhood infrastructure including services, and relationships with other areas, although it is not clear what critical thresholds are and if they have been passed, or how many neighbourhoods.

It is also possible that future neighbourhood policy may have an effect on neighbourhood effects. From 2010, the UK's new neighbourhood policy, with a coalition government attempting to reduce public expenditure, constitutes another experiment. This includes the withdrawal of most traditional area based initiatives alongside spending cuts and attempts to devolve decision-making through 'localism'. These changes have the *potential* to change the characteristics of neighbourhoods which are seen to create neighbourhood effects, including population composition, neighbourhood infrastructure including services and the relationship with other areas. Thus it is possible that policies will lead to new, altered or increased neighbourhood effects.

## References

Adler, M., & Ziglio, E. (1996). *Gazing into the oracle: The Delphi method and its application to social science and public health*. Bristol: Jessica Kingsley.

Bailey, N., Barnes, H., Livingston, M., & Mclennan, D. (2013). Understanding neighbourhood population dynamics for neighbourhood effects research: A review of recent evidence and data source developments. In M. van Ham, D. Manley, N. Bailey, L. Simpson, & D. Maclennan (Eds.), *Understanding neighbourhood dynamics: New insights for neighbourhood effects research* (chap. 2, pp. 23–42). Dordrecht: Springer.

Bolster, A., Burgess, S., Johnston, R., Jones, K., Propper, C., & Sarker, R. (2007). Neighbourhoods, households and income dynamics. A semi-parametric investigation of neighbourhood effects. *Journal of Economic Geography, 7*(1), 356–386.

Bond, L., Sautkina, E., & Kearns, A. (2010). Mixed messages about mixed tenure: Do reviews tell the real story? *Housing Studies, 26*(1), 69–91.

Brunton-Smith, I., & Jackson, J. (2013). Urban fear and its roots in place. In V. Ceccato (Ed.), *Urban fabric of crime and fear*. London: Springer.

Cheshire, P. C. (2007). *Segregated neighbourhoods and mixed communities: A critical analysis*. New York: Joseph Rowntree Foundation.

Cheshire, P. (2012). Are mixed community policies evidence based? A review of research on neighbourhood effects. In M. van Ham, D. Manley, N. Bailey, L. Simpson, & D. Maclennan (Eds.), *Neighbourhood effects research: New perspectives* (chap. 12, pp. 267–294). Dordrecht: Springer.

Connolly, S., O'Reilly, D., Rosato, M., & Cardwell, C. (2011). Area of residence and alcohol-related mortality risk: A five-year follow-up study. *Addiction, 106*(1), 84–92.

Dabinett, G. (2001). *A review of the evidence base for regeneration policy and practice*. London: DETR.

Dearden, L., & Sibieta, L. (2010). Ethnic inequalities in child outcomes. In K. Hansen, H. Joshi, & S. Dex (Eds.), *Children of the 21st century: The first 5 years* (pp. 169–184). Bristol: Policy Press.

Evans, M. (2004). Introduction: Is policy transfer a rational process? In M. Evans (Ed.), *Policy transfer in global perspective* (pp. 1–9). Aldershot: Ashgate.

Feinstein, L., & Duckworth, K. (2006). *Development in the early years: Its importance for school performance and adult outcomes Wider Benefits of Learning* (Research Report No. 20). London: Centre for Research on the Wider Benefits of Learning.

Feinstein, L., Lupton, R., Hammond, C., Mujtaba, T., & Sorhaindo, A. with Tunstall, R., Richards, M., Kuh, D., & Jonson, J. (2008). *The public value of social housing: A longitudinal analysis of the relationship of housing and life chances*. London: University of London, Centre for Research on the Wider Benefits of Learning, Institute of Education.

Fieldhouse, E., & Cutts, D. (2008). Mobilisation or marginalisation? Neighbourhood effects on Muslim electoral registration in Britain in 2001. *Political Studies, 56*(2), 333–354.

Galster, G. (2007). Should policymakers strive for neighbourhood social mix? An analysis of the Western European evidence base. *Housing Studies, 22*(4), 523–545.

Galster, G. (2012). The mechanism(s) of neighbourhood effects: Theories, evidence and policy implications. In M. van Ham, D. Manley, N. Bailey, L. Simpson, & D. Maclennan (Eds.), *Neighbourhood effects research: New perspectives* (chap. 2, pp. 23–56). Dordrecht: Springer.

Hedman, L., & van Ham, M. (2012). Understanding neighbourhood effects: Selection bias and residential mobility. In M. van Ham, D. Manley, N. Bailey, L. Simpson, & D. Maclennan (Eds.), *Neighbourhood effects research: New perspectives* (chap. 4, pp. 79–101). Dordrecht: Springer.

Treasury, H. M. (2003). *The Green Book: Appraisal and evaluation in central government. Treasury Guidance*. London: TSO.

Jones, E. (2010). Parental relationships and parenting. In K. Hansen, H. Joshi, & S. Dex (Eds.), *Children of the 21st century: The first 5 years* (pp. 53–75). Bristol: Policy Press.

Linstone, H. A., & Turoff, M. (1975). *The Delphi method*. Reading: Addison-Wesley.

Lupton, R., & Kneale, D. (2012). Theorising and measuring place in neighbourhood effects: The example of teenage parenthood in England. In M. van Ham, D. Manley, N. Bailey, L. Simpson, & D. Maclennan (Eds.), *Neighbourhood effects research: New perspectives* (chap. 6, pp. 121–146). Dordrecht: Springer.

Lupton, R., Tunstall, R., Sigle-Rushton, W., Obolenskaya, P., Sabates, R., Meschi, E., et al. (2009). *Growing up in social housing in Great Britain: The experience of four generations*. London: Tenant Services Authority/JRF/Scottish Government.

Manley, D., & van Ham, M. (2012). Occupational mobility and neighbourhood effects. In M. van Ham, D. Manley, N. Bailey, L. Simpson, & D. Maclennan (Eds.), *Neighbourhood effects research: New perspectives* (chap. 7, pp. 147–174). Dordrecht: Springer.

Mason, P., Kearns, A., & Bond, L. (2011). Neighbourhood walking and regeneration in deprived communities. *Health & Place, 17*(3), 727–737.

Meen, G., Nygard, C., & Meen, J. (2013). The causes of long-term neighbourhood change. In M. van Ham, D. Manley, N. Bailey, L. Simpson, & D. Maclennan (Eds.), *Understanding neighbourhood dynamics: New insights for neighbourhood effects research* (chap. 3, pp. 46–62). Dordrecht: Springer.

Mullins, D. (2006). Exploring change in the housing association sector using the Delphi method. *Housing Studies, 21*(2), 227–251.

Neighbourhood Effects. (2013). Neighbourhood effects seminar website. www.neighbourhoodeffects.org.

Pawson, R. (2006). *Evidence-based policy: A realist perspective*. New York: Sage.

Robinson, D. (2010). The neighbourhood effects of new immigration. *Environment and Planning A, 42*(10), 2451–2466.

Schoon, I., Cheng, H., & Jones, E. (2010). Resilience in children's development. In K. Hansen, H. Joshi, & S. Dex (Eds.), *Children of the 21st century: The first 5 years* (pp. 235–248). Bristol: Policy Press.

Small, M. L., & Feldman, J. (2012). Ethnographic evidence, heterogeneity and neighborhood effects after moving to opportunity. In M. van Ham, D. Manley, N. Bailey, L. Simpson, & D.

Maclennan (Eds.), *Neighbourhood effects research: New perspectives* (chap. 3, pp. 57–78). Dordrecht: Springer.

Social Research Association. (2003). *Ethical guidelines.* http://www.thesra.org.uk/documents/pdfs/ethics03.pdf. Accessed Mar 2011.

Tunstall, R., Lupton, R., Kneale, D., & Jenkins, A. (2011a). *Teenage housing and neighbourhoods and adult outcomes: Evidence from the 1970 cohort study* (Case Paper No. 64). London: CASE. http://sticerd.lse.ac.uk/dps/case/cr/CASEreport64.pdf.

Tunstall, R., Lupton, R., Kneale, D., & Jenkins, A. (2011b). *Growing up in social housing in the new millennium: Housing, neighbourhoods and early outcomes for children born in 2000* (Case Paper No. 143). London: CASE. http://sticerd.lse.ac.uk/dps/case/cp/CASEpaper143Tunstall.pdf.

Tunstall, R., Lupton, R., Green, A., Watmough, S., & Bates, K. (2012). *The task for disadvantaged young job seekers: A job in itself?* York: Joseph Rowntree Foundation.

van Ham, M., & Manley, D. (2010). The effect of neighbourhood housing tenure mix on labour market outcomes: A longitudinal investigation of neighbourhood effects. *Journal of Economic Geography, 10*(2), 257–282.

van Ham, M., Manley, D., Bailey, N., Simpson, L., & Maclennan, D. (2012). Introduction. In M. van Ham, D. Manley, N. Bailey, L. Simpson, & D. Maclennan (Eds.), *Neighbourhood effects research: New perspectives* (pp. 1–22). Dordrecht: Springer.

van Ham, M., Manley, D., Bailey, N., Simpson, L., & Maclennan, D. (2013). Understanding neighbourhood dynamics: New insights for neighbourhood effects research. In M. van Ham, D. Manley, N. Bailey, L. Simpson, & D. Maclennan (Eds.), *Understanding neighbourhood dynamics: New insights for neighbourhood effects research* (pp. 1–22). Dordrecht: Springer.

Volker, B., Mollenhorst, G., & Schutjens, V. (2013). Neighbourhood social capital and residential mobility. In M. van Ham, D. Manley, N. Bailey, L. Simpson, & D. Maclennan (Eds.), *Understanding neighbourhood dynamics: New insights for neighbourhood effects research* (chap. 7, pp. 139–160). Dordrecht: Springer.

www.neighbourhood.org.uk, including all presentations in this series available on the site.

# Chapter 10
# Neighbourhood Based Policies in the Netherlands: Counteracting Neighbourhood Effects?

Gideon Bolt and Ronald van Kempen

## Introduction

There is a long tradition of neighbourhood based policies in the Netherlands, but it is only since the 1990s that this policy has had a strong focus on changing the population composition of neighbourhoods. Until the 1990s area based policies were focused on the physical upgrading of neighbourhoods. However, evaluations of urban renewal policies in the 1980s revealed that the physical investments did not solve the social and economic problems in most urban renewal areas. As a consequence, both the national government and the governments of the big cities came to the conclusion that the concentration of low income groups was the root of the problem (Beaumont et al. 2003). One of their objections to the concentration of poverty was the fear that it might have a negative effect on the social mobility opportunities of residents, due to the absence of positive role models and a lack of opportunities to build social capital. However, the most dominant motivation for social mixing policies was not the fear for negative neighbourhood effects, but the idea that the liveability of the neighbourhoods (in terms of social cohesion, nuisances and safety) could be improved by breaking-up the concentration of poverty (Bolt and Van Kempen 2011).

While social mix was initially mainly defined in terms of income, the negative effects of ethnic segregation came more and more to the fore since the beginning of the twenty-first century (Van Kempen and Bolt 2009). Residential concentration of ethnic minorities is seen as major impediment to their integration by the national government, as concentration (supposedly) leads to fewer social contacts with native Dutch and less identification with the Dutch society (see, for example, Ministry of

G. Bolt (✉) • R. van Kempen
Faculty of Geosciences, Utrecht University, P.O. Box 80.115,
3508 TC Utrecht, The Netherlands
e-mail: g.s.bolt@uu.nl; r.vankempen@uu.nl

Justice 2005). Nevertheless, the measures that are taken to counter segregation are not so much focused on ethnic mix (due to anti-discrimination legislation), but on income mix. In this paper we aim to answer the following question: Are the instruments that are brought into action against concentrations of low incomes effective in reducing income and ethnic segregation?

Before we answer this question we will give a brief overview of neighbourhood based policy in the Netherlands since the 1980s. This is followed by an evaluation of the empirical underpinnings of the current mixing policies. In several policy documents, the anti-segregationist viewpoint is defended by refereeing to academic studies that give evidence of negative neighbourhood effects in the Netherlands, especially with respect to the integration of ethnic minorities. A critical analysis of these studies reveals that the evidence base is very weak.

The core of the paper is the evaluation of policies aimed at the deconcentration of low incomes (and indirectly at the deconcentration of minority ethnic groups). Like in many other European countries, there are basically two ways in the Netherland to combat income segregation (Bolt 2009). First, the spatial dispersion of low income can be addressed through housing allocation policy. In 2005 a law has been implemented which gives municipalities the opportunity to exclude people who cannot financially support themselves from the rental housing market in so-called problem areas (almost by definition low income neighbourhoods). We will show that the law has mainly a symbolic value. There is hardly an effect on income segregation, let alone ethnic segregation.

The second type of desegregation policy is focused at housing diversification in poor areas. An analysis of population developments in diversified areas reveals that there are indeed indications that both income and ethnic segregation are declining. However, we also show that the desegregationist effects of new housing in poor areas are not sufficient to compensate for the segregationist effects of new housing in other areas. Moreover, we see clear signs of reconcentration in other neighbourhoods when we take a close look at the dispersal patterns of households that are forced to move out diversified neighbourhood.

In the epilogue we conclude that the effectiveness of desegregation policy is disappointing, at least from the point of view of policymakers. Moreover, we argue that other housing policies (which are not intended to affect segregation) might actually have the (unintended) consequence that income and ethnic segregation will increase in the future.

## Neighbourhood Based Policy in the Netherlands

The Netherlands have a long tradition of policies aimed at the renewal of urban neighbourhoods. In the 1970s and 1980s the focus was mainly on improving the physical quality of the neighbourhood by means of demolition of derelict housing. Social problems and problems associated with concentration and segregation were hardly mentioned. Urban renewal was based on the principle of 'building for the neighbourhood residents'. They were given the right to be re-housed in the same area. This was made possible by building mainly in the social rented sector. Due to the extensive governments subsidies for the construction of social housing at the time, rent levels could be kept at a low level. Because the urban renewal areas already contained an over-representation of low-income households, the urban renewal policies did not result in a change in the population composition of neighbourhoods (Beaumont et al. 2003).

From the second half of the 1980s onwards, policymakers became more aware of the fact that physical renewal alone was insufficient to solve the social and economic problems in neighbourhoods and that it was equally important to address economic, social, cultural and environmental factors. In the years to follow, the Ministry of Housing, Spatial Planning and the Environment increasingly realized that market oriented policies might lead to spatial concentrations of low-income households in general, and poor ethnic minority households specifically. These ideas resulted in the policy outlined in the 1997 White Paper on urban restructuring. The objective of the White Paper was clearly stated as achieving a mixed population: "Although there are no extreme concentrations of vulnerable groups, there are certain neighbourhoods where problems prevail. There is a chance that this will lead to mutually reinforcing processes of dilapidation in parts of the cities. In several neighbourhoods liveability and safety are under pressure … In some neighbourhoods where one-sidedness can occur or already dominates, increasing the diversity of the housing stock … can facilitate physical, social and cultural improvement of living and working environments in these neighbourhoods" (Ministry of Housing, Spatial Planning and the Environment 1997, p. 70).

The policy mainly focused on the neighbourhoods built between 1945 and 1965. Concretely, the policy of urban restructuring sought to diminish socio-economic spatial segregation and concentration of low-income groups through interventions in the housing stock: the upgrading and sale of social rented dwellings, selective demolition, and the construction of more expensive dwellings (Ministry of Housing, Spatial Planning and the Environment 1997; Van Beckhoven and Van Kempen 2003). By aiming to retain middle-class households, this policy intended to promote the social and economic vitality of the city by reducing the unemployment rate, increasing liveability, public safety, and entrepreneurship in the worst neighbourhoods (Van Beckhoven and Van Kempen 2003).

Since 2001 the debate on the social mix changed in tone. Until that point the policy discourse had been focused on income mix. After 2001, the problems associated with spatial concentrations of minority ethnic groups were featured explicitly, for instance in the influential Memorandum of a commission that investigated the effectiveness of Dutch integration policy. Parallel to the development in the UK (Robinson 2005), there was a clear shift towards an assimilation discourse. The idea that minority ethnic groups should adapt to prevailing culture became dominant and segregation was seen as a barrier to this assimilation process. Through the 1990s the residential segregation of minority ethnic groups had mainly been interpreted as an indication that there is ethnic inequality in the access to housing (Bolt and Van Kempen 2002). Since 2001, however, the focus has been more and more on the supposedly negative consequences of residential segregation on the integration of minority ethnic groups. In the Yearly Memorandum on Integration Policy (Ministry of Justice 2005, p. 19, own translation) it is stated that "[…] Concentration is especially disadvantageous for integration because it results in an accumulation of social problems which may eventuate in a state of affairs that is very hard to handle […]. Concentration is also disadvantageous because it makes the ethnic dividing lines more visible in a more concentrated way. This harms the image of ethnic minorities […]. Finally, concentration is particularly disadvantageous for the possibilities for meeting and contacts between persons from different origin groups […] the diminishing contacts with native Dutch indirectly influence the social chances of ethnic minorities".

At the beginning of 2007, a new government was installed in the Netherlands. This coalition government of Christian and Social Democrats and a smaller Christian party placed area-based policy prominently on its main agenda: "'Working together, living together', is the slogan of the new Dutch government. 'Living together' takes place to a significant extent in districts. The street, the district and the neighbourhood, in addition to people's homes and workplaces, form the environment in which daily life is lived. The quality of that living environment and the manner in which we interact within it therefore largely determines the quality of a society as a whole" (Website of the Ministry of Housing, Spatial Planning and the Environment 2009).

One of the changes in the government structure, compared with earlier periods, was the appointment of a new Minister of Housing, Districts and Integration. Significant is that the field of "Integration" was now incorporated in the Ministry of Housing and Spatial Planning, implying that integration can be seen (at least partially) as a spatial problem.

Even in the spring of 2007, this Minister announced that her policy on district improvement would be aimed at only 40 'priority neighbourhoods' in Dutch cities. This policy would be aimed at transforming problematic urban districts into areas in which a diversity of people would like to live. In her first White Paper ("From Attention Neighbourhoods to Strong Neighbourhoods"; Ministry of Housing, Spatial Planning and the Environment 2007), the new Minister stated clearly that, in the 40 areas she had selected for her new policy of creating (within a period of 8–10 years) strong neighbourhoods, the problems were derived to a large extent from the population composition of these areas. At the start of the memorandum she states that the selected areas have an over-representation of deprived households. It is also stated that these areas have an over-representation of non-western minority ethnic groups. Middle-income groups increasingly leave these areas, while at the same time low-income households and those with few chances on the labour market move in. This trend has a very negative consequence for people in these areas: "Dropout from school, a deteriorated and one-sided living environment offering limited opportunities for social contacts, high level of (youth) employment, a limited integration of newcomers, emancipation, particularly of non-western women, staying behind, a lack of employment opportunities […] As a consequence, the role of the city as emancipation machine for lower income groups and ethnic minorities is put under pressure" (Ministry of Housing, Spatial Planning and the Environment 2007, p. 3).

Two years later, The Minister put significant emphasis on the role of neighbourhoods as a means for the integration of ethnic minorities in the 'Integration letter': "By doing things together in district and neighbourhood, citizens will become closer to each other, differences will be less threatening and there will even be more room for diversity. Segregation hampers that. Next to that, it stops the exchange of knowledge about Dutch society and it appears if there is no necessity anymore to command the Dutch language well [...] The Netherlands should not be a country of parallel communities, but should be a country of equal opportunities for everyone" (Ministry of Housing, Spatial Planning and the Environment 2009, p. 11/12).[1]

Since 2010, a new coalition government of Christian Democrats, Liberals with support of the very right-wing Geert Wilders' party (PVV, Partij voor de Vrijheid, which translates as Party for Freedom), has been in charge. This result is that there is much less focus on urban and neighbourhood policies. The Ministry of Housing, Spatial Planning and the Environment was completely abolished. The prominent neighbourhood policy was not headed by a specific Minister anymore and was replaced to the Ministry of Interior Affairs. It was not a coincidence that since this move almost nothing was heard any more of this policy. While under the former Cabinet, the urban policies were in the newspapers almost every week (positively and negatively), under the new Cabinet it seemed to have disappeared completely. The lack of a specific Minister undoubtedly plays a role here. The present Minister of Interior Affairs does not have housing and neighbourhoods on her media agenda. In the present housing policy, the emphasis is now on stimulating home ownership by reducing the transfer tax rate and by introducing the *Right to Buy* (Ministry of Interior Affairs 2011a). According to a new law, housing associations are required to offer 75 % of their housing stock for sale. However, it is unlikely that this law will actually be implemented. There are some legal complications as housing associations are private organisations and it is questionable whether they can be forced to sell their property.[2]

Investments in cities and neighbourhood are greatly reduced by the right-wing government. After 2014 the central government will not invest in urban renewal anymore. The central budget for the 40 attention areas is also reduced to zero. That does not mean that neighbourhood policy is abolished, but the initiative is left more to local governments and to the citizens. In the Vision on Housing (Ministry of Interior Affairs 2011a) several references are made to Big Society, a second concept that is borrowed from the British Conservatives (after the Right to Buy whereby tenants in social housing could elect to purchase their property, often with substantial discounts).

There is also a clear break with the previous government with regard to integration policies. In its vision on integration, the right wing coalition agitates against 'multiculturalism' and 'cultural relativism' and stresses tough measures, like stricter

---

[1] The fear of parallel communities is also theme that also plays a role in the discourse on integration in other European countries, like the UK (Phillips 2006) or Germany (Gruner 2010).

[2] Moreover, the break-up of the coalition (April, 2012) will probably lead to the withdrawal of the law.

immigration rules and letting immigrants pay for the integrations courses that they are forced to attend. However, the government shares with the previous government the idea that segregation is an important factor in the lack of integration of ethnic minorities: "Too many children grow up in an environment in dysfunctional families in an environment where unemployment, debts, school dropout, and criminal behaviour are the order of the day" (Ministry of Interior Affairs 2011b, p. 1).

The vision of the present and previous governments that segregation is a barrier to integration is based on a limited number of empirical studies, which are referred to in several policy documents. These studies (supposedly) show that there is a negative effect of ethnic concentration on the identification with the Netherlands (Havekes and Uunk 2008) and on social ties with native Dutch (Van der Laan Bouma-Doff 2007; Gijsberts and Dagevos 2007b). In the next section, we will discuss the strength of the empirical basis for the anti-segregation policies.

## Questioning the Empirical Basis for Desegregation Policies

### *Measuring Social Ties*

Social ties with native Dutch are measured in a very crude way. The dependent variable in the study of Van der Laan Bouma-Doff (2007) is based on the question "Do you sometimes associate with (White) Dutch people in your spare time? (Yes/No)". Gijsberts and Dagevos (2007a), using the same dataset (Social Position and Use of Provisions by Ethnic Minorities 2002), chose a slightly different dependent variable, based on the question: "Do you engage in your free time mainly with members of your own ethnic group or with members of the Dutch population?". On the basis of a multilevel regression analysis, these studies aim to test (implicitly) the social isolation thesis, put forward in Wilson's 1987 classic, *The Truly Disadvantaged*. The social isolation thesis consists of two parts: (1) the absence of resource-rich people in high-poverty neighbourhoods makes it difficult for the poor to form relationships with resource-rich people. (2) This reduces the ability of poor households to achieve social and economic mobility.

The studies mentioned above test the first part of the thesis in a disputable way, as the dependent variable does not give much insight in the quality and quantity of social ties. Next to that, it is (implicitly) assumed that native Dutch are resource-rich and members of minority ethnic group are resource-poor. Another problem is that the second part of Wilson's thesis (social isolation hampers social mobility) is not tested. Neighbourhood effects are about the role of neighbourhoods in social mobility (see for an overview Van Ham et al. 2012, 2013), and social isolation is the most often mentioned explanation for neighbourhood effects to occur. Therefore, the crucial question is whether the lack of contacts with native Dutch has a negative effect on social mobility. In another paper, Doff (2010) looks at the effects of ethnic concentration on labour market participation for four different ethnic groups in the

Netherlands. She only finds a statistical effect of concentration in the analysis on Moroccans. More important for the testing of the social isolation thesis is that she included a contact variable in her analysis (this time, she used the operationalisation of Gijsberts and Dagevos 2007a). This variable turned out to have no effect on labour market participation, which confirms our impression that this measurement does not tell anything about the quality of someone's social network.

## *Methodological Shortcomings*

There are several shortcomings in the studies policy documents refer to as underpinning of desegregation policy (Buck 2001; Galster 2005). One of these problems is the *omitted-variable problem*. In many cases, the effect of ethnic concentration on some measure of integration is estimated, without accounting for all other relevant variables (at both the individual and the neighbourhood level) that may help to explain the integration of minority ethnic groups. This problem applies also to the studies that are quoted by the Dutch government to underpin the vision that ethnic concentration has a negative effect on the integration of minorities (Gijsberts and Dagevos 2007a; Van der Laan Bouma-Doff 2007). As explained above, these studies found a negative effect of ethnic concentration on contacts with native Dutch, but they did not control for other neighbourhood characteristics. Boschman (2012) shows that the effect of neighbourhood composition on interethnic contacts disappears when other neighbourhood characteristics are taken into account. Even more problematic is that it is usually not possible to account for the *problem of selection*. People are not randomly assigned across neighbourhoods, but they select their neighbourhood. This implies that there is a bias from unobservable characteristics of individuals that influence both the level of integration and the choice of the neighbourhood. For example, an immigrant who is not interested in having any friends outside his own ethnic community is likely to live in an immigrant dense neighbourhood *and* has no members of the native majority in his circle of friends. If an analysis subsequently concludes that living in an ethnic neighbourhood has a negative effect on the likelihood of developing bridging ties, then the association may be spurious and not causal: The unobserved variable – reluctance to associate with members of the native majority – accounts both for the increased probability to live in an ethnic neighbourhood and the decreased likelihood to have a diverse social network.

## *Overemphasizing the Role of the Neighbourhood*

Like the authors mentioned above, Van Eijk (2010a) also tested the validity of Wilson's isolation thesis by examining the role of the neighbourhood in increasing the chances of contact with resource-rich people (operationalized as people with a high level of education). She collected in-depth information on the composition of social networks in three Rotterdam neighbourhoods. The neighbourhoods were selected on the basis of socioeconomic composition: one neighbourhood (Blijdorp) is rich, a second (Hillesluis) is poor and the third (Cool) is a socioeconomically mixed neighbourhood. The analysis shows that the neighbourhood does not play the expected role in the composition of networks. Residents in the poor neighbourhood have the fewest resource-rich ties and also the fewest number of ties with people outside the neighbourhood. However, this result can be explained by householders' socioeconomic position and not by the characteristics of their surrounding neighbourhood. The limited role of the neighbourhood is also illustrated by the fact that, on average, only 17 % of the network members of respondents live in the surrounding neighbourhood. Moreover, half of these ties were not formed in a neighbourhood setting. Many respondents' ties with colleagues or family members that live in the same neighbourhoods were not formed *because* connected people lived in the same neighbourhood.

On the basis of this outcome, it is questionable whether the neighbourhood can be viewed as a meaningful framework through which integration can be achieved. Nevertheless, it is a view that is still very strongly supported in the international literature on ethnic segregation, regardless of the question whether researchers are emphasizing positive or negative aspects of living in ethnic neighbourhoods (Bolt et al. 2010). The assumption that the ethnic make-up of where one lives plays a determining role in one's social interactions seems to get increasingly outdated in modern societies, which are characterised by the increasing spatial scale of social relations (Bolt et al. 1998). Zelinsky and Lee (1998) argue that ethnic groups are increasingly forming 'communities without propinquity'. They adopt the term 'heterolocalism' to denote the trend that recent populations of migrants are characterised by a dispersed pattern of residential location, while maintaining a strong internal social cohesion, by means of modern transport and communication networks. A specific form of 'heterolocalism' is transnationalism, a concept that has received wide attention across sociology, geography, and urban studies (Faist 2000). According to the definition of Portes and colleagues (1999) transnationalism refers to occupations and activities that require regular and sustained social contacts over time across national borders for their implementation. For some scholars transnational ties and local ties are mutually exclusive, which implies that transnationalism prevents immigrants from adapting to their new places of residence (Çağlar 2001). However, on the basis of a research on the identities of Turks in the German city of Duisburg, Ehrkamp (2005) concludes that transnational ties enable local attachment rather than prevent it (cf. Kivisto 2001). By transforming their neighbourhood into Turkish space, they create a new place for belonging. More fundamentally, Ehrkamp (2005) dissociates herself from the traditional literature in which place is conceptualized as containerised space. To come to a better understanding of the interrelationship between social relations and space, places should be conceptualized as nodes within transport and communication networks (Massey 1994; Castells 2000). Therefore, ethnic communities are "best conceived of as webs that extend across city-regions with nodes in areas of residential concentration, as opposed to entities that rely on residential propinquity to preserve cultural difference" (Drever 2004, p. 1437).

## Desegregation by Means of Housing Allocation

The regulation of the influx of low-income groups is the most heavily debated anti-segregation measure in the Netherlands. Following the initiative of the city of Rotterdam, a national law has been passed, which gives cities the opportunity to ask permission to the Ministry of Housing, Spatial Planning and the Environment to implement a divergent allocation policy for certain deprived parts of the cities. The new law (entitled 'Special Measures for Urban Issues', nicknamed the 'Rotterdam Act') was implemented in January 2005 and allows municipal governments to exclude people who depend on social security (apart from social security for the elderly) and cannot financially support themselves (Rotterdam, 2006), and who have not lived in the municipal region in the preceding 6 years, from the rental housing market in so-called problem areas (Van Eijk 2010b). The realization of this law can be attributed to the electoral triumph in 2001 of anti-immigrant party Leefbaar Rotterdam. While the trigger for the Rotterdam-law was the increasing ethnic concentration in Rotterdam, the law is not directly addressing the minority ethnic groups (Bolt 2004). The core problem is defined as the concentration of disadvantaged people and not the concentration of immigrants. In this way, the sting is somewhat removed from political discussion, but this redefinition of the problem is not credible. All measures to regulate the city population are substantiated with data on the spatial concentration of the 'attention groups', namely Surinamese, Antillean, Cape Verdeans, Turks, Moroccans and immigrants from other poor countries. Not once are figures on socio-economically disadvantaged households presented (Bolt 2004; Van Eijk 2010b). This contradiction is 'solved' by using the following semantic formulae: "… Ethnicity or descent is not the main issue. It is the relative wealth and socio-economic position of newcomers and the opportunities in the city for social mobility. In short, the colour is not the problem, but the problem does have a colour" (Municipality of Rotterdam 2003, p. 12).

Before the legal application of the law, Rotterdam carried out an experiment in three neighbourhoods. House seekers with an income of less than 120 % of the minimum wage were excluded from the rental market in these areas. The evaluation by the Municipality showed that the share of low-income households moving in social rented dwellings decreased from 79 % before the experiment to 37 % after the experiment. This might seem like a spectacular change, but Ouwehand and Doff (2007) stress that the decrease of 42 % corresponds with an absolute difference of 9 households(!). With the implementation of the Rotterdam law, the allocation criterion was not income anymore, but dependence on social security. Next to that, the number of designated areas was expanded in Rotterdam. Nevertheless, the actual effect of the 'Rotterdam-law' on segregation is probably very limited. Although all bigger cities (>100.000) inhabitants could apply the law, Rotterdam is still the only city that has done that. It has been estimated that the policy only affects a relative small group of households on benefits, as it is not possible to exclude house-seekers that have lived for at least six years in greater Rotterdam. Yearly about 300 house-seekers are not allowed to move into the designated areas on a total of 600,000 house-seekers. The number of affordable dwellings that is accessible for this group is reduced by about 1 % as a consequence of the restrictive policy (Gijsberts and Dagevos 2007b). As a consequence, the Rotterdam law has only a very limited influence on segregation levels.

While no other city in the Netherlands has applied the Rotterdam law, there are more examples of cities where lower income households are excluded from certain housing blocks or complexes, like Culemborg (Gijsberts and Dagevos 2007b) and The Hague (Kokx and Van Kempen 2010). The difference with Rotterdam is that the policy is focused at a very low spatial scale. The similarity with Rotterdam is that having a low income is explicitly related to problems with liveability (Kokx and Van Kempen 2010). In other words, exclusion of low incomes is used as a strategy to improve the manageability of the neighbourhood (cf. Goodchild and Cole 2001).

## Adaptation of the Housing Stock

### *Changing Population Composition in Urban Restructuring Neighbourhoods*

In their evaluation of urban restructuring in Dutch cities, Wittebrood and Permentier (2011) assessed the effect of physical interventions on the population composition of neighbourhoods. They developed a data set that contains the changes in the housing stock (2000–2006) of all neighbourhoods in the 31 biggest cities in the Netherlands. The data revealed that substantial physical change had taken place in 100 neighbourhoods (from a total of 655). In 64 neighbourhoods urban restructuring had taken place with the specific intent of changing the social composition of the population through replacing part of the social housing stock with owner-occupied dwellings. To obtain a valid estimate of the impact of the physical intervention, a quasi-experimental approach was chosen. For each of the 64 neighbourhoods ('experimental neighbourhoods') a similar one was sought where no intervention had taken place, but with otherwise similar characteristics ('control neighbourhoods'). In 18 cases, it was not possible to find a match. These unmatched neighbourhoods were classified as 'unique neighbourhoods'.

Neighbourhoods selected for urban restructuring are characterised by a strong concentration of ethnic minorities and low income households. In 2000 about 30 % of the households in the experimental neighbourhoods were in the lowest income quintile (Table 10.1). The concentration of low income households has slightly increased in the period 2000–2006. Although there was an even stronger increase in the control neighbourhoods, where the proportion of low income households rose by 2.5 %, it still has to be concluded that restructuring apparently did not help to reduce the concentration of low incomes. A similar conclusion can be drawn with respect to the ethnic concentration. The proportion of ethnic minorities is increasing in urban restructuring neighbourhoods, albeit at a smaller rate than in

**Table 10.1** Population changes in urban restructuring neighbourhoods

|  | Unique neighbourhoods (n=18) | Experimental neighbourhoods (n=46) | Control neighbourhoods (n=46) |
|---|---|---|---|
| % low income households 2000 | 35.2 | 29.8 | 30.9 |
| % low income households 2006 | 36.6 | 30.2 | 33.4 |
| Difference 2000–2006 | 1.4 | 0.4 | 2.5 |
| % ethnic minorities 2000 | 40.7 | 25.2 | 25.7 |
| % ethnic minorities 2006 | 47.0 | 30.0 | 33.3 |
| Difference 2000–2006 | 6.3 | 4.8 | 7.6 |

*Source*: Wittebrood and Permentier (2011)

similar neighbourhoods where urban restructuring did not take place. This leads to the conclusion that urban restructuring did not reduce the concentration of low incomes and ethnic minorities.

## *Displacement and Segregation*

Evaluations of urban restructuring in the Netherlands usually deal only with developments within the urban restructuring neighbourhoods themselves. However, looking only at developments within the restructuring neighbourhoods themselves does not yield a complete picture of the effects of urban restructuring. To assess the impact on segregation, it is crucial to understand the mobility behaviour of displaced households.

On the basis of a national housing survey, Bolt and colleagues (2009) made a comparison between the neighbourhood choices of displaced movers and regular movers. Compared with other movers with otherwise similar characteristics, displaced households are less likely to move to neighbourhoods with a lower share of low-income households or minority ethnic groups. The reasons vary: the choice for a new dwelling often has to be made under duress, affordable vacant dwellings cannot be found everywhere, and adequate knowledge of housing opportunities in and around the city is not available. Moreover, the priority status of displaced households often only applies when they opt for a similar dwelling as their current one.

Displaced households in the Netherlands are offered a 'certificate of urgency' which gives them priority over others seeking a dwelling. If a regular social housing applicant and a displaced household apply for the same dwelling, the dwelling is usually allocated to the latter. However, having a priority rating does not lead to unlimited opportunities: displaced households often can only apply for a dwelling with roughly the same characteristics as the previous one in terms of number of rooms or type of dwelling. Thus it is unlikely that they can obtain a more desirable dwelling in a more affluent neighbourhood, since their 'certificate of urgency' would not be valid there and they cannot compete successfully for such a dwelling on the basis of waiting time. Other home seekers can afford to wait longer for a desirable vacant dwelling. In sum, the windows of opportunity are narrow for those who are being displaced and as a result, displaced households move to neighbourhoods with a similar population composition as the areas they are leaving behind (Posthumus et al. 2013). In other words; maybe concentrations of low-income households are broken in the targeted areas, but there is a strong suggestion that new concentrations emerge elsewhere.

On the basis of a research in three Dutch cities (The Hague, Utrecht and Leiden) Bolt and Van Kempen (2010a) found that there were segregation tendencies even *within* the displaced households group. Higher income groups are less likely to move to a poorer neighbourhood than low income groups are. For households in the lowest income category (earning less than €1,000 a month), the average household income in their destination neighbourhood is €1,800. For the medium and highest income categories these figures are €2,300 and €3,900 respectively. Nevertheless, even households in the lowest income category move, on average, to substantially more affluent neighbourhoods.

Belonging to a minority ethnic group reduces one's propensity to move to a more affluent neighbourhood as well as the likelihood of moving to a neighbourhood with a lower proportion of minorities. While native Dutch households move to neighbourhoods with a substantially lower proportion of ethnic minorities (The percentage of minorities is, on average, 6 % lower in the new neighbourhood) non-Dutch households move to comparable neighbourhoods in terms of ethnic makeup (with only a 1 % difference between the neighbourhoods of origin and destination). In other words, the sorting process amongst displaced households is not different from the sorting process amongst regular movers (Bolt et al. 2008), which leads the conclusion that urban restructuring cannot be seen as an effective tool to reduce ethnic segregation.

## *Effects of Green Field Developments on Segregation*

While social mix is an important goal of urban restructuring, this theme plays a much more modest role in the development of green field locations.[3] While 60 % of the housing associations regard creating a 'balanced' population composition as an important motive for their investment in urban restructuring neighbourhoods, only one fifth indicates that social mix is a relevant issue for their activities at greenfield locations (De Kam and Needham 2003).

This is remarkable as most newly built dwellings are not built within restructuring neighbourhoods, but on large scale new housing estates, mostly at the outskirts of the city (Jókövi et al. 2006). They were built to attract the relatively wealthy: compared to the existing urban housing stock, the newly built housing stock has a relatively large share of owner-occupied, expensive and single-family dwellings

---

[3]The term greenfield location refers to undeveloped land in a city or rural area, commonly used for agriculture. The advantage of developing greenfield areas is that there is no need to remodel or demolish an existing structure. In the Netherlands, most greenfield locations that are being developed can be found at the outskirts of the city.

**Table 10.2** Segregation indices in Rotterdam, The Hague and Utrecht

|  |  | Rotterdam | The Hague | Utrecht |
|---|---|---|---|---|
| SI lowest income quintile | 1999 | 13.3 | 17.3 | 11.5 |
|  | 2005 | 14.4 | 21.3 | 13.6 |
|  | Change | 1.1 | 4.0 | 2.1 |
| SI highest income quintile | 1999 | 28.4 | 31.0 | 23.2 |
|  | 2005 | 28.4 | 36.7 | 26.5 |
|  | Change | 0.0 | 5.7 | 3.3 |
| SI Turks and Moroccans | 1999 | 48.2 | 51.6 | 40.8 |
|  | 2005 | 43.6 | 52.1 | 43.1 |
|  | Change | −4.6 | 0.5 | 2.3 |

*Source*: Bolt and Van Kempen (2010b)

**Table 10.3** Change SI 1999–2005, by neighbourhood type

|  | New housing estates | Priority neighbourhoods | Other neighbourhoods |
|---|---|---|---|
| SI lowest income quintile |  |  |  |
| Rotterdam | 0.3 | 0.5 | 0.3 |
| The Hague | 2.8 | 1.0 | 0.1 |
| Utrecht | 2.6 | −0.4 | 0.0 |
| SI highest income quintile |  |  |  |
| Rotterdam | 0.9 | 0.2 | −1.1 |
| The Hague | 5.8 | 1.5 | −1.6 |
| Utrecht | 3.8 | 2.0 | −2.4 |
| SI Turks and Moroccans |  |  |  |
| Rotterdam | 0.3 | −4.3 | −0.7 |
| The Hague | 1.7 | 0.0 | −1.2 |
| Utrecht | 1.2 | 2.4 | −1.3 |

*Source*: Bolt and Van Kempen (2010b)

(De Jong et al. 2008). On the basis of American research, it can be expected that this fosters the segregation process. New housing fuels the income sorting process by making it easier for higher status households to move out of lower status neighbourhoods (Dwyer 2007; Yang and Jargowsky 2006).

This is confirmed when we compare the segregation trends of The Hague and Utrecht (where, respectively, 9.6 and 7.4 % of residents live in new housing estates[4]) and Rotterdam (where only 0.5 % of the residents live at new housing estate). While income segregation is increasing in Utrecht and The Hague, it has remained stable in Rotterdam (Table 10.2).

Table 10.3 describes the contribution of different types of neighbourhoods to the development of the segregation index over the years 1999–2005. A neighbourhood

---

[4] A new housing estate is defined as a neighbourhood were at least 1,000 dwellings haven been built in the period 1999–2004 and where at least 80 % of the dwellings are built in that period.

contributes to the segregation index if low-income households in that neighbourhood are under- or overrepresented compared to the city level average. The larger the under- or overrepresentation and the larger the number of inhabitants in the neighbourhood, the larger the contribution of the neighbourhood to the segregation index. New housing estates had (almost) no inhabitants in 1999 and therefore did not contribute to the segregation index in that year, therefore they will always have a positive effect on the development of the segregation index. Priority neighbourhoods[5] and other neighbourhoods already contributed to the segregation index in 1999, a negative effect on the development of the segregation index indicates a decrease in the over- or underrepresentation of low-income households and/or a relative decline of the number of residents.

The sharp increase in income segregation in the city of The Hague can largely be attributed to the developments in the new housing estates. There has been a strong increase in the number of inhabitants in these neighbourhoods and low-income households are underrepresented here. Besides that, also the contribution of priority neighbourhoods to the segregation index is relatively high. This indicates that, despite the process of demolition and restructuring in these neighbourhoods, the overrepresentation of low-income households in these neighbourhoods has further increased.

In the city of Utrecht the strong increase in the number of inhabitants in new housing estates, and the underrepresentation of low-income groups in these neighbourhoods, has led to an increase in the contribution of these neighbourhoods to the segregation index. In the city of Utrecht, however, the contribution of the priority neighbourhoods to the segregation index has decreased. The share of low-income households in these neighbourhoods increased less than in The Hague, and the number of inhabitants decreased. Therefore in total the segregation index did not increase as much as in The Hague. In Rotterdam income segregation did barely increase. The extensive restructuring in the priority neighbourhoods has succeeded in keeping middle- and higher-income households within these neighbourhoods.

There are also clear differences between the cities with respect to the development of ethnic segregation. Rotterdam is the only city of the three where the segregation of Turks and Moroccans has decreased. This is mainly caused by the decreasing population size in priority neighbourhoods, which is partly due to the process of urban restructuring. The increase of ethnic segregation in Utrecht and The Hague can to a large extent be attributed to the population developments in new housing estates. Turks and Moroccans are underrepresented there, in The Hague even more than in Utrecht. The overrepresentation of Turks and Moroccans in priority neighbourhoods has increased in both The Hague and Utrecht, but in The Hague the relative number of inhabitants has decreased. Therefore in Utrecht the effect of priority neighbourhoods to the segregation index has increased while in The Hague the effect of these neighbourhoods has remained the same. The contribution of other neighbourhoods to the segregation index decreased both in Utrecht and in The Hague, due to the decrease in the (relative) number of inhabitants of these neighbourhoods.

---

[5]Priority neighbourhoods are the 40 areas that were assigned by the Minister of Housing in 2007.

The conclusion is that building dwellings on new housing estates has led to an increase in both income and ethnic segregation in Utrecht and the Hague. Policy investments in priority neighbourhoods have not been effective in counterbalancing these segregationist tendencies.

## Epilogue

Desegregation policies in the Netherlands are not very effective in reducing income or ethnic segregation. Ethnic segregation can only be addressed indirectly by trying to reduce income segregation, but there is not a very strong link between both types of segregation (Bolt et al. 2008). Income segregation is targeted by a combination of housing allocation policy and restructuring of the housing stock. The 'Rotterdam law' has been widely discussed in the Netherlands, but has not been applied outside Rotterdam yet. Even within Rotterdam, there seems to be only a limited effect on income segregation. Urban restructuring is applied in all major cities. There seems to be an effect of changing the composition of the housing stock on income and ethnic segregation, provided that renters are replaced owner occupiers. However, this effect is rather limited and is (at least partly) nullified by the tendency of displaced households to move to other poor and/or immigrant dense areas.

While area based policies have a limited downward effect on segregation patterns, sectoral housing policy may have (unintentionally) the opposite effect. We have for instance seen that building at green field sites leads to greater segregation in the city as a whole. Another example is the introduction of choice based letting (CBL) to give people more choice in where and how they live. This system is by far the most common housing allocation method in the Netherlands (Van Daalen and Van der Land 2008). Although CBL is often criticized for strengthening ethnic segregation tendencies, there is no academic research in the Netherlands that links segregation to housing allocation systems. However, Manley and Van Ham (2011) have shown for the UK that CBL does indeed increase the level of segregation, which can be attributed to the strong tendency of ethnic minorities to move into the least desirable neighbourhoods.

Recently, the present (right-wing) government has decided that the households earning more than €43,000 will have to pay extra rent in the future. The idea is that it is a misappropriation of subsidies that many middle-class households still live in socially rented dwellings. The attempts to reduce this so-called distortion, known informally as 'silting', will reinforce the trend that the social rented sector gradually becomes the exclusive domain of low-income households (Van Ham et al. 2006). The *Right to Buy* policy will probably have the same effect if the Dutch government decide to follow the same path as the UK (Jones and Murie 2006). On top of that, there will be an effect of the agreement of the previous Dutch government with the European Commission on the question how to deal with the state aid for the housing associations' sector (Priemus and Gruis 2011). In order to get a 'level playing field' of competition between social and public housing providers, it is decided that at least 90 % of socially rented dwellings should now be allocated to low income groups (earning less than €33,614). The current discussion in the Netherlands is about the implications of this policy for median income groups. In some regions of the Netherlands, there is a lack of housing opportunities for these groups, since eligibility criteria for mortgage loans have been made more stringent too (RLI 2011). The implications for residential segregation, however, do not play a part in the discussion, although the measure has already lead to lower investments of housing associations in social and tenure mix, as such restructuring activities cannot be financed with guarantees from the EC (Priemus and Gruis 2011). Moreover, the measure will further increase residualisation of the social rented sector and, consequently, income segregation. This will not necessarily translate into a higher level of ethnic segregation, as the marginalization of the social sector in terms of income does not go hand in hand with a greater orientation of minority ethnic groups on the social rented sector (Van Ham et al. 2006).

It has to be concluded that there is a puzzling paradox in the Dutch debate on anti-segregation policy. On the one hand, there is a lot of discussion about measures that should reduce segregation (restructuring policy, Rotterdam law), although their effect is limited. On the other hand there is lack of attention for the segregationist effects of other policy measures (planning of new housing estates, *Right to Buy*, limiting the access of median incomes to social housing) that were not enforced with the aim to affect segregation.

# References

Beaumont, J., Burgers, J., Dekker, K., Dukes, T., Musterd, S., Staring, R., & van Kempen, R. (2003). Urban policies in the Netherlands. In P. de Decker, J. Vranken, J. Beaumont, & I. Van Nieuwenhuyze (Eds.), *On the origins of urban development programmes in nine European countries* (pp. 119–137). Antwerpen/Apeldoorn: Garant Publishers.
Bolt, G. (2004). Over spreidingsbeleid en drijfzand [On dispersal policy and quicksand]. *Migrantenstudies, 20*(2), 60–73.
Bolt, G. (2009). Combating residential segregation of ethnic minorities in European cities. *Journal of Housing and the Built Environment, 24*(4), 397–405.
Bolt, G., & Van Kempen, R. (2002). *Wonen in multiculturele steden* [Living in multicultural cities]. The Hague: Ministry of Housing, Spatial Planning and the Environment.
Bolt, G., & van Kempen, R. (2010a). Dispersal patterns of households who are forced to move: Desegregation by demolition: A case study of Dutch cities. *Housing Studies, 25*(2), 159–180.
Bolt, G., & van Kempen, R. (2010b). Segregatie: recente ontwikkelingen in zes stadsgewesten [Segregation: recent developments in six urban regions]. In F. van Dam et al. (Eds.), *Nieuwbouw, verhuizingen en segregatie – Effecten van nieuwbouw op de bevolkingssamenstelling van stadswijken* (pp. 89–106). Den Haag/Bilthoven: Planbureau voor de Leefomgeving.
Bolt, G., & van Kempen, R. (2011). Succesful mixing? Effects of urban restructuring policies in Dutch neighbourhoods. *Tijdschrift voor Economische en Sociale Geografie, 102*(3), 361–368.
Bolt, G., Burgers, J., & van Kempen, R. (1998). On the social significance of spatial location; spatial segregation and social inclusion. *Netherlands Journal of Housing and the Built Environment, 13*(1), 83–95.
Bolt, G., Van Kempen, R., & Van Ham, M. (2008). Minority ethnic groups in the Dutch housing market: spatial segregation, relocation dynamics and housing policy. *Urban Studies, 45*(7), 1359–1384.

Bolt, G., van Kempen, R., & van Weesep, J. (2009). After urban restructuring: relocations and segregation in Dutch cities. *Tijdschrift voor Economische en Sociale Geografie, 100*(4), 502–518.

Bolt, G., Özüekren, A. S., & Phillips, D. (2010). Linking integration and residential segregation. *Journal of Ethnic and Migration Studies, 36*(2), 169–186.

Boschman, S. (2012). Residential segregation and interethnic contact in the Netherlands. *Urban Studies, 49*(2), 353–367.

Buck, N. (2001). Identifying neighbourhood effects on social exclusion. *Urban Studies, 38*(12), 2251–2275.

Çağlar, A. S. (2001). Constraining metaphors and the transnationalisation of spaces in Berlin. *Journal of Ethnic and Migration Studies, 27*(4), 601–613.

Castells, M. (2000). *The rise of the network society.* Oxford: Blackwell.

De Jong, A., Van der Broek, L., Declerck, S., Klaver, S., & Vernooij, F. (2008). *Regionale woningmarktgebieden, verschillen en overeenkomsten* [Regional housing markets, differences and similarities]. Rotterdam/Den Haag: NAi/RPB.

De Kam, G., & Needham, B. (2003). *Een hele opgave. Over sociale cohesie als motief bij stedelijke herstructurering* [A big challenge. Social cohesion as motive for urban restructuring]. Den Haag/Utrecht: DGW/NETHUR.

Doff, W. (2010). *Puzzling neighbourhood effects: Spatial selection, ethnic concentration and neighbourhood impacts.* Amsterdam: Ios Press.

Drever, A. I. (2004). Separate spaces, separate outcomes? Neighbourhood impacts on minorities in Germany. *Urban Studies, 41*(8), 1423–1439.

Dwyer, R. E. (2007). Expanding homes and Increasing Inequalities; U.S. Housing Development and the Residential Segregation of the Affluent. *Social Problems, 54*(1), 23–46.

Ehrkamp, P. (2005). Placing identities: Transnational practices and local attachments of Turkish immigrants in Germany. *Journal of Ethnic and Migration Studies, 31*(2), 345–364.

Faist, T. (2000). *The volume and dynamics of international migration and transnational social spaces.* Oxford: Oxford University Press.

Galster, G. C. (2005). *Neighbourhood mix, social opportunities, and the policy challenges of an increasingly diverse Amsterdam.* Amsterdam: Wibaut Lecture, Institute for Metropolitan and International Development.

Gijsberts, M., & Dagevos, J. (2007a). The socio-cultural integration of ethnic minorities in the Netherlands: Identifying neighbourhoods effects on multiple integration outcomes. *Housing Studies, 22*(5), 805–831.

Gijsberts, M., & Dagevos, J. (2007b). *Interventies voor integratie. Het tegengaan van etnische concentratie en het bevorderen van interetnisch contact* [Interventions for integration. Combating ethnic concentration and stimulating interethnic contact]. Den Haag: Sociaal en Cultureel Planbureau.

Goodchild, B., & Cole, I. (2001). Social balance and mixed neighbourhoods in Britain since 1979: A review of discourse and practice in social housing. *Environment and Planning D: Society and Space, 19*(1), 103–121.

Gruner, S. (2010). 'The others don't want …'. Small-scale segregation: Hegemonic public discourses and racial boundaries in German neighbourhoods. *Journal of Ethnic and Migration Studies, 36*(2), 275–292.

Havekes, E. A., & Uunk, W. (2008). Identificatie in context. Het effect van de etnische samenstelling van de buurt op de identificatie van allochtonen met Nederlanders [Identification in context. The effect of ethnic composition of the neighbourhood on the identification of ethnic minorities with native Dutch]. *Mens en Maatschappij, 83*(4), 376–393.

Jókövi, M., Boon, C., & Filius, F. (2006). *Woningproductie ten tijde van VINEX; een verkenning* [Housing production during VINEX; an exploration]. Rotterdam/Den Haag: NAi/RPB.

Jones, C., & Murie, A. (2006). *The right to buy: Analysis and evaluation of a housing policy.* Oxford: Wiley Blackwell Publishing.

Kivisto, P. (2001). Theorizing transnational immigration: A critical review of current efforts. *Ethnic and Racial Studies, 24*(4), 549–577.

Kokx, A., & van Kempen, R. (2010). A fact is a fact, but perception is reality: stakeholders' perceptions and urban policies in the process of urban restructuring. *Environment and Planning C: Government and Policy, 28*, 335–348.

Manley, D., & van Ham, M. (2011). Choice-based letting, ethnicity and segregation in England. *Urban Studies, 48*(14), 3125–3143.

Massey, D. (1994). *Space, place, and gender.* Minneapolis: University of Minnesota Press.

Ministry of Housing, Spatial Planning and the Environment. (1997). *Nota stedelijke vernieuwing* [Memorandum on Urban Renewal]. The Hague: Ministry of Housing, Spatial Planning and the Environment.

Ministry of Housing, Spatial Planning and the Environment. (2007). *Actieplan Krachtwijken: van aandachtswijk naar krachtwijk* [Action plan strong neighbourhoods: From attention neighbourhoods to strong neighbourhoods]. The Hague: Ministry of Housing, Spatial Planning and the Environment.

Ministry of Housing, Spatial Planning and the Environment. (2009). *Integratiebrief* [Integration Letter]. The Hague: Ministry of Housing, Spatial Planning and the Environment.

Ministry of Interior Affairs. (2011a). *Woonvisie* [Vision on Housing]. The Hague: Ministry of Interior Affairs.

Ministry of Interior Affairs. (2011b). *Integratie, binding en burgerschap* [Integration, binding and citizenship]. The Hague: Ministry of Interior Affairs.

Ministry of Justice. (2005). *Jaarnota integratiebeleid 2005* [Yearly memorandum on integration policy]. The Hague: Ministry of Justice.

Municipality of Rotterdam. (2003). *Rotterdam zet door: op weg naar een stad in balans* [Rotterdam presses on: on the way to a city in balance]. Rotterdam: Gemeente Rotterdam.

Ouwehand, A.L., Van der Laan Bouma-Doff, W. (2007). *Excluding disadvantaged households into Rotterdam neighbourhoods; equitable, efficient or revanchist?* Paper for the ENHR conference on Sustainable Urban Areas, Rotterdam, 25–28 June.

Phillips, D. (2006). Parallel lives? Challenging discourses of British Muslim self-segregation. *Environment and Planning D: Society and Space, 24*(1), 25–40.

Portes, A., Guarnizo, L. E., & Landolt, P. (1999). The study of transnationalism: Pitfalls and promise of an emergent research field. *Ethnic and Racial Studies, 22*(2), 217–237.

Posthumus, H., Bolt, G., & Van Kempen, R. (2013). Urban restructuring, displaced households and neighbourhood change: Results from three Dutch cities. In M. Van Ham, D. Manley, N. Bailey, L. Simpson, & D. Maclennan (Eds.), *Understanding neighbourhood dynamics. New insights for neighbourhood effects research* (pp. 87–109). Dordrecht: Springer.

Priemus, H., & Gruis, V. (2011). Social housing and illegal state aid: The agreement between European commission and Dutch government. *International Journal of Housing Policy., 11*(1), 89–104.

RLI. (2011). *Open doors, closed doors: median-income groups and the housing market.* The Hague: The Councils for the Environment and Infrastructure (RLI).

Robinson, D. (2005). The search for community cohesion: Key themes and dominant concepts of the public policy agenda. *Urban Studies, 42*(8), 1411–1427.

Van Beckhoven, E., & Van Kempen, R. (2003). Social effects of urban restructuring: a case study in Amsterdam and Utrecht, the Netherlands. *Housing Studies, 18*(6), 853–875.

Van Daalen, G., & van der Land, M. (2008). Next steps in choice-based letting in the Dutch social housing sector. *International Journal of Housing Policy, 8*(3), 317–328.

Van der Laan Bouma-Doff, W. (2007). Confined contact: Residential segregation and ethnic bridges in the Netherlands. *Urban Studies, 44*(5/6), 997–1017.

Van der Laan Bouma-Doff, W. (2008). Concentrating on participation: Ethnic Concentration and Labour Market Participation of Four Ethnic Groups. *Schmollers Jahrbuch/Journal of Applied Social Science Studies, 128*(1), 153–173.

Van Eijk, G. (2010a). *Unequal networks: Spatial segregation, relationships and inequality in the city.* Amsterdam: Ios Press.

Van Eijk, G. (2010b). Exclusionary policies are not just about the 'neoliberal city': A critique of theorie of urban revanchism and the case of Rotterdam. *International Journal if Urban and Regional Research, 34*(4), 820–834.

Van Ham, M., van Kempen, R., & van Weesep, J. (2006). The changing role of the Dutch social rented sector. *Journal of Housing and the Built Environment, 21*(3), 315–335.

Van Ham, M., Manley, D., Bailey, N., Simpson, L., & Maclennan, D. (2012). Introduction. In M. van Ham, D. Manley, N. Bailey, L. Simpson, & D. Maclennan (Eds.), *Neighbourhood effects research: New perspectives* (pp. 1–22). Dordrecht: Springer.

Van Ham, M., Manley, D., Bailey, N., Simpson, L., & Maclennan, D. (2013). Understanding neighbourhood dynamics: new insights for neighbourhood effects research. In M. van Ham, D. Manley, N. Bailey, L. Simpson, & D. Maclennan (Eds.), *Understanding neighbourhood dynamics: new insights for neighbourhood effects research* (pp. 1–22). Dordrecht: Springer.

Van Kempen, R., & Bolt, G. (2009). Social cohesion, social mix, and urban policies in the Netherlands. *Journal of Housing and the Built Environment, 24*(4), 457–475.

Wilson, W. J. (1987). *The truly disadvantaged: The inner city, the underclass and public policy*. Chicago: The University of Chicago Press.

Wittebrood, K., & Permentier, M. (2011). *Wonen, wijken and interventies. Krachwijkenbeleid in perspectief* [Housing, neighbourhood and interventions. The 'empowered neighbourhoods' policy in perspective]. Den Haag: Sociaal en Cultureel Planbureau.

Yang, R., & Jargowsky, P. A. (2006). Suburban development and economic segregation in the 1990s. *Journal of Urban Affairs, 28*(3), 253–273.

Zelinsky, W., & Lee, B. A. (1998). Heterolocalism: An alternative model of the socialspatial behaviour of immigrant ethnic communities. *International Journal of Population Geography, 4*(4), 281–298.

# Chapter 11
# U.S. Assisted Housing Programs and Poverty Deconcentration: A Critical Geographic Review

George C. Galster

## Introduction

American housing policymakers have been confronted with the geographic implications of their strategies for almost half a century.[1] Following the end of legal segregation of public housing with the Executive Order of 1964 and the publication of the Kerner Commission report in 1968 in the aftermath of the prior four summers' urban civil disruptions, the federal government began to grapple with the possibility that where they were supplying housing assistance was perhaps contributing to the poverty problem more than its solution (Goering 1986; Galster 2008). Both public housing and assistance provided to privately owned developments for low-income tenants began to come under criticism by scholars (e.g., Rainwater 1970) and federal courts (e.g., Gautreaux case; see Polikoff 2006) for their role in creating and maintaining ghettos. This geographic analytical focus gained academic if not policy salience with the publication of Wilson's The Truly Disadvantaged (1987) and my introduction of and formal conceptualization of the term "geography of opportunity" (Galster and Killen 1995). Over two decades of ever-intensifying inter-disciplinary research and policy discussion on "neighbourhood effects" followed (see van Ham et al. 2012, 2013).

Rising concerns by the courts, scholars, and activists over the personal and social costs arising from concentrating low-income (typically minority) households in urban neighbourhoods with high proportions of similarly disadvantaged households prompted several types of programmatic responses by federal government housing policymakers (Popkin et al. 2000; Goetz 2003; McClure 2006, 2008). Arguably, the

---

[1] For historical overviews and details on current federal housing policy and programs, see Galster (2008), Katz and Turner (2008), Khadduri and Wilkins (2008), Schwartz (2010), and Landis and McClure (2010).

G.C. Galster (✉)
Department of Urban Studies and Planning, Wayne State University, Detroit, MI, USA
e-mail: aa3571@wayne.edu

earliest was an attempt by the U.S. Department of Housing and Urban Development (HUD) to redirect the vestiges of new housing construction and acquisition under the public housing program toward small-scale sites outside of neighbourhoods of concentrated disadvantage beginning in the late 1960s and early 1970s (Hogan 1996). The second was HUD's increasing emphasis on attaching housing assistance to the needy tenant instead of to a dwelling unit, beginning formally with the creation of tenant-based housing assistance certificates in Section 8 of the 1974 Housing and Community Development Act. Since the inception of the "Section 8" (re-titled Housing Choice Voucher, HCV, in the 2000s) program, there have been a few changes in program administrative rules[2] and experiments with providing pre-move assistance and counseling to subsidized tenants in an effort to encourage them to use their voucher to move to lower-poverty neighbourhoods offering superior quality of life and opportunities. By 1994 the conditions of some public housing estates had grown so dire that HUD initiated a third response: the HOPE VI (Housing Opportunities for People Everywhere) program. The notion was to demolish or rehabilitate the worst public housing estates, ultimately replacing them with mixed-income (often mixed-tenure) developments. Original low-income residents would either inhabit the affordable units on the redeveloped sites or would be helped to move elsewhere with tenant-based housing assistance or conventional or scattered-site public housing.[3]

At the outset I should make it clear that the deconcentration of poverty has never been a major, consistently pursued goal of federal housing policy, nor have HUD programs or administrative rules been comprehensively and systematically oriented toward achieving this goal. Indeed, the federal effort at poverty deconcentration could be described as token, fragmented, and reluctant. Scattered-site public housing was rarely adopted by the local housing authorities that manage public housing and HCV programs, and often only under the impetus of a court order. This initiative never represented more than a tiny share of public housing units nationwide (Hogan 1996). Though there have been several small poverty deconcentration demonstration programs involving HCVs (Schwartz 2010), they have involved only a few dozen local housing authorities representing a small share of all HCVs. In addition, HCVs with stipulations for deconcentration have frequently been required as

---

[2] These new "portability" rules allowed HCV holders to use the assistance outside of the jurisdiction of the local public housing authority issuing the voucher. However, as explained below, local authorities often undermined these rules.

[3] During this period there were also several changes to existing housing program rules that encouraged deconcentration. First, the HUD rule that required local housing authorities to replace every demolished public housing unit with another one somewhere in the jurisdiction, was replaced with a rule allowing a HCV to substitute for the lost unit. Second, HUD allowed a wider range of incomes to qualify for public housing, while simultaneously placing more households with very low incomes into the HCV program instead of traditional public housing concentrations. Finally, as HUD's affordability restrictions on many under-maintained privately owned and operated rental developments originally subsidized under the Section 8 New Construction/Rehab, Section 236, or other site-based federal assistance programs expired they permitted the "vouchering out" of their low-income tenantry instead of rehabilitating the site (Varady and Walker 2000).

elements of court-ordered public housing desegregation decrees, not because of HUD initiatives (Popkin et al. 2003). Arguably, the HOPE VI program was more motivated by an urgent political need to defuse Congressional Republican efforts to abolish HUD than by an overarching commitment to deconcentrate poverty. Moreover, federal housing efforts are bureaucratically fragmented. By far, the major current program for the construction of affordable housing in the U.S. is the Low Income Housing Tax Credit (LIHTC) program, yet this is administered by the Treasury Department, not HUD. As explained below, this program's rules are schizophrenic regarding poverty deconcentration.

In this paper I do not challenge the notion that deconcentrating poverty and reducing future concentrations of poverty is a worthy goal of federal housing policy (cf. Cisneros 1996; Galster 2002; Goetz 2003; Arthurson, Chap. 12, this volume). Nor in this paper do I raise the thorny questions of whether deconcentration ultimately benefits the low-income households who may be involved (cf. Goering and Feins 2003; DeLuca and Dayton 2009; Galster 2013), the host communities that may become more diverse as a consequence of these programs (cf. Galster et al. 2003), or the communities from which the poor move (cf. Galster 2003). Rather, here I take a distinctly geographic perspective and consider the degree to which these aforementioned federal housing programs succeeded in opening up a wider variety of spatial opportunities for low-income households to live in lower-poverty, less minority concentrated neighbourhoods.[4] I then address what individual, structural, and administrative forces may have influenced the success of these programs in this geographic regard. Finally, I consider what spatial lessons the U.S. experience with deconcentration strategies offers to an international audience.

I rely upon secondary analyses of studies of the aforementioned four types of federal deconcentration programs. As such, I am constrained in my operationalization of neighbourhood indicators and bases for comparison.[5] I thus am unable (with rare exceptions) to explore the degree to which these programs have fostered deconcentration along lines of improving access of low-income households to appropriate employment, superior school districts, or other dimensions of opportunity that are often weakly measured by neighbourhood poverty rates and minority population percentages that have been traditionally employed in research. Similarly, I am often unable to compare geographic outcomes against alternative benchmarks: locations of recipients pre- vs. post-program participation, locations of recipients across programs, and locations of comparable households who are not recipients.

---

[4] This paper does not explore other, non-federal programs aimed at deconcentrating poverty that are initiated by some states, counties and cities. These include inclusionary zoning requirements for new, private housing developments and gentrification "circuit-breakers" that provide sustained housing affordability in revitalizing neighborhoods. For more on these options, see Levy et al. (2006), Pendall (2008), and Schuetz et al. (2011).

[5] In every study utilized, "neighborhood" is operationalized as a census tract: a census Bureau-defined area of about 4,000 inhabitants that is delineated to be as homogeneous as possible and bounded by clear topographical or human-made features. I therefore use census tract and neighborhood as synonyms here.

I also note as introduction that virtually all extant research is descriptive or quasi-experimental in its design; the exception is the Moving To Opportunity (MTO) demonstration, which used a random assignment design. Because there is a great deal of administratively-based and recipient-based selection into programs, claims about the independent causal impact of a program on recipient location cannot be made, with the possible exception of MTO. Moreover, because studies typically report locations only of those who succeed in participating in the given program (i.e., either passed the screening for site-based projects or successfully leased an apartment through the HCV program), the full program effects can be overstated (Clark 2005).[6] To complicate matters still further, there is a great deal of functional interdependence among the programs. HCV holders often reside in LIHTC developments. HOPE VI projects rely on HCVs to relocate most of their original tenants. Comparisons among randomly assigned groups in the Moving To Opportunity (MTO) demonstration were confounded by the fact that many control group households were later required to leave their public housing unit due to HOPE VI initiatives. Court-ordered public housing desegregation mandates (such as in Baltimore) also offered them HCVs for relocation that placed geographic limitations on their use. Thus, the independent geographic impact of a particular program on the locations of subsidy recipients may be camouflaged behind the forthcoming statistics showing their geographic patterns.

Despite these caveats, some clear and important findings can be discerned. I proceed by analyzing the geographic patterns of participants in each of the aforementioned strands of federal policy—scattered-site public housing, HCVs, LIHTC, and HOPE VI—and compare these patterns to other low-income renters not receiving subsidies and across programs to the extent feasible. I turn next to characteristics of the low-income participants, the structural aspects of local housing markets, and housing program administrative rules that may influence the geographic outcomes of the programs. Finally, I draw lessons from this analysis for policymakers in other nations who may wish to pursue their own deconcentration strategies.

## Scattered-Site Public Housing

Local Public Housing Authorities (PHAs) that develop and manage public housing in the American system were encouraged by U.S. Department of Housing and Urban Development (HUD) beginning in the 1970s to develop more housing on a "scattered-site" basis. This typically was operationalized as the construction and/or acquisition of low-density buildings with fewer than 15 units per site in locations that where not disproportionately minority-occupied (Hogan 1996). This strategy was not widely adopted across the nation, and cross-PHA variations in the density and locations of "scattered sites" were huge.

---

[6]In other words, they report only "treatment on treated" results, not "intent to treat" results.

Beginning in the early 1980s and continuing for two decades, the impetus for scattered-site public housing was primarily provided by the federal courts. In dozens of locales across the country, PHAs and HUD were sued by minority public housing tenants claiming a variety of discriminatory and segregationist actions (Julian and Daniel 1990). All these cases were resolved with court-ordered deconcentration efforts, most involving a combination of scattered-site public housing and HCVs issued to plaintiffs (Popkin et al. 2003).

With very little research attention focused on it, the most recent data on scattered-site public housing come from a 1994 survey of selected large- and medium-sized PHAs. It found that in the large PHAs such housing represented 8 % of all PHA units, and were scattered with 6.2 units per site, on average. The comparable figures for the medium-sized PHAs were: 9.5 % share with 4.7 units per site (Hogan 1996). Though the case studies revealed the popularity of scattered-site compared to conventional public housing on the part of program administrators, tenants, and the general public, a near-elimination of funding for new public housing development of any sort (especially after the advent of HOPE VI) relegated this strategy to a marginal or "boutique" status.

Nevertheless, the 1994 survey provided some suggestive evidence from nine case study sites about where scattered-site units were located. A unit-weighted average of reported data (Hogan 1996: Tables 3–6, 3–7) reveals the following percentages of scattered-site units in census tracts with 1990 higher-than-citywide: median income (28.5 %); poverty rates (58.8 %); minority occupancy rates (61.4 %); and high school graduation rates (43.5 %). These figures suggest that the sampled scattered-site public housing developments were located in better neighbourhoods than conventional public housing, but nevertheless were in neighbourhoods with higher rates of poverty, minority occupancy, and high-school leaving than the average for their cities. However, there is so much variation in neighbourhood characteristics of scatted-site developments both within and across cities, so these averages should be interpreted with considerable caution.

Hogan (1996) also investigated two special cases of Chicago (IL) and Yonkers (NY), which were required to build scattered-sites in response to public housing desegregation court decisions; see Table 11.1. In both cases, the scattered-sites were located in census tracts that had substantially higher incomes and lower poverty rates, unemployment rates, and rates of black occupancy, compared to the conventional public housing developments that previously were their only public housing options. Other generalizations are more difficult. In Yonkers the scattered sites were in places with more white and fewer Hispanic residents and more with college degrees; not so for Chicago. Moreover, it is clear that the scattered sites in Yonkers were considerably more advantaged locales in multiple dimensions than scattered sites in Chicago. Thus, while it is clear that in general scattered site public housing programs have offered superior neighbourhood environments for low-income tenants compared to conventional, large-scale, concentrated public housing developments, the gain achieved is contextualized by local market structures, including frequent neighbourhood opposition.

**Table 11.1** Comparison of neighbourhood characteristics in Chicago and Yonkers, by scattered-site and conventional public housing developments (participants are black and Hispanic)

|  | Chicago | | Yonkers | |
|---|---|---|---|---|
|  | Scattered-sites | Concentrated conventional[a] | Scattered-sites | Concentrated conventional |
| White (%) | 18 | 15 | 91 | 19 |
| Black (%) | 27 | 83 | 2 | 55 |
| Hispanic (%) | 52 | 1 | 4 | 24 |
| Poor (%) | 35 | 67 | 4 | 26 |
| Median family income ($) | 19,817 | 11,948 | 53,646 | 26,660 |
| Unemployed (%) | 17 | 34 | 4 | 12 |
| Age 25+ with college degree (%) | 10 | 9 | 26 | 14 |

*Source*: Hogan (1996: Tables 5–6, 6–1)
[a]Four largest Chicago Housing Authority public housing developments subject to lawsuit

## Tenant-Based Rental Assistance: Housing Choice Vouchers

Since its inception in 1974, tenant-based rental assistance had followed the general formula that the qualifying household must contribute a share of its income (currently 30 %) toward rent of an apartment that meets certain quality standards and whose landlord agrees to sign a minimum 1-year lease with the tenant and the PHA administering the voucher. The value of the subsidy is computed as the difference between metropolitan area's "fair market rent" (currently the 40th percentile of rents for the given apartment size the family qualified for, based on a recent survey of that entire metro area) and the tenant contribution. In its early "certificate" incarnation, the program required the tenant to find apartments at or below fair market rent. This was modified in the current "voucher" version, though the value of the subsidy was not increased if the tenant chose to occupy a more expensive apartment; very few can afford this extra expense. Once issued a voucher, the tenant has usually 3 or more months depending on the PHA to locate an apartment with a landlord willing to participate in the HCV program, complete requisite lease paperwork and have the apartment inspected.[7] As of 2009, almost a third of all households receiving federally financed housing assistance received their benefit through a HCV (Schwartz 2010: Table 1.1)

In principle, the augmented purchasing power provided by the HCV should reduce the financial constraints on low-income households' ability to occupy apartments in lower-poverty, lower minority concentration neighbourhoods. Those who hoped the HCV program would produce different spatial outcomes for low-income households were sobered, however, by the findings of the Experimental Housing

---

[7]Nationally about 30 % of all those issued HCVs cannot lease up within the required period and forfeit their vouchers (Grigsby and Bourassa 2004).

Allowance of the late 1970s (Cronin and Rasmussen 1981) and the first wave of city-specific case studies after program rollout (e.g., Hartung and Henig 1997; Newman and Schnare 1997; Turner 1998b). They showed that most HCV holders went to neighbourhoods that indeed were slightly less poor and minority-occupied, but still were relatively poor, segregated places compared to the generic neighbourhood. Moreover, many HCV users did not move at all, preferring instead to reduce their rent contributions to their current landlord.[8]

Subsequent national studies of HCV geographic outcomes provided a more nuanced portrait but did not alter its fundamental contours. Pendall (2000) compared the neighbourhood circumstances of a 1998 nationwide sample of HCV holders and low-income renters who received no assistance. He found that those with HCVs were only 75 % as likely to live in distressed neighbourhoods, on average. There were substantial variations, however, across metropolitan areas and race of HCV holder (with blacks being much more likely to use their vouchers in distressed neighbourhoods). McClure (2006) found in fiscal year 2002 that HCV holders experienced modestly lower average neighbourhood poverty rates than all very low-income renters (those earning less than 50 % of the metro area median income) in central cities (23.2 % vs. 24.4 %). Yet, the opposite relationship proved true in the suburbs (13.5 % vs. 12.1 %), producing only a small difference nationwide (18.9 % vs. 19.8 %). In a companion study (2008), McClure found that only 26 % of HCV holders resided in census tracts with less than 10 % poverty rates. This figure was one only percentage point higher than the average poverty rate in the locations of unsubsidized renters in the same income bracket as most HCV tenants (i.e., under 30 % of metro area median income). The performance of minority HCV holders was even worse in this regard: only 17 % of black and 19 % of Hispanic HCV holders resided in neighbourhoods with under 10 % poverty rates. In the only study to examine cross-housing program comparative safety characteristics of neighbourhoods, Lens and colleagues (2011) recently found that HCV holders in ten large cities, resided in neighbourhoods with lower crime rates than those in place-based assisted housing, on average.

It should be noted that comparing HCV holders to other low-income renters obscures some important unobservable differences between the groups, so the aforementioned differences (or lack thereof) might be due purely to selection bias of who takes up HCV and/or succeeds in leasing up. This possibility was tested explicitly in a random assignment experiment (Patterson et al. 2004), but this did nothing to shake the central conclusion reached above: use of a voucher resulted in only small improvements of neighbourhood conditions on many dimensions.

A different basis of comparison—longitudinal changes in households' locations before and after receipt of a HCV—paints a similar portrait.[9] Feins and Patterson (2005) conducted the most comprehensive longitudinal analysis using a national

---
[8] Finkel and Buron's (2001) study of 48 housing authorities showed that 21 % of HCV holders leased in place.

[9] Results depend, however, on which metro area is being studied and whether the HCV holders move to the suburbs from the city; see, e.g., Finkel and Buron (2001) and Varady and Walker (2003a; 2003b).

**Table 11.2** Comparison of neighbourhood characteristics of HCV participants' pre-program, initial program, and second program locations (national sample 1995–2002; all ethnic groups)

| Neighbourhood characteristic (2000) | Pre-program location | 1st program location | 2nd program location |
|---|---|---|---|
| Poor (%) | 18.4 | 20.6 | 19.5 |
| Receiving public assistance (%) | 6.4 | 7.7 | 6.7 |
| Families w/ children w/ female head (%) | 28.9 | 33.8 | 33.0 |
| High school dropouts (%) | 15.7 | 17.0 | 16.4 |
| Unemployed (%) | 8.2 | 8.9 | 8.6 |
| Males participating in labor force (%) | 67.7 | 68.5 | 68.5 |
| Females participating in labor force (%) | 55.5 | 56.2 | 56.7 |
| Families with no employees (%) | 14.6 | 14.7 | 14.1 |
| Households w/ income 2 X poverty (%) | 60.4 | 56.4 | 57.9 |
| Adults w/ some college education (%) | 20.2 | 20.3 | 21.0 |
| Adults w/ college degrees (%) | 23.4 | 21.0 | 21.8 |
| Housing units owner-occupied (%) | 59.0 | 53.0 | 55.1 |
| Population African-American (%) | 22.1 | 27.1 | 26.8 |
| Population Hispanic (%) | 13.6 | 13.8 | 14.4 |
| Population non-Hispanic White (%) | 59.3 | 54.3 | 53.5 |

*Source*: Feins and Patterson (2005: Exhibits 3, 4)

sample of those entering the HCV program during 1995–2002. They discovered that the trajectory of moves was not into significantly better neighbourhoods (measured on many characteristics) over time; see Table 11.2. Several things are significant from Table 11.2. First, by most indicators the post-HCV neighbourhoods were inferior to the pre-HCV ones. Second, the second neighbourhood occupied by HCV holders was generally inferior to the first neighbourhood they occupied with their HCV. Third, the average differences in either direction were small.[10]

Feins and Patterson's (2005) multivariate models showed an interesting geographic aspect, however. Moving greater distances with HCVs was associated with lower neighbourhood poverty rates and greater neighbourhood owner-occupancy rates. For example, those moving 1–5 miles saw an average 1 % point decrease in neighbourhood poverty rate, whereas those (few recipients) moving over 10 miles saw at least double that reduction.[11]

These results all suggest that merely increasing the effective affordability of decent-quality vacant apartments via a HCV is insufficient to get much average improvement in the geographic outcomes for program participants compared to comparable renters who are not subsidized. But how much of this is related to distance of initial move and

---

[10] Neither of these findings are surprising given the large share of recipients who did not move after receipt of a HCV.

[11] All of these studies' conclusions must be interpreted carefully because a non-trivial number of HCV holders live in units supplied under the auspices of the LIHTC program (Williamson et al. 2009). The functional overlap between this program and the HCV program and its implications will be described more fully below.

longevity since departure from original neighbourhood? What happens to outcomes if stronger or weaker constraints on geography are imposed? What happens if more mobility assistance and counseling is provided to HCV recipients? Three special programs involving HCVs provide some answers to these questions: Gautreaux phases I and II and the Moving To Opportunity (MTO) demonstration; for a good overview of these three programs and their results, see Goering and Feins (2003) and Duncan and Zuberi (2006).

## *The Gautreaux Phase I and II Programs*

In 1966 Dorothy Gautreaux, representing the class of black residents of Chicago Housing Authority (CHA) public housing projects, filed suit against CHA and HUD, alleging a variety of discriminatory practices. After extended court battles, the Supreme Court found in her favour (Rubinowitz and Rosenbaum 2000). The first court-mandated remedy provided 7,000 HCVs issued from 1976 to the late 1990s to black CHA tenants and waiting list candidates, and mandated extensive pre-move counselling and moving assistance for participants provided by a local non-profit fair housing organization. Initially the court required that all such HVCs be used in suburbs with less than 30 % black populations. Roughly four-fifths of the participants were ultimately placed in such suburbs. However, as the Chicago rental market tightened in particular years, some families were permitted to be placed in Chicago City neighbourhoods that were considerably less advantaged and had higher minority shares than 30 %, but were deemed "revitalizing" (Rubinowitz and Rosenbaum 2000).

Recent follow-up investigations of "Gautreaux I" revealed that movers to suburbs indeed succeeded in getting into and staying in much safer, whiter neighbourhoods with better schools than the neighbourhoods they left (Keels et al. 2005; DeLuca and Rosenbaum 2003).[12] Origin communities were on average 42 % poor and 83 % black, whereas most suburban relocatees at the time of survey 15–20 years after initial move lived in neighbourhoods that were 16 % poor and 48 % black, on average. However, over time they tended to move (or have their neighbourhoods change) in such a way that some of the initial drop in the percentage of black residents in the neighbourhood was erased. Even more impressive was the durability of these gains for the second generation (Keels et al. 2005). Both original heads of households (mostly mothers) who moved and their children who moved with them but were adults by the time of follow-up research showed impressive persistence of residential environments; see top panel of Table 11.3.

There was a supplementary phase of the Gautreaux litigation (commonly called Gautreaux II) that commenced in 2002, which provided another few hundred more

---

[12] A similar finding regarding the superiority of suburban compared to city destinations emerged from Goetz's (2003) evaluation of HCV users involved in the court-ordered Minneapolis public housing desegregation case. These results must be interpreted with caution, however, as both were based quasi-experimental evaluation designs and thus selection bias affects the results.

**Table 11.3** Comparison of origin and subsequent neighbourhoods for suburban participants in Gautreaux I and II programs (all participants are black)

| Program and neighbourhood characteristic (various years) | Origin | 1st placement | Mother's current[a] | Children's current[a] |
|---|---|---|---|---|
| Gautreaux I program | | | | |
| Poor in neighbourhood (%) | 42 | 17 | 16 | 18 |
| Black in neighbourhood (%) | 83 | 28 | 48 | 44 |
| Gautreaux II program | | | | |
| Poor in neighbourhood (%) | 49 | 13 | 27 | N/A |
| Black in neighbourhood (%) | 80 | 11 | 61 | N/A |

*Source*: Duncan and Zuberi (2006: Figures 2–5)
*N/A* not available
[a]Mothers are those originally placed; Children are the adult children of these mothers

HCVs (Pashup et al. 2005). Unlike the first phase, these HCVs had to be used in neighbourhoods that were less than 30 % black-occupied and less than 23.49 % (the city-wide average) poor. Compared to Gautreaux I suburban movers, the second-phase movers evinced larger initial reductions in neighbourhood percentages of poverty and black residents, but a much stronger erosion of these contextual gains over a shorter period; see the bottom panel of Table 11.3.[13]

## *The Moving To Opportunity Demonstration*

The Congressionally authorized MTO demonstration operating from 1994 to the late 1990s also employed HCVs but differed from the Gautreaux programs in many ways (Goering and Feins 2003; de Souza Briggs et al. 2010a, b). It is an experimental research effort undertaken in five metropolitan areas, not an effort to redress past discrimination in one city. It was established as a classic random assignment experiment, wherein families with children living in public housing complexes in highly disadvantaged neighbourhoods who volunteered to participate were randomly assigned to one of three groups and then tracked over a now nearly 20-year span. One was the control; one was given a HCV with neither restrictions nor mobility assistance (like the generic HCV program); the third ("experimental" group) was given a HCV that only could be used in a neighbourhood with less than a 10 % poverty rate but also provided mobility assistance by a local non-profit agency. Importantly, the experimental group (like all Section 8 tenants) was only required to remain in their initial, low-poverty neighbourhood for 1 year.

---

[13]A similar finding emerged in new analysis of black Baltimore public housing tenants who volunteered to move with HCVs to low-poverty (<10 %), low-minority (<30 %), low assisted housing (<5 %) neighbourhoods pursuant to a recent court-mandated desegregation decree (DeLuca and Rosenblatt 2011). Over a third moved within the first 3 years after the mandated 1-year tenure in such target neighbourhoods, and when they did so their destination neighbourhoods increased on average from 23 to 62 % black-occupied and from 8 to 16 % poverty rates.

**Table 11.4** Comparison of neighbourhood characteristics, by MTO control group, experimental movers, and HCV movers (5 years after random assignment; primarily black with some Hispanic participants)

| Neighbourhood characteristic (2000) | Control group | Experimental movers | HCV movers |
|---|---|---|---|
| Poor (%) | 38.6 | 27.4 | 28.3 |
| Families w/ children w/ 2 parents (%) | 38.5 | 52.7 | 46.1 |
| Employed (%) | 81.0 | 88.5 | 86.2 |
| Households w/ income 2 X poverty (%) | 37.4 | 59.2 | 47.4 |
| Adults w/ some education beyond HS (%) | 30.7 | 43.5 | 37.7 |
| Adults w/ college degrees (%) | 15.1 | 23.1 | 18.3 |
| Housing units owner-occupied (%) | 23.0 | 43.1 | 33.1 |
| Population non-Hispanic White (%) | 12.4 | 22.0 | 12.4 |

*Source*: Orr et al. (2003) Exhibits 2.8, 2.10; note: only statistically significant regression-adjusted differences between control and other group are shown

Feins (2003) analyzed the initial moves of participants. She found a 35 %-point reduction in average neighbourhood percentage poor for experimental movers and a 23 %-point reduction for HCV holders, in contrast to the comparison group. Much more modest reductions (8–10 % points) were observed for average neighbourhood percentages of black population. But these early gaps significantly narrowed by the time of the interim study 5 years after initial assignment (Orr et al. 2003; Clark 2005); see Table 11.4. This narrowing occurred because: (1) many experimental households moved to higher-poverty, higher-percentage black neighbourhoods after the first and second year of assignment to a low-poverty neighbourhood; (2) many neighbourhoods initially occupied witnessed rising trends in poverty; and (3) many control households moved out of public housing, often due to their fears of crime and gangs or the demolition of their projects due to HOPE VI or court-ordered desegregation plans (Clark 2005).

Nevertheless, as Table 11.4 shows, on every quantitative measure of neighbourhood employed, both the HCV group and the experimental group attained a superior neighbourhood environment than the control group 5 years after initial assignment. The same can be said when the battery of participant-assessed characteristics of neighbourhoods is considered; see Table 11.5. However, the gaps in either objective or subjective measures of neighbourhood were considerably narrowed when HCV and experimental groups were contrasted, though the latter resided in superior environments in every aspect except household member victimization.

Subsequent qualitative investigations of movers in MTO turned up some additional insights about altered neighbourhood conditions. de Souza Briggs and colleagues (2010a, b) concluded that, in addition to notable gains in mental and physical health, the major environmental gains experienced by the experimental group were gender-specific. Girls in experimental families gained substantially from the reduced stress associated with enhanced personal security in their new locations. In particular, they were removed from rampant predatory behaviours, including gangs who pressured them into early sex. A less salutary outcome was also revealed: 70 % of experimental household children were attending same school district as originally

**Table 11.5** Comparison of MTO participants' assessments of neighbourhood, by control group, experimental movers, and HCV movers (5 years after random assignment; primarily black with some Hispanic participants)

| Neighbourhood characteristic (various years) | Control group | Experimental movers | HCV movers |
|---|---|---|---|
| Satisfied w/ neighbourhood (%) | 47.5 | 76.8 | 65.5 |
| Feeling safe at night (%) | 54.9 | 85.2 | 70.5 |
| w/ Litter/graffiti/abandoned buildings (%) | 70.4 | 46.8 | 57.7 |
| w/ Public drinking/groups hanging out (%) | 69.5 | 33.5 | 52.9 |
| w/ Police not responding (%) | 33.7 | 7.7 | 18.0 |
| w/ Person in household victimized by crime during past 6 months (%) | 20.9 | 12.4 | 12.0 |

*Source*: Orr et al. (2003) Exhibit 3.5; note: only statistically significant regression-adjusted differences between control and other group are shown

(Orr et al. 2003). This was produced by a combination of short-distance moves, open (non-geographically based) enrolment policies of local schools, and parents who did not wish to disrupt children's social and school networks.

## *Other Efforts at Deconcentrating HCV Holders*

As suggested in the introduction, there were additional efforts initiated by the federal courts or by HUD to use HCVs to significantly alter the geographic outcomes of HCV recipients. Besides the famous Gautreaux case noted above, there were over a dozen other PHA racial desegregation case settlements begun in the 1980s and 1990s that used HCVs (Goetz 2003; Popkin et al. 2003). Two other HUD-initiated programs of the 1990s tried to change the geography of HCV use, though their efforts have never been evaluated systematically (Schwartz 2010). The Regional Opportunity Counselling Program was established in 1997 in 16 metropolitan areas. It tried to build collaborations between central city and suburban housing authorities designed to increase residential options for HCV holders by overcoming bureaucratic barriers and offering counselling assistance. The Vacancy Consolidation Program, targeted at public housing developments slated for demolition in 15 housing authorities, also provided encouragement and counselling for relocates using HCVs. Neither imposed any requirements on the types of neighbourhood that recipients must select.

## Housing Development Through the Low-Income Housing Tax Credit Program

The LIHTC program was created as part of the Tax Reform Act of 1986, which, among other things, removed substantial tax incentives for investments in rental housing development. The LIHTC is administered by the Department of Treasury (not HUD), which grants a per capita value of tax credit allocations to each state's designated housing agency. Subject to broad guidelines, each state develops criteria for awarding these tax credits and holds annual competitions among prospective developers for projects designed with a minimum share of "affordable" units.[14] Developers awarded the credits sell them (at prices that reflect a variety of market and institutional conditions) to large companies seeking tax relief via a secondary market specifically designed for this purpose, thereby providing equity to the development. Designated LIHTC units must remain affordable for 15 years (Schwartz 2010). The subsides involved are not as deep in prior site-based federal assisted housing programs and rents are not set at 30 % of income, and thus the clientele targeted by the LIHTC program (typically earning 50–60 % of metro area median incomes) is not as low-income as typically served by the public housing or HCV programs (Khadduri and Wilkins 2008). As of 2009, the LIHTC program was assisting 21.5 % of all households receiving federally financed housing aid (Schwartz 2010: Table 1.1) and represents the largest contemporary producer of affordable housing.

Several early studies of the program revealed that there was a clear tendency for most LIHTC units to be built in areas of higher-than-average poverty and minority concentrations (Newman and Schnare 1997; Roisman 1998; Freeman 2004; Rohe and Freeman 2001), especially if they were located in central cities. McClure (2006) found, however, that as the LIHTC program has evolved it has increasingly developed units in the suburbs; in the most recent year analyzed (2002) almost equal shares went to central cities and suburbs. After examining construction produced over all the first 15 years of the program, McClure (2006) showed that 29 % of all LIHTC units were built in neighbourhoods with less than 10 % poverty rates, and only 8.5 % were built in those with higher than 40 % poverty rates. Nevertheless, these figures were virtually identical to those for all renters earning less than 50 % of the metro area median income (27 and 9 %, respectively), suggesting that little deconcentration was achieved by LIHTC developments.. The same conclusion was reached by McClure (2008) in a follow-up investigation of the locations of the 80,000 LIHTC units put into service nationally during 2002. He found that only 32 % were located in neighbourhoods with less than 10 % poverty rates. Moreover, this figure was 2 % points lower than the average poverty rate in the locations of unsubsidized renters in the same income bracket as most LIHTC tenants. Interpretation of these statistics must be done with caution, however, as we know nothing about LIHTC occupants' origins. We therefore have no idea if occupants in the suburban, low-poverty area LIHTC projects were primarily relocatees from poor, central city neighbourhoods or suburbanites.

Moreover, the independent geographic impact of the LIHTC program is particularly difficult to assess due to the functional overlap between this program and the

---

[14]To be eligible to apply for the program, developments must have a minimum of 20 % of the units renting for no more than 30 % of a figure equaling 50 % of the metropolitan area's median family income or, equivalently, a minimum of 40 % of the units renting for no more than 30 of 60 % of the metropolitan area's median family income (Schwartz 2010).

HCV program. States have often encouraged LIHTC developers to lease to HCV holders because otherwise very low-income unsubsidized renters would often be unable to afford rents in these projects. Williamson and colleagues (2009) discovered that 63 % of Florida LIHTC developments housed HCV holders, an average of 9 % of the tenantry per development, constituting a whopping 16 % of all the HCV holders in the state. Thus, it is likely that the siting of LIHTC units affects opportunities for a non-trivial number of HCV holders and, conversely, the impact of the LIHTC program is strongly influenced by the presence of the HCV option. It is also significant that while only 12 % of LIHTC units were located in Florida concentrated poverty neighbourhoods, 30 % of all voucher-holders living in LIHTC units resided in such neighbourhoods. The authors conclude for Florida that LIHTC contributed to concentration of disadvantage directly via their siting and indirectly by drawing disproportionate numbers of voucher holders to these distressed neighbourhoods. We do not know the extent to which this conclusion can be generalized.

## Mixed-Income Redevelopment of Former Public Housing Estates Through the HOPE VI Program

Initiated in 1994 (in the same statute as MTO), the sixth program within the Housing Opportunities for People Everywhere (HOPE) programs was saddled with a host of oft-conflicting goals which partly frustrated the objective of poverty deconcentration (Turbov 2006). The program called for the revitalization of "severely distressed" public housing sites (characterized by physical decay, high vacancies, drugs, gangs and violence) through locally developed PHA-private developer/financier partnerships. These partnerships competed for HUD grant allocations, which helped finance the demolition and rehabilitation of public housing units, the construction of new units on site, the temporary relocation of displaced tenants, and the provision of HCVs to displaced tenants who were unwilling or unable to return to the redeveloped sites. Though there are vast differences in the features of the redeveloped sites (Popkin et al. 2004), a universal feature is a mixture of public housing tenants with those of higher incomes and, often, some owner-occupants. In all cases there has been a net reduction in the number of units for public housing tenants on site. HOPE VI was discontinued in 2006, though new variants are now being proposed under HUD's "Neighbourhoods of Choice" rubric. All totaled, HOPE VI demolished about 150,000 dilapidated public housing units in 224 different projects nationwide (Landis and McClure 2010).

The net geographic impact of the HOPE VI program is a complex amalgam of both who ends up residing on the redeveloped sites and what happens to those who were displaced. Among the last group, some moved to other public housing projects, some were able to use HCVs (and go to other communities comparable to those of other HCV holders), and some were not and had to fend in the private rental market. The national HOPE VI tracking study found that after the first eight years of the program only 19 % of original residents were living on the redeveloped sites,

**Table 11.6** Comparison of HOPE VI participants' assessments of neighbourhood, by type of assistance and location (primarily black with some Hispanic participants) All figures as percentages

| Neighbourhood characteristic (2000) | Baseline HOPE VI | Non-movers[a] | Other PH develop. | HCV holder | Unassisted rental |
|---|---|---|---|---|---|
| Drug selling | 77 | 72 | 45 | 23 | 17 |
| Shooting/violence | 66 | 48 | 32 | 11 | 21 |
| Feel safe outside home at night | 55 | 57 | 68 | 83 | 74 |

*Source*: Buron (2004) national HOPE VI tracking study of eight sites
[a]Non-movers are those temporarily relocated on-site until redevelopment was completed

29 % were in other public housing, 33 % were using HCVs, and 18 % had left housing assistance (Popkin et al. 2004), proportions that roughly matched those obtained in a slightly earlier study of a different sample of 73 projects (Kingsley et al. 2003).

The most detailed information about the geographic outcomes associated with residents of PHA sites that were redeveloped under HOPE VI has been provided in Buron's (2004) study of eight longitudinal case study sites. He found that comparing initial conditions on-site to those at first HCV relocation residence, the average neighbourhood poverty rate dropped from 40 to 28 %, but the average share of minority residents only dropped from 92 to 87 %. The main gain was in residents' perceptions of safety: reports of "big problems with…:" "shootings and violence" fell from 67 to 20 %, "people selling drugs" fell from 77 to 30 %, "gangs" fell from 49 to 17 %, and "people being attacked or robbed" fell from 25 to 9 %. However, these effects were distinctive according to whether the relocatees moved to another public housing unit on the site being redeveloped, or moved off-site without assistance, with a HCV, or into another public housing development elsewhere; see Table 11.6. Nevertheless, relocatees on average experienced neighbourhoods that they perceived as much safer than the original HOPE VI sites before redevelopment (cf. Popkin and Cove 2007).

Kingsley and colleagues (2003) studied geographic outcomes for movers from all 73 HOPE VI sites as of 2000. They found that 31 % used HCV, 49 % went to other public housing and 20 % moved elsewhere without assistance. On average, relocatees moved 3.9 miles from their original HOPE VI site, saw a decline in their neighbourhood poverty rates from 61 to 27 % and a decline in their neighbourhood percent minority from 88 to 68 %. Relocatees using HCVs followed roughly comparable patterns of clustering as those in the generic HCV program, though relocates were slightly more likely (13 % vs. 10 %) to cluster in tracts that already had 10 % HCV households or more.

Buron and colleagues (2007) also found that most HOPE VI relocatees using HCVs saw a large improvement in their neighbourhood quality in terms of poverty rates and safety, compared to their former public housing estates. However, these relocatees not only faced the normal challenges as general HCV holders but also the extra adjustments associated with moving out of public housing (e.g., being responsible for timely utilities and rent payments). The fact that there were no additional program funds allocated within the HOPE VI program to counsel and assist such HCV-using relocates move to substantially lower poverty, lower minority neighbourhoods implicitly suggests their lack of serious commitment to the deconcentration goal. Thus, there should be some concern over the sustainability of these initial gains by HOPE VI relocatees using HCVs.

Over the past decade, the Chicago Housing Authority (CHA) has been the national leader in transforming most of its public housing estates in HOPE VI-like manner, though using additional sources of funds beside this federal program. As such it has been the object of intense study that, among other things,[15] has reinforced some of the worrisome prospects concerning geographic sustainability noted above. Popkin and colleagues (2000) found that former CHA public housing residents had different and more severe needs that inhibited their successful leasing up of apartments with HCVs than generic HCV holders. For example, because of rampant gang activity and drug markets, it was common for CHA relocatees to have at least one member of their family with a criminal history, rendering them both disqualified to return to CHA units on the redeveloped site and easy to deny by private landlords in the rental market. Many CHA relocatees also had severe personal challenges (e.g., mental and physical disabilities; responsibilities for many children) that made it difficult to find appropriate private rental dwellings and successfully lease up. Most had no experience in searching for housing, interfacing effectively with landlords or, once housed, behaving appropriately as a private tenant (such as paying utility bills or allowing informal occupancy arrangements in violation of leases). But the problems were not only confined to relocatees. Popkin and colleagues (2000) found in that CHA residents who lived on the original sites but now occupied the redeveloped sites were having trouble complying with the new, tough lease requirements because they were not getting the supportive services they needed. This evidence speaks to the minimal successes that HOPE VI has had in substantially increasing housing opportunities for former public housing residents in non-poor environments.

## Comparing Geographic Outcomes Across U.S. Assisted Housing Programs

The prior analysis has relied upon studies that have essentially made within-program comparisons of geographic outcomes for participants. Here I turn to the handful of studies that used common bases to compare outcomes across programs. An introductory note of caution in interpreting the following results is in order because of the functional overlaps between many programs. HCV holders may have been moved under the auspices of the generic program (while many stayed and leased up in place), the HOPE VI program, or the "vouchering out" of tenants in privately

---

[15] For example, Jacob (2004) found that children of CHA relocatees using HCVs did not get substantially improved experiences of school quality.

**Table 11.7** Comparison of participants' neighbourhood characteristics, by Federal Housing Assistance Program

| Federal program | % of units[a] | Neighbourhood condition | | In 1990 |
|---|---|---|---|---|
| | | Poor (%) | Minority (%) | Renter-occupied (%) |
| Total | 100 | 26 | 45 | 66 |
| Section 8 Voucher/Certificate | 30 | 20 | 41 | 60 |
| Public housing | 25 | 36 | 59 | 74 |
| Section 8 Moderate Rehabilitation | 2 | 29 | 53 | 70 |
| Section 8 New/Substantial Rehab. | 19 | 21 | 34 | 64 |
| Section 236 | 9 | 21 | 40 | 67 |
| Other site-based assistance | 8 | 28 | 55 | 68 |
| Low-Income Housing Tax Credit | 7 | 21 | 37 | 60 |

*Source*: Pendall (2000) based on national HUD databases and U.S. Census data
[a]Receiving federal subsidy and occupied as of 1998

owned subsidized apartments developed under the Section 8 or Section 236 programs noted below. Moreover, some number of HCV holders reside in LIHTC units.

In the first cross-program comparative study of geographic outcomes of an older set of programs, Newman and Schnare (1997) found during the early 1990s that only 15 % of HCV holders resided in neighbourhoods with 30 % or higher poverty rates, which compared favourably to 54 % of residents in public housing and 23 % of residents in privately owned, HUD-subsidized projects.[16] Only 5 % of HCV holders resided in neighbourhoods of over 40 % poverty, compared to 36 % of residents in public housing and 13 % of residents in privately owned, HUD-subsidized projects. These data form a pattern that has often been observed subsequently: among those receiving U.S. federal housing subsidies, public housing residents generally live in the most-disadvantaged neighbourhoods, followed by residents in other-site-based assistance programs, followed by HCV holders residing in the private rental market.

Pendall (2000) provided a more comprehensive, cross-program comparative study of geographic outcomes; see Table 11.7. His figures showed that, on average, residents of public housing (25 % of all assisted tenants as of 1998) experienced the highest rates of neighbourhood poverty (36 %), minority occupancy (59 %), and renter occupancy (74 %). Mean neighbourhood features of units produced by the Section 8 Moderate Rehabilitation and miscellaneous site-based programs (10 % of all assisted tenants) ranked closely behind. Next in rank, with virtually identical conditions for residents, came locations associated with the LIHTC program (7 % of the total), Section 236 new construction subsidy program (9 % of the total), Section 8 New or Substantial Rehabilitation program (19 % of total), and HCV program (30 % of total). This cluster of both site-based and tenant-based subsidy programs had their average tenants occupying neighbourhoods that were: 20–21 % poor, 34–41 % minority, and 60–67 % renter-occupied. Compared to public

---

[16]Newman and Schnare (1997) did not consider the LIHTC program that had begun just before the study.

**Table 11.8** Comparison of participants' neighbourhood poverty rates, by federal housing assistance program and area where assistance used

| Federal program | Neighbourhood Poverty rate | | In 2005 |
|---|---|---|---|
| | Suburbs (%) | Central City (%) | Total[c] (%) |
| Section 8 Voucher/Certificate[a] | 14 | 23 | 19 |
| Low-Income Housing Tax Credit[b] | 13 | 26 | 19 |
| Renter households in poverty | 15 | 27 | 22 |
| Renter households < 50 % AMI | 12 | 24 | 20 |

*Source*: McClure (2006: Table 1) based on national HUD databases and U.S. Census data
*AMI* metropolitan area median income
[a]Used during fiscal year 2002
[b]Placed in service 1987–2002
[c]Includes non-metropolitan areas

housing, these differentials were greatest in the aspect of neighbourhood poverty rates and least in the aspect of neighbourhood renter-occupancy rates.

Important geographic nuance to this portrait has been provided by McClure (2006), DeFillippis and Wyly (2008), and Talen and Koschinsky (2011). In his nation-wide study, McClure (2006) discovered that neighbourhood poverty rate differentials between the HCV and LIHTC programs and compared to unsubsidized, lower-income renters depended upon whether a central city or suburban location was considered; see Table 11.8. In all cases the average neighbourhood poverty rates experienced by program participants and generic low-income renters were roughly twice as high in the central cities. However, whereas in the central cities the HCV holders' mean neighbourhood poverty rate was 3 % points less than residents in LIHTC developments, the reverse was true (by 1 % point) in the suburbs. Nevertheless, in both geographic contexts the HCV holders only slightly "out-performed" what McClure used as their unsubsidized comparison group (renter households in poverty) and LIHTC residents slightly "under-performed" what McClure used as their unsubsidized comparison group (renter households earning less than half the metro area median income). From a different perspective, the performance of the LIHTC program in expanding options in the suburbs appears more favourable. McClure (2006) showed that if only suburban destinations are considered, a substantially higher share of all units provided by the LIHTC program go to low-poverty (0–10 %) neighbourhoods compared to shares of HCV holders (50 % vs. 43 %). Nevertheless, Table 11.8 makes it clear that, on average, neither the LIHTC nor the HCV program operating in either central cities or suburbs produces a substantially different distribution of low-income households by neighbourhood poverty rates than what is produced by comparable unsubsidized renters in the market place.

DeFillippis and Wyly's (2008) study of New York City revealed that HCV holders were not more likely than residents in subsidized housing (supported by either the federal government and/or the city itself) to live in lower-poverty or less minority-concentrated neighbourhoods. They concluded that, especially in tight housing markets (partially made so by rent control in the case of New York) and markets undergoing much gentrification (such as New York), preserving the site-based subsidized housing stock was more important for poverty deconcentration efforts than more vouchers. But this is debatable as the sole goal.

Talen and Koschinsky (2011) found in Chicago that although on average HCV holders experienced less poverty and minority concentrations in their neighbourhoods than residents of site-based assisted units, there was an important distinction related to degree of concentration of assistance in the neighbourhood. In areas with high concentrations of either HCV or assisted sites, the above relationship was reversed, and the households in site-based assisted developments lived in more advantaged neighbourhoods. This suggests that when forces lead to high concentrations of HCV holders in particular neighbourhoods it often erases their potential locational advantages. I explore this potential endogeneity further below.

## Explanations for Findings: Individual, Structural, and Administrative Rules

Taken at face value, the aforementioned studies lead to the following conclusions:

1. Residents of U.S. public housing on average reside in significantly more disadvantaged neighbourhoods compared to participants in any other assisted housing program and most other low income renters.
2. Residents of other types of site-based assisted housing programs (particularly LIHTC) do not, however, reside in significantly different residential environments than tenant-based HCV holders
3. HCV holders fare somewhat better in neighbourhood poverty rates than equivalent households who do not receive subsidies, but the comparative differences are even smaller when the LIHTC program is compared to its equivalent private renter standard.
4. HCV holders in general do not substantially improve their neighbourhood circumstances with subsequent moves; indeed if their initial move was (perhaps with counselling assistance) to a low-income, predominantly white neighbourhood, their subsequent moves were to higher-poverty, higher-minority share neighbourhoods.

Why do these patterns emerge? There is little debate regarding conclusion 1. Due to their construction at large scales and high densities, their explicit history of racial segregation, their historical evolution to house only the neediest, and the concomitant negative spill over effects on their environs, traditional American public housing has almost tautologically resulted in concentrations of disadvantage (Hirsch 1983; Goering 1986; Julian and Daniel 1990; Massey and Kanaiaupuni 1993; Schill and Wachter 1995; Coulibaly et al. 1998).

There is more contention over the sources of conclusions 2., 3., and 4. There are three not mutually exclusive but distinct sets of arguments here: the "individualist," "structuralist," and "program rules" explanations. The individualist view focuses on characteristics of the program participants that influence how they use their HCV geographically, such as preferences, housing search patterns, social networks, personal psychological and intellectual resources, family responsibilities, criminal histories, and the like (e.g., Clark 2005, 2008). The structuralist view emphasizes geographic constraints imposed on program participants by the operation of metropolitan housing markets (such as low vacancy rates, racial discrimination, and selective participation of landlords in the HCV program) and public transportation systems (e.g., de Souza Briggs et al. 2010b; DeLuca and Rosenblatt 2011). The program rules view emphasizes the constraining impacts of the regulations associated with the major housing programs in question and their administration in local housing markets.

Unfortunately, some research findings do not help us sort out the individualist and structuralist explanations for the geographic performance of HCV holders because they are consistent with both. For example, a common consensual finding emerges from a variety of investigations of the geographic impact of the HCV program: ethnic-racial differences. Inferior outcomes were uniformly associated with minority ethnic status, even after controlling for other characteristics. Being black was especially associated with reaping small geographic gains from use of HCV (Hartung and Henig 1997; Newman and Schnare 1997; Turner 1998a, b; Pendall 2000; Basolo and Nguyen 2005; McClure 2008). This, of course, could be the result of all the (unmeasured) individualist factors above that are correlated with race, and/or racial discrimination in rental markets. As another example, Turner (1998b) found that in most metropolitan areas she studied there were greater shares of below-FMR units in low-poverty areas than shares of HCV holders residing there, suggesting something either about the search patterns and/or preferences of the HCV holders and/or the willingness of landlords in such areas to participate in the voucher program.[17] However, other research clearly offers support to elements of all three positions; I discuss these next.

## *Evidence Supporting the Individualist Position*

There is convincing evidence that low-income households in general and HCV holders in particular are deeply embedded in highly localized social networks. This "bonding social capital" can provide invaluable sources of support (money, child-care, in-kind assistance) and information, though this also sometimes comes with a burdensome set of responsibilities (de Souza Briggs et al. 2010a, b). These networks can provide a geographic centre of gravity for residents who are granted

---

[17]Similarly intriguing but ambiguous evidence has been gleaned from other programs as well. Buron (2004) notes that many HOPE VI relocatees moved to public housing that was nearly as distressed as the ones from which they left. He could not attribute the reasons but speculated on a combination of preferences, inability to qualify for private housing, lack of time to find alternatives, housing market constraints, or lack of knowledgeable and conscientious relocation assistance.

HCVs in two ways: spatially biased information and a need for proximity for assistance family can offer. HCV holders without counselling support typically activate local networks to help them locate a place to use their voucher (Deluca et al. 2011). Unfortunately, given the constrained geographic scope of these networks and other search strategies (such as personally looking for "For Rent" signs), few vacant apartments in advantaged neighbourhoods are uncovered. Even if they have information about rental opportunities in advantaged neighbourhoods, HCV holders may not wish to move there if the distance to their kin and friendship networks and institutional ties seems prohibitive. A particularly poignant if unique example was provided by Goetz (2003), who observed Asian immigrant residents of a Minneapolis public housing project vigorously opposed moving in compliance with a court-ordered desegregation decree and, when issued a HCV and forced to leave, stayed very close. Moreover, with little first-hand or second-hand information about alternative neighbourhoods, new HCV holders often have limited bases upon which to compare them and make more globally informed choices. de Souza Briggs and colleagues (2010a, b) determined that many MTO movers focused mainly on avoiding danger, not moving to "places of opportunity." But some also moved to escape predatory relatives and neighbours. They did not choose what might be perceived as much better options because they had never experienced them and thus did not know what they were missing; "information poverty" they called it. In their extensive, open-ended interviews with low-income black households in Baltimore and Mobile AL, Deluca and colleagues (2011) found little salience of "neighbourhood" in the residential choices of their interviewees, other than a desire for relative safety at the small geographic scale around the dwelling; dwelling characteristics dominated the selection process.

Of course, lack of information with geographic breadth is not a pure individualist trait but likely is reflective of housing market structure, as amplified below. Furthermore, evidence on moving destinations of HCV holders should not be taken as proof of "revealed preference." Mobility reflects a variety of structural constraints, including limited information and limited housing options, not just preferences. This point gains powerful nuance with recent discoveries by Deluca and colleagues (2011). Remarkably, they found that 70 % cited reasons for last move that were beyond their control, what the authors called "reactive moves."[18] Dwelling unit failure was the most frequently cited cause of mobility (25 %). Reactive moves must occur in a matter of weeks to avoid homelessness, so expediency is salient. Not surprisingly under these circumstances, search processes rely upon "leads" from family and friend networks and seeking nearby "for rent" signs being paramount, with highly localized moves aimed at securing decent dwellings (not necessarily decent neighbourhoods) being the result. Though family ties were activated by necessity during these reactive housing searches, many did not express a desire to do so or to retain close contacts. Similarly,

---

[18]This is consistent with Fairchild and Tucker (1982), who found that blacks were much more likely than whites to experience events that would trigger involuntary moves, such as evictions, intolerable housing quality breakdowns, and domestic violence.

though low-income black households typically moved to places with high proportions of black neighbours, it was not because they in any way "preferred" that racial composition, contrary to the conclusions of Clark (2008).

## Evidence Supporting the Structuralist Position

Many studies clearly indicate various types of housing market structural barriers that influence where HVC holders reside. One set relates to the availability of vacant, appropriately priced rental units located in low-poverty, low minority concentration neighbourhoods whose landlords are willing to participate in the HCV program. Pendall's (2000) regressions based on a nationwide sample of HCV holders showed that structure of local housing market, specifically the supply of rental housing in non-distressed and distressed tracts, was a key determinant of the share of HCV holders living in distressed tracts, controlling for metro area poverty and racial composition. Turner's (1998b) study of HCV holders in six metro areas found substantial differences in outcomes depending on local housing market conditions. In a few metropolitan areas the combination of good locations for public and other site-based assisted housing and tight private rental markets led HCV holders to underperform their site-based counterparts in terms of poverty and minority neighbourhood indicators. These conclusions were echoed in the subsequent study by Finkel and Buron (2001) involving more metropolitan areas. The MTO research documented substantial differences in the ability of experimental and generic HCV groups to lease an apartment due to the relative paucity of units available in low-poverty neighbourhoods (Orr et al. 2003). Even with their counselling, the experimental group's lease up rate was 14 % points lower than the HCV participants,' though much higher than the success rate in Gautreaux (Shroder 2003). Subsequent qualitative research has further emphasized how difficult it was for MTO experimental households to find housing in low-poverty neighbourhoods, even with the assistance of counselors (de Souza Briggs et al. 2010a, b). Finally, tight rental markets in the Chicago suburbs forced the Gautreaux I program to modify its desegregation criteria (Rubinowitz and Rosenbaum 2000), as noted above.

A closely related structural bias induces HCV use in disadvantaged neighbourhoods because landlords often eagerly recruit HCV holders there. In these neighbourhoods private landlords are more likely faced with high vacancies and respond by aggressively marketing their units to voucher holders (Galster et al. 1999), especially at local housing authorities (Deluca et al. 2011).

Taken holistically, the studies discussed in the preceding two paragraphs clearly indicate synergistic biases in the way housing markets operate to limit geographic opportunities for lower-income households, whether they have HCVs or not. Areas of opportunity are often inaccessible to HCV holders because they are too expensive, have few vacant rental units, and/or have few landlords willing to participate in the program, all precisely because they are areas in high demand by more affluent segments of the housing market. Simultaneously, areas of disadvantage have all the opposite factors at work: lower rents, many vacant rental units, and landlords so desperate for tenants that they actively recruit HCV holders. Thus, the same market forces that produce income group segregation remain powerful determinants of the geography of HCV holder residence.

What's more, racial differences in structural barriers are also apparent. Basolo and Nguyen's (2005) study of HCV users in a large Southern California PHA found that HCV users' perceptions of barriers to mobility were primarily "too few homes to rent" and "landlord will not rent to Section 8." Huge racial differences in responses were revealed, as blacks and to a lesser degree Hispanics were more likely to cite these obstacles than whites, with the incidence of black responses at least 22 % points higher than whites'. Finally, the most recent national investigation of racial discrimination that employed carefully matched "testers" revealed substantial rates of differential access afforded black and Hispanic apartment seekers (Turner et al. 2002). Although the study did not have the testers use HCVs, we can presume that such discriminatory barriers based on race would be of relevance to the outcomes of black and Hispanic HCV users. If a landlord wished to discriminate illegally against a minority HCV holder, it is an easy and virtually undetectable subterfuge to merely decline on the legal basis of unwillingness to participate in the HCV program.

A final structural barrier is that many holders of a new HCV cannot lease up because they lack the requisite savings to cover the costs of moving, security deposits, and other fees associated with acquiring a new apartment (Popkin and Cunningham 1999). Though there may be means of covering these costs through special emergency grant funds or charitable contributions, such often do not arrive in time before the voucher lease up period expires (Marr 2005).

Though all these structuralist scenarios provide plausible explanations for the modest geographic performance of HCV holders in any given cross-section, they are less persuasive explanations for the erosion of geographic advantages over subsequent HCV-supported moves, as observed by Feins and Patterson (2005) and especially in the MTO demonstration (Orr et al. 2003; Clark 2005; Turney et al. 2006). Here, different sorts of structural barriers must be brought to bear. First, once in an advantaged neighbourhood (perhaps through the assistance of mobility counsellors), HCV holders are often forced to move because the landlord is unwilling to continue participating in the program. After examining the geographic patterns of HCV holders who received their vouchers as part of a public housing desegregation suit in Baltimore, Deluca and Rosenblatt (2011) found that nearly half of those who had moved after 4 years from their original site (in a low-poverty, low-minority, low-assisted household neighbourhood) were forced to do so because landlords refused to continue participation. This echoed results from MTO (Orr et al. 2003), where the two most frequently cited reasons by experimental group households for leaving their first, low-poverty neighbourhood were "leasing problems" (22 %) and "problems with landlords" (20 %). Second, HCV holders in advantaged neighbourhoods may face harassment or more subtle forms of discrimination and ostracism (based on their class and/or racial status) that makes them uncomfortable and desirous of more diverse environs. Even more neutrally, they are unlikely to form close social bonds with their new neighbours or get deeply involved with new institutions, thus a subtle sense of alienation may remain (de Souza Briggs et al. 2010b). Third, these locations may raise insurmountable challenges to negotiate the spatial mismatch of home, work, socialization, and childcare, especially if the HCV holder lacks an automobile (de Souza Briggs et al. 2010a, b). MTO families that initially moved to low-poverty neighbourhoods often excessively distanced themselves from pre-existing job and social networks and eventually felt compelled to move closer to the urban core (Turney et al. 2006; de Souza Briggs et al. 2010a, b). Once any or all of these reasons trigger a move from the initial location, all the individualist and structuralist factors noted above return to play and produce the observed pattern of mobility into somewhat less-advantaged places, closer to the original, disadvantaged neighbourhood.

## *Evidence on Assisted Housing Program Rules and Administration*

There are several fundamental elements in the design and administration of HCVs that limit their efficacy as a vehicle for deconcentrating poverty among recipients. First, the aforementioned asymmetry in landlords' willingness to participate in the program (less in more desirable neighbourhoods and vice versa) would be rendered moot if all landlords were required to participate in the program, but this compulsion is contrary to current rules. Second, HCV Fair Market Rents (FMRs) have been consistently lowered since the inception of the program, from 50th to 45th to the current 40th percentiles of the metropolitan-wide distributions of rental units for the particular category of bedroom in question. Thus, the purchasing power of the HCV has been eroded and thus the regions over which recipients can afford to use it have shrunk. Third, the fundamental nature of the FMR creates an economic incentive for recipients toward HCV use in more disadvantaged neighbourhoods. Since the value of the HCV is based on metropolitan-wide rent distributions, a HCV holder can find cheaper accommodations in more disadvantaged submarkets within the region. Some PHAs permit reductions in tenant out-of-pocket contributions to rent if they can lease such below-FMR apartments, thus unwittingly providing an incentive for choosing disadvantaged neighbourhoods. Fourth, the time limitation of how quickly recipients must lease-up encourages them to settle on neighbourhoods with which they are already most familiar. Fifth, qualitative evidence suggests that some PHAs discourage those to whom they issue vouchers from using them outside of their jurisdiction (Marr 2005). Finally, the operation of the LIHTC program that works to recruit HCV holders to developments in more distressed neighbourhoods has been alluded to earlier (Williamson et al. 2009).

Why the LIHTC program does not generally out-perform the unsubsidized rental market in providing neighbourhood options for low-income households can be explained by program rules that encourage development of exclusively low-income projects in disadvantaged neighbourhoods. From the outset, the LIHTC program structure has not favoured poverty deconcentration, though some had hoped that this program could more easily overcome political opposition from suburbs than other subsidized housing vehicles. First, Treasury Department rules require that states favour applications for developments in "qualified census tracts" (QCTs) that are part of "comprehensive redevelopment initiatives." QCTs are neighbourhoods wherein half or more of the residents have incomes below 60 % of the metro area median family income and the poverty rate is 25 % or more. Once granted to QCTs, the associated developers get 30 % more tax credits than would otherwise have been allocated. The QCTs ideally are areas that will be redeveloped for middle- or upper-income private housing, whereupon the location of a LIHTC development might create a more diverse community in the long run than otherwise would have been the case. Unfortunately, this "creation of an island of affordability in a sea of gentrification" has happened rarely; the gentrification has typically not materialized. Second, Treasury Department rules stipulate that at least one-tenth of all credits be allocated to non-profit developers; currently over a fifth are so allocated nationwide, on average. Most non-profit housing developers in the U.S. are community development corporations that are based in disadvantaged neighbourhoods, have substantial governance representation from these neighbourhoods, and focus on providing affordable housing to these places. Not surprisingly, these first two rules have resulted in a disproportionate number of LIHTC projects being built in areas that have remained primarily low-income and minority-occupied. Third, because program rules typically favour developments that provide larger proportions of affordable units, there is little incentive for developers to design mixed-income projects that would enhance economic diversity at the small geographic scale. Finally, within the federal guidelines there is latitude for states to value various aspects of applications for tax credits, including geographic criteria, and there have been few state schemes that that heavily favoured LIHTC developments that provide opportunities for deconcentration.

## Policy Implications for U.S. and European Contexts

### *Proposals to Better Deconcentrate Poverty in the U.S.*

For almost a quarter century there have been discussions of wide-ranging policy reform proposals aimed at (among other things) improving the geographic distribution of U.S. federal housing assistance, (see: Goering 1986; Goering et al. 1995; Turner 1998a; Turner and Williams 1998; Popkin and Cunningham 1999; Katz and Turner 2001, 2008; Pendall 2000; Achtenberg 2002; Galster et al. 2003; Grigsby and Bourassa 2004; Khadduri 2005; Popkin et al. 2004, 2005; McClure 2008; Khadduri and Wilkins 2008; Pendall 2008; de Souza Briggs et al. 2010a, b; Landis and McClure 2010). There seems to be an emerging consensus that what is required is a broad palette of reforms involving both supply-side (dwelling-based) and demand-side (tenant-based) housing strategies (tailored to the particulars of the local metropolitan market), plus complementary non-housing strategies. The suggested reform proposals have included:

Demand-Side Housing Assistance (HCV) proposals:

- Direct leasing and brokerage for connecting HVC holders to market-rate rental housing and LIHTC developments in good neighbourhoods
- Financial incentives to HCV holders and potential HCV landlords in desirable areas, such as raising Fair Market Rent levels there
- Intensified pre-move mobility counselling and aid, coupled with post-move follow-up, support, and assistance when necessary
- PHA performance incentives rewarding those who help HCV holders move outside disadvantaged neighbourhoods and promote a more effective use of inter-PHA portability of HCVs
- End PHA administration of vouchers and contract to non-profit organizations with metro-wide coverage
- Prohibitions on the use of HCVs in certain neighbourhoods/requirements that they can only be used in more "opportunity rich" neighbourhoods
- Requirements for all landlords to participate in HCV program upon request
- Making comparative school performance data more available to HCV parents
- Beefed-up fair housing enforcement aimed at users of HCVs who are minority and families with children

Supply-Side Housing Assistance (scattered-site public housing, LIHTC, HOPE VI) proposals[19]:

- Changing rules of LITC allocations to discourage development in poor neighbourhoods and to create more income mixing within developments
- Changing the basis for state allocations of tax credits from per capita to favor those state with tighter housing markets
- Limitations on where developments can be sited ("neighbourhood impaction standards") to avoid concentrations of low-income or assisted households
- Preserving affordable housing in gentrifying areas, perhaps by offering tax abatements or freezing assessed values for property tax purposes
- Empowering metropolitan planning organizations to tie receipt of federal grants to suburban jurisdictions with their creation of "fair share" assisted housing development

Non-housing proposals:

- Car vouchers to help navigate the tricky transportation requirements for home-work-childcare-church transitions
- Attaching child-care vouchers and training assistance to housing assistance
- Holistic matching of housing and other supportive welfare and educational services across agencies

---

[19] Another policy option here is inclusionary zoning for new, privately developed complexes, though in the U.S. this has been devolved to the state and local governments so I do not list it here among federal reform proposals.

## *Potential Lessons for Western Europe and Further Afield*

What can housing policymakers in Western Europe learn from the past U.S. experience and this panoply of proposals? Before addressing this question I must emphasize that: (1) fundamental differences in U.S. and Western European housing systems make the origins of the poverty deconcentration problem and its resolution distinctive; (2) Western European housing systems cannot be generalized without perils of oversimplification; and thus (3) detailed policy recommendations for Western Europe based on American experience should not be made.[20]

The first point is essential and bears amplification. On the demand-side of the equation, the Western European challenges related to poverty deconcentration are considerably less severe than in the U.S. Because tenant-based housing allowances are an entitlement and after-tax income distributions are much more compressed in Western Europe, there will be less severe neighbourhood sorting according to income transpiring through market processes. On the supply-side, there are several factors that also make the issue considerably different in Western Europe. These include a large social rental sector encompassing a wide range of incomes, centralized or regionalized planning systems that can exert direct control over where this social housing is located and how it is interspersed with market-rate dwellings, universal participation of private market landlords in housing allowance schemes[21] and, frequently, some form of rent restrictions and/or a relatively small and underdeveloped private rental sector. Because of these differences, the fundamental distinction in the origins of the problem of poverty concentration is that it is primarily market-driven in the case of the U.S. and state-driven in the case of Western Europe.

The fundamental difference in the nature of the potential solutions to the problem is that the U.S. has indirect and relative weak policy levers while Western Europe has the opposite. At its core, federal housing programs designed to assist lower-income households represent only 5.4 % of the entire U.S. housing stock and assists only a third of eligible households as of 2009 (Schwartz 2010). Thus, even if these programs were systematically designed to deconcentrate poverty (which, as I argued above, they are not) it is arguable that they would have only a modest impact on the geography of disadvantage. Moreover, U.S. federal policy is delivered against a backdrop of fragmented local governments that typically lack regional coordination for deconcentration efforts. Finally, the politics surrounding the deconcentration of poverty in the U.S. is indelibly stained with racism that arguably has constrained the aggressiveness with which any such initiatives could be pursued. Though such racial-ethnic issues certainly are present in Western Europe, I do not believe that they have attained the degree of longstanding cultural and political salience as they have in the U.S.

---

[20] As an illustration of these points, see Priemus and colleagues (2005).
[21] Because tenants receive the rental allowance directly, landlords do not contract directly with a local housing authority for part of the rent payment as in the U.S. and thus do not have the option of not participating.

Despite these fundamental trans-Atlantic differences, I think some broad lessons from the American experience do have international traction. The efficacy of tenant-based (demand-side) efforts to deconcentrate poverty will be inversely related to the:

- Tightness of the local metropolitan housing market in neighbourhoods that represent desirable destinations for assisted tenants
- Extent that concentrated low-income households constitute racial-ethnic-immigrant minorities and the private housing market is balkanized by discriminatory barriers
- Strength of local social ties among concentrated low-income households and the density of location-specific institutions purveying cultural capital to these communities
- The degree of safety and other aspects about quality of residential life and public services available in neighbourhoods occupied disproportionately by the poor that would make them less likely to seek alternative locations

The efficacy of dwelling-based (supply-side) efforts to deconcentrate poverty will be directly related to the:

- Regulatory powers granted to local public planning and housing development authorities to plan regionally in a dispersed manner to avoid concentrations de novo
- Geographic area over which these powers may be exercised
- Tightness of the local metropolitan housing market overall that will limit the ability of higher-income households to avoid living in mixed-income neighbourhoods
- The nature of neighbourhood-based facilities and services, including schooling quality and jobs programs

## *Two Final Policy Caveats*

In closing this policy discussion two caveats are in order. First, while there may be a general consensus on both sides of the Atlantic that concentrations of poverty are bad things that should be avoided to the extent feasible, the counterfactual is rarely specified in any detail. Neighbourhood diversity is hard to both define and make operational in practice. Five thorny practical issues arise in particular (Tunstall and Fenton 2006: 25–26; Kleinhans 2004; Galster 2013):

- *Composition*: On what basis(es) are we mixing people in the deconcentrated neighbourhood alternative: ethnicity, race, religion, immigrant status, income, housing tenure...all, or some of the above?
- *Concentration*: What is the desired amount of mixing in question? Which amounts of which groups comprise the ideal mix, or are minimally required to produce the desired outcomes?
- *Scale*: Over what level(s) of geography should the relevant mix be measured? Does mixing at different spatial scales yield different outcomes?
- *Distance*: How far away from the prior area of concentrated deprivation should low-income households be moved to achieve a more socially mixed

neighbourhood?[22] And how is distance managed for low income residents who need services only offered in central places?
- *Speed*: How rapidly do such programs need to be brought up to scale? Are demonstrations helpful in the transition?

Many different combinations of the above elements characterize different aspects of poverty deconcentration policies in different national contexts, though often not explicitly. Indeed, the counterfactual to concentrated poverty – "social mix" – is an intrinsically vague, slippery term; it is typically used to mean different things by different people. Planners and policymakers must be more precise and explicit in specifying the parameters of these five aspects of social mix before they can recommend specific policies and practices to deconcentrate poverty.

The second caveat relates to the efficacy of assisted housing policy in general to radically change the socioeconomic opportunities of low-income households and their families by changing their geographic contexts. I think the evidence is clear that for certain families in certain contexts the differences supplied by geography alone can be substantial. However, for many low-income residents of concentrated poverty neighbourhoods it will take more than changing their location given their durable and potentially constraining connections to social networks based in disadvantaged neighbourhoods and a variety of personal attributes that will continue to limit their upward mobility unless addressed directly (de Souza Briggs et al. 2010a, b; Goetz 2010). This is clearly recognized by the current administration at HUD, as embodied in the principles of their Choice Neighbourhoods program. Several local housing authorities at this writing are experimenting with new collaborations between local educators and other service providers to comprehensively and holistically help subsidized households improve their economic and social prospects (U.S. Department of Housing and Urban Development 2011).

## Conclusions

At a descriptive level, three conclusions can be drawn regarding the impact of U.S. federal assisted housing policy on deconcentrating poverty. First, residents of U.S. public housing on average reside in significantly more disadvantaged neighbourhoods compared to participants in any other assisted housing program. Second, residents of other types of site-based assisted housing programs (particularly LIHTC) do not reside in significantly different residential environments than tenant-based HCV holders. HCV holders fare somewhat better in neighbourhood poverty rates than equivalent households who do not receive subsidies, but the comparative differences are even smaller when the LIHTC program is compared to equivalent private renters.

---

[22] Much U.S. evidence suggests that moving from concentrations of poverty had little salutary impact on households unless the destination is far distant from the original neighbourhood; see Rubinowitz and Rosenbaum (2000), Goetz (2003), and Feins and Patterson (2005).

Third, HCV holders do not substantially improve their neighbourhood circumstances with subsequent moves; indeed if their initial move is (perhaps with counselling assistance) to a low-income, predominantly white neighbourhood, their subsequent moves are generally to higher-poverty, higher-minority share neighbourhoods. In other words, the public housing program historically intensified the concentration of poverty and subsequent demand-side and supply-side policies have had relatively little impact in improving the neighbourhood conditions of recipients.

Why these subsequent policy efforts over several decades have produced so little deconcentration of poverty is subject to considerable debate. Arguments involving the characteristics of residents of concentrated poverty neighbourhoods (such a binding local social ties), arguments citing structural barriers of many sorts in the housing market, and arguments involving the rules and administration of HCV and LIHTC programs all have merit. It is extremely difficult to quantify precisely the relative contributions of these three strands of argument. Thus, I believe that all "pure" explanations should be rejected in favour of some middle-ground position.

What should be done in the U.S. to enhance the efficacy of assisted housing programs to deconcentrate poverty has been the longstanding object of vigorous debate. Some amalgam of supply-side and demand side reforms, coupled with non-housing strategies hold most promise. The U.S. experience in this regard offers several broad lessons to housing policymakers in Western Europe, even though there are vast differences in the origins and policy options available for addressing concentrated poverty.

Scholars should recognize how challenging it is to measure precisely the independent causal impacts on the residential geography of recipients emanating from specific programs providing federal housing assistance in the United States. There is selection bias in terms of program participation, with distinctly different clienteles participating in the various programs. There may be substantial functional overlaps and interrelationships among the programs. Though experimentally designed research holds promise in sorting out some of these confounding biases, it is costly. Perhaps most importantly, the major studies describing the geography of housing assistance assume that the locations and mobility behaviours of other households that define aggregate neighbourhood characteristics are unaffected by the geographic decisions of assisted households or developers of site-based assisted housing. Clearly this is untrue in some circumstances, as we know that over-concentrations of assisted housing can lead to endogenous neighbourhood reactions (Galster et al. 1999, 2003, 2008; Varady and Walker 2003a, b). Some interesting efforts to model holistically these dynamic neighbourhood mobility interrelationships among assisted and unassisted households are being undertaken at this writing by Owens (2011). Not only are the effects of current programs hard to discern, but the tools to incrementally build poverty deconcentration and neighbourhood mixing programs are in its infancy. All these realms offer fertile areas for future scholarship that are likely to yield important insights for policymakers interested in altering the geography of opportunity.

# References

Achtenberg, E. (2002). *Stemming the tide: A handbook on preserving subsidized multifamily housing*. New York: Local Initiatives Support Corporation.

Basolo, V., & Nguyen, M. (2005). Does mobility matter? An analysis of housing voucher holders' neighbourhood conditions by race and ethnicity. *Housing Policy Debate, 16*(3/4), 297–324.

Buron, L. (2004). *An improved living environment? Neighbourhood outcomes for HOPE VI relocatees*. Washington, D.C.: The Urban Institute. http://www.urban.org/publications/311059.html.

Buron, L., Levy, D., & Gallagher, M. (2007). *Housing choice vouchers: How HOPE VI families fared in the private market*. Washington, D.C.: The Urban Institute. http://www.urban.org/publications/311487.html.

Cisneros, H. (1996). Regionalism: The new geography of opportunity. *National Civic Review, 85*(2), 35–48.

Clark, W. A. V. (2005). Intervening in the residential mobility process: Neighbourhood outcomes for Low-income populations. *PNAS, 102*(43), 15307–15312.

Clark, W. A. V. (2008). Re-examining the moving to opportunity study and its contribution to changing the distribution of poverty and ethnic concentration. *Demography, 45*(3), 515–535.

Coulibaly, M., Green, R., & James, D. (1998). *Segregation in federally subsidized low-income housing in the United States*. Westwood: Praeger.

Cronin, F. J., & Rasmussen, D. W. (1981). Mobility. In M. Struyk & M. Bendick (Eds.), *Housing vouchers for the poor: Lessons from a national experiment* (pp. 107–128). Washington, D.C.: Urban Institute Press.

de Souza Briggs, X., Comey, J., & Weismann, G. (2010a). Struggling to stay out of high-poverty neighbourhoods: housing choice and locations in moving to opportunity first decade. *Housing Policy Debate, 20*(3), 383–427.

de Souza Briggs, X., Popkin, S., & Goering, J. (2010b). *Moving to opportunity: The story of an American experiment to fight ghetto poverty*. Oxford/New York: Oxford University Press.

DeFilippis, J., & Wyly, E. (2008). Running to stand still through the looking glass with federally subsidized housing in New York City. *Urban Affairs Review, 43*(6), 777–816.

DeLuca, S., & Dayton, E. (2009). Switching social contexts: The effects of housing mobility and school choice programs on youth outcomes. *Annual Review of Sociology, 35*, 457–491.

Deluca, S., & Rosenbaum, J. (2003). If low-income blacks are given a chance to live in white neighbourhoods, will they stay? Examining mobility patterns in a quasi-experimental program with administrative data. *Housing Policy Debate, 14*(3), 305–346.

Deluca, S., & Rosenblatt, P. (2011). *Walking away from the wire: Residential mobility and opportunity in Baltimore*. Unpublished paper, Department of Sociology, Johns Hopkins University.

Deluca, S., Rosenblatt, P., & Wood, H. (2011). *Why poor people move (and where they go): Residential mobility, selection and stratification*. Unpublished paper, Department of Sociology, Johns Hopkins University.

Duncan, G., & Zuberi, A. (2006). Mobility lessons from Gautreaux and moving to opportunity. *Northwestern Journal of Law and Social Policy, 1*(1), 110–126.

Fairchild, H. H., & Tucker, B. M. (1982). Black residential mobility: Trends and characteristics. *Journal of Social Issues, 38*, 51–74.

Feins, J. (2003). A cross-site analysis of MTO's locational impacts. In J. Goering & J. Feins (Eds.), *Choosing a better life? Evaluating the moving to opportunity social experiment* (pp. 81–116). Washington, D.C.: Urban Institute Press.

Feins, J., & Patterson, R. (2005). Geographic mobility in the Housing Choice Voucher Program, 1995–2002. *Cityscape, 8*(2), 21–48.

Finkel, M., & Buron, L. (2001). *Study on Section 8 voucher success rates* (Quantitative study of success rates in metropolitan areas, Vol. 1). Washington, D.C.: Abt Associates for U.S. Department of Housing and Urban Development.

Freeman, L. (2004). *Siting affordable housing: Location and trends of low-income Housing Tax Credit Developments in the 1990s*. Washington, D.C.: Brookings Institution Center on Urban and Metropolitan Policy. Available at: http://www.brookings.edu/reports/2004/04metropolitan policy_freeman.aspx.

Galster, G. (2002). An economic efficiency analysis of deconcentrating poverty populations. *Journal of Housing Economics, 11*(4), 303–329.

Galster, G. (2003). The effects of MTO on sending and receiving neighborhoods. In J. Goering & J. Feins (Eds.), *Choosing a better life? Evaluating the moving to opportunity social experiment* (pp. 365–382). Washington, DC: Urban Institute Press.

Galster, G. (2008). Scholarship on U.S. housing planning and policy: The evolving topography since 1968. *Journal of the American Planning Association, 74*(1), 1–12.

Galster, G. (2013). Neighbourhood social mix: Theory, evidence, and implications for policy and planning. In N. Carmon & S. Fainstein (Eds.), *Planning as if people mattered*. Philadelphia: University of Pennsylvania Press.

Galster, G., & Killen, S. (1995). The geography of metro-politan opportunity: A reconnaissance and conceptual framework. *Housing Policy Debate, 6*(1), 7–44.

Galster, G., Tatian, P., & Smith, R. (1999). The impact of neighbors who use Section 8 certificates on property values. *Housing Policy Debate, 10*(4), 879–917.

Galster, G., Tatian, P., Santiago, A., Pettit, K., & Smith, R. (2003). *Why NOT in my back yard? The neighbourhood impacts of assisted housing*. New Brunswick: Rutgers University/Center for Urban Policy Research Press.

Galster, G., Cutsinger, J., & Malega, R. (2008). The costs of concentrated poverty: Neighbourhood property markets and the dynamics of decline. In N. Retsinas & E. Belsky (Eds.), *Revisiting rental housing: Policies, programs, and priorities* (pp. 93–113). Washington, D.C.: Brookings Institution Press.

Goering, J. (Ed.). (1986). *Housing desegregation and federal policy*. Chapel Hill: University of North Carolina Press.

Goering, J., & Feins, J. (Eds.). (2003). *Choosing a better life? Evaluating the moving to opportunity social experiment*. Washington, D.C.: Urban Institute Press.

Goering, J., et al. (1995). *Promoting housing choice in HUD's rental assistance programs: A report to Congress*. Washington, D.C.: U.S. Department of Housing and Urban Development.

Goetz, E. (2003). *Clearing the way: Deconcentrating the poor in urban America*. Washington, D.C.: The Urban Institute Press.

Goetz, E. (2010). Better neighbourhoods, better outcomes? Explaining relocation outcomes in HOPE VI. *Cityscape, 12*(1), 5–32.

Grigsby, W., & Bourassa, S. (2004). Section 8: The time for fundamental program change. *Housing Policy Debate, 15*(4), 805–834.

Hartung, J., & Henig, J. (1997). Housing vouchers and certificates as a vehicle for deconcentrating the poor: Evidence from the Washington, DC, metropolitan area. *Urban Affairs Review, 32*(3), 403–419.

Hirsch, A. (1983). *Making the Second Ghetto: Race and housing in Chicago 1940–1960*. Cambridge: Cambridge University Press.

Hogan, J. (1996). *Scattered-site public housing: Characteristics and consequences*. Washington, D.C.: US Department of Housing and Urban Development.

Jacob, B. (2004). Public housing, housing vouchers and student achievement: evidence from public housing demolitions in Chicago. *American Economic Review, 94*(1), 233–258.

Julian, E., & Daniel, M. (1990). Separate and unequal: The root and branch of public housing segregation. *Clearinghouse Review, 23*, 666–688.

Katz, B., & Turner, M. (2001). Who should run the housing voucher program? a reform proposal. *Housing Policy Debate, 12*(2), 239–262.

Katz, B., & Turner, M. (2008). Rethinking U.S. rental housing policy: A new blueprint for federal, state and local action. In N. Retsinas & E. Belsky (Eds.), *Rethinking rental housing: Policies, programs and priorities* (pp. 319–358). Washington, D.C.: Brookings Institution.

Keels, M., Duncan, G., Deluca, S., Mendenhall, R., & Rosenbaum, J. (2005). Fifteen years later: Can residential mobility programs provide a permanent escape from neighbourhood crime and poverty? *Demography, 42*(1), 51–73.

Khadduri, J. (2005). Comment on Basolo & Nguyen, 'Does mobility matter?'. *Housing Policy Debate, 16*(3–4), 325–334.

Khadduri, J., & Wilkins, C. (2008). Designing subsidized rental housing programs: What have we learned? In N. Retsinas & E. Belsky (Eds.), *Rethinking rental housing: Policies, programs and priorities* (pp. 161–190). Washington, D.C.: Brookings Institution.

Kingsley, G. T., Johnson, J., & Pettit, K. L. S. (2003). Patterns of Section 8 relocation in the HOPE VI program. *Journal of Urban Affairs, 25*(4), 427–447.

Kleinhans, R. (2004). Social implications of housing diversification in urban renewal: A review of recent literature. *Journal of Housing and the Built Environment, 19*(4), 367–390.

Landis, J., & McClure, K. (2010). Rethinking federal housing policy. *Journal of the American Planning Association, 76*(3), 319–348.

Lens, M. C., Gould Ellen, I., & O'Regan, K. (2011). Do vouchers help low-income households live in safer neighbourhoods? *Cityscape, 13*(3), 135–160.

Levy, D., Comey, J., & Padilla, S. (2006). *Keeping the neighbourhood affordable*. Washington, D.C.: The Urban Institute.

Marr, M. (2005). Mitigating apprehension about Section 8 vouchers. *Housing Policy Debate, 16*(1), 85–112.

Massey, D., & Kanaiaupuni, S. (1993). Public housing and the concentration of poverty. *Social Science Quarterly, 74*(1), 109–122.

McClure, K. (2006). The Low-Income Housing Tax Credit program goes mainstream and moves to the suburbs. *Housing Policy Debate, 17*(3), 419–446.

McClure, K. (2008). Deconcentrating poverty with housing programs. *Journal of the American Planning Association, 74*(1), 90–99.

Newman, S., & Schnare, A. (1997). 'And a suitable living environment': the failure of housing programs to deliver on neighbourhood quality. *Housing Policy Debate, 8*(4), 703–741.

Orr, L., Feins, J., Jacob, R., Beechcroft, E., Sanbonmatsu, L., Katz, L., Liebman, J., & Kling, J. (2003). *Moving to opportunity for fair housing demonstration: Interim impacts evaluation*. Washington, D.C.: U.S. Department of Housing and Urban Development.

Owens, A. (2011). *Neighbourhood poverty and the changing geography of subsidized housing*. Unpublished doctoral dissertation, Department of Sociology, Harvard University, Cambridge.

Pashup, J., Edin, K., Duncan, G., & Burke, K. (2005). Participation in a residential mobility program from the client's perspective: Findings from Gautreaux Two. *Housing Policy Debate, 16*(3–4), 361–392.

Patterson, R., & 7 others. (2004). *Evaluation of the Welfare to Work Voucher Program. Report prepared by Abt Associates and QED Group*. Washington, D.C.: U.S. Department of Housing and Urban Development.

Pendall, R. (2000). Why voucher holder and certificate users live in distressed neighbourhoods. *Housing Policy Debate, 11*(4), 881–910.

Pendall, R. (2008). From hurdles to bridges: Local land use regulations and the pursuit of affordable rental housing. In N. Retsinas & E. Belsky (Eds.), *Rethinking rental housing: Policies, programs and priorities* (pp. 225–274). Washington, D.C.: Brookings Institution.

Polikoff, A. (2006). *Waiting for Gautreaux*. Evanston: Northwestern University Press.

Popkin, S., & Cove, E. (2007). *Safety is the most important thing: How HOPE VI helped families*. Washington, D.C.: Urban Institute Policy Brief. http://www.urban.org/publications/311486.html.

Popkin, S., & Cunningham, M. (1999). *CHAC Section 8 program: Barriers to successful leasing up*. Washington, D.C.: The Urban Institute.

Popkin, S., Buron, L., Levy, D., & Cunningham, M. (2000). The Gautreaux legacy: What might mixed-income and dispersal strategies mean for the poorest public housing tenants? *Housing Policy Debate, 11*(4), 911–942.

Popkin, S., Galster, G., Temkin, K., Herbig, C., Levy, D., & Richter, E. (2003). Obstacles to desegregating public housing: Lessons learned from implementing eight consent decrees. *Journal of Policy Analysis and Management, 22*(2), 179–200.

Popkin, S., Katz, B., Cunningham, M., Brown, K., Gustafson, J., & Turner, M. (2004). *A decade of HOPE VI: Research findings and policy challenges*. Washington, DC: The Urban Institute and The Brookings Institution. http://urban.org/uploadedPDF/411002HOPEVI.pdf

Popkin, S., Cunningham, M., & Burt, M. (2005). Public housing transformation and the had-to-house. *Housing Policy Debate, 16*(1), 1–24.

Priemus, H., Kemp, P., & Varady, D. (2005). Housing Vouchers in the U.S., Great Britain, and the Netherlands: Current issues and future perspectives. *Housing Policy Debate, 16*(3–4), 575–609.

Rainwater, L. (1970). *Behind Ghetto Walls*. Chicago: Aldine.

Rohe, W., & Freeman, L. (2001). Assisted housing and residential segregation: The role of race and ethnicity in the siting of assisted housing developments. *Journal of the American Planning Association, 67*(3), 279–292.

Roisman, F. (1998). Mandates unsatisfied: The Low Income Housing Tax Credit program and the civil rights laws. *University of Miami Law Review, 52*, 1011–1050.

Rubinowitz, L., & Rosenbaum, J. (2000). *Crossing the class and color lines: From public housing to white suburbia*. Chicago: University of Chicago Press.

Schill, M., & Wachter, S. (1995). The spatial bias of federal housing law and policy. *University of Pennsylvania Law Review, 143*(5), 1285–1342.

Schuetz, J., Meltzer, R., & Been, V. (2011). Silver bullet or trojan horse? The effects of inclusionary zoning on local housing markets. *Urban Studies, 48*(2), 273–296.

Schwartz, A. (2010). *Housing policy in the United States* (2nd ed.). Oxford: Routledge.

Shroder, M. (2003). Locational constrain, housing counseling and successful lease-up. In J. Goering & J. Feins (Eds.), *Choosing a better life? Evaluating the moving to opportunity social experiment* (pp. 59–80). Washington, D.C.: Urban Institute Press.

Talen, E., & Koschinsky, J. (2011). Is subsidized housing in sustainable neighbourhoods? *Housing Policy Debate, 21*(1), 1–28.

Tunstall, R., & Fenton, A. (2006). *In the mix: A review of mixed income, mixed tenure and mixed communities*. York: Joseph Rowntree Foundation, English Partnerships, and the Housing Corporation.

Turbov, M. (2006). Public housing redevelopment as a tool for revitalizing neighbourhoods: How and why did it happen and what have we learned? *Northwestern Journal of Law and Social Policy, 1*(1), 167–201.

Turner, M. (1998a). Moving out of poverty: Expanding mobility and choice through tenant-based housing assistance. *Housing Policy Debate, 9*(2), 373–394.

Turner, M. (1998b). *Affirmatively furthering fair housing: Neighbourhood outcomes for tenant-based assistance in six metropolitan areas*. Washington, D.C.: Urban Institute.

Turner, M., & Williams, K. (1998). *Housing mobility: Realizing the promise*. Washington, D.C.: Urban Institute.

Turner, M., Ross, S., Galster, G., & Yinger, J. (2002, June). *Discrimination in Metropolitan Housing Markets: National results from phase I of HDS 2000*. Washington, D.C.: Urban Institute Report (6977) to U.S. Department of Housing and Urban Development.

Turney, K., Clampet-Lundquist, S., Edin, K., Kling, J., & Duncan, G. (2006). Neighbourhood effects on barriers to employment: Results from a randomized housing mobility experiment in Baltimore. *Brookings-Wharton Papers on Urban Affairs, 2006*, 137–187.

U.S. Department of Housing and Community Development. (2011). Choice neighbourhoods: History and HOPE. *Evidence Matters* (Winter), 1–7.

van Ham, M., Manley, D., Bailey, N., Simpson, L., & Maclennan, D. (2012). Introduction. In M. van Ham, D. Manley, N. Bailey, L. Simpson, & D. Maclennan (Eds.), *Neighbourhood effects research: New perspectives* (pp. 1–22). Dordrecht: Springer.

van Ham, M., Manley, D., Bailey, N., Simpson, L., & Maclennan, D. (2013). Understanding neighbourhood dynamics: New insights for neighbourhood effects research. In M. van Ham, D. Manley, N. Bailey, L. Simpson, & D. Maclennan (Eds.), *Understanding neighbourhood dynamics: New insights for neighbourhood effects research* (pp. 1–22). Dordrecht: Springer.

Varady, D., & Walker, C. (2000). Vouchering out distressed subsidized developments: Does moving lead to improvements in housing and neighbourhood conditions? *Housing Policy Debate, 11*(1), 115–162.

Varady, D., & Walker, C. (2003a). Using housing vouchers to move to the suburbs: How do families fare? *Housing Policy Debate, 14*(3), 347–382.

Varady, D., & Walker, C. (2003b). Using housing vouchers to move to the suburbs: The Alameda County, California experience. *Urban Affairs Review, 39*(2), 143–180.

Williamson, A., Smith, M., & Strambi-Kramer, M. (2009). Housing choice Vouchers, the Low-Income Housing Tax Credit, and the Federal Poverty Deconcentration Goal. *Urban Affairs Review, 45*(1), 119–132.

Wilson, W. J. (1987). *The truly disadvantaged: The Inner City, the Underclass and Public Policy.* Chicago: The University of Chicago Press.

# Chapter 12
# Neighbourhood Effects and Social Cohesion: Exploring the Evidence in Australian Urban Renewal Policies

**Kathy Arthurson**

## Introduction

Social housing estates constructed in Australia in the post Second World War period are perceived as increasingly problematic for policy makers in the contemporary era. The concentrations of residents experiencing issues of poverty, unemployment and behavioural issues are exacerbated by the targeting of access only to high need groups that include ex-prisoners, people with substance abuse, and mental health issues. Originally these groups were ineligible to access social housing as when it was initially conceptualised in Australia the target group was low income working class families. Nevertheless the client group only forms part of the present-day challenges for housing authorities that manage the housing. The housing assets are aging, often run down, concentrated in particular neighbourhoods, largely on the fringe areas of cities, and uniform in design and style, which differentiates the estates from adjacent neighbourhoods of private housing. They were built swiftly to meet demand so the houses are often of poor quality construction. In recent years there has been civic disorder experienced on some estates, including Rosemeadow, Macquarie Fields and Redfern social housing estates in Sydney, New South Wales. These incidents reflect the results of social and economic change, including high levels of unemployment, in particular experienced by disenfranchised youth. The media has given extensive and often sensationalist coverage to these incidents prompting concerns that threats to social cohesion are emanating from and linked to neighbourhoods typified by high concentrations of social housing (Arthurson 2004). This emphasis was evident in explaining the rationale underlying the decision taken to demolish one of the problematic estates, Villawood (East Fairfield) estate in the late 1990s and replace it totally with private housing. The New South Wales Minister

K. Arthurson (✉)
Faculty of Health Sciences, Southgate Institute for Health, Society
and Equity, Flinders University, Bedford Park, SA, Australia
e-mail: kathy.arthurson@flinders.edu.au

for Urban Affairs and Planning (1997: p. 2; 1998) evoked images of Villawood as a homogeneous social housing estate with "systemic social and criminal problems", and private housing was proposed as the solution to "overcome the problems of the past to provide a safe and secure environment for new families and future generations". From this perspective spatial concentration of social housing represents a threat to social cohesion. There is no denying that there are problems involving crime and violence on some social housing estates but these depictions are not the only realities in the day-to-day lives of residents and are not applicable to all estates or indeed across all parts of estates.

While the example of Villawood illustrates an extreme policy reaction to addressing neighbourhood based problems in areas of concentrated social housing and implied threats to social cohesion, the key contemporary policy solution largely involves implementing estate renewal projects incorporating changes to housing tenure and social mix to create more heterogeneous estates. The discussion commences with outlining Australian estate renewal approaches and questions whether US style 'neighbourhood effects' policies that are often drawn on in the debates are relevant in the Australian context. Then the chapter explores the empirical evidence linking mixed communities to social cohesion. Finally attention turns to the current research findings, which examine some of the linkages made between social mix policies and social cohesion from the point of view of residents of three reordered mixed tenure neighbourhoods.

## Australian Renewal Approaches: Addressing Neighbourhood Problems or Neighbourhood Effects?

The first attempts at renewal in areas of concentrated social housing in Australia were conducted in the mid 1980s. Projects focused on physical renewal and demolition of housing to address structural problems, rather than social problems on the estates. The initial projects consisted of building new social housing on site through subdividing pre-existing large backyards into a number of smaller allotments thereby increasing the density of housing (Arthurson 2002). This approach was revised as recognition increased that there were others issues to contend with in the neighbourhoods beyond the age, physical condition and related high maintenance expenditure requirements of the first generation post Second World War housing. Social housing tenants were increasingly characterised by large numbers of low income and single parent families and blue collar workers affected by processes of industry restructuring, which resulted in higher levels of unemployment and lower rates of workforce participation within the estates compared to adjoining neighbourhoods. The situation is typified by Mitchell Park one of the case study estates for the empirical research reported later in this chapter. In 1995, the resident unemployment rate was 16 % compared to an average of 12 % in the metropolitan area of Adelaide, 25 % of residents were sole parent families with 57 % on low annual incomes (Proctor 1995; SA Better Cities 11 Steering Committee 1995). As concern and awareness about the extent of the problems grew, instead of continuing to increase the concentration of social housing in the neighbourhoods through demolition and urban infill, attention turned to diversifying social mix as part of estate renewal initiatives. Social mix is commonly used to refer to the level of socioeconomic variance of residents, measured by housing tenure within a particular spatially delineated area, age range or ethnic mix of residents. While it can represent all of the above features in the context of urban renewal, policy makers tend to use the two terms social mix and housing tenure mix interchangeably (Arthurson 2002).

The key characteristic taken into account in selecting estates for latter-day regeneration initiatives was the large concentrations of residents experiencing socioeconomic disadvantage (Fulcher 1999). In most Australian States 'community renewal' type programs have been initiated. The most recent responses embrace 'whole of government' approaches to addressing social exclusion, arguing that the complexity and interconnected nature of estate residents' problems require solutions that are much broader than changes to physical infrastructure and housing carried out independently of other social concerns. Whole of government models envisage working in collaboration and partnership with a range of government and non-government agencies including, but not restricted to the fields of education, health and police to facilitate integrated service delivery at the local neighbourhood level (Arthurson 2003). Additional activities to making improvements to housing and physical environments include initiatives to: coordinate service provision at the local level; provide employment opportunities for residents; partnerships to empower communities; and diversifying housing tenure and social mix within the regeneration area to break down the concentrations of disadvantaged residents (Arthurson 2012). Changes to social mix are achieved through reordering the mix of housing dwellings in terms of quality, size and tenure type, involving demolition of obsolete housing, building new housing for sale on the private market to attract home buyers into the area and permanently relocating some social housing tenants to other neighbourhoods.

Underlying social mix policies is the idea that concentrated social housing leads to 'problem neighbourhoods' and that for disadvantaged people their issues are compounded simply from living amongst other like minded, socioeconomically similar, underprivileged people. The implications of these contentions are that it is considered more beneficial for disadvantaged social housing tenants to reside in mixed tenure neighbourhoods, with greater social balance, than what are envisaged as homogeneous neighbourhoods made up of concentrated social housing households. While this idea appears similar to the concept of 'neighbourhood effects' identified in the US, as discussed shortly the different social and political context in Australia makes its application less relevant in this milieu. Nevertheless, policy makers argue that there are numerous benefits from rebalancing social mix, or in effect thinning out spatial concentrations of social housing tenants. The anticipated benefits include: lowering area-based stigma; role modelling effects from more advantaged neighbours; broadening social interactions for disadvantaged groups that lead to benefits such as employment opportunities; and creating more inclusive communities with enhanced social cohesion (Arthurson 2002; Wood 2003). It is this latter aspiration of policy makers to "integrate the estates into the wider community," and increase or maintain levels of "social cohesion" and "community integration" within the neighbourhoods (Arthurson 2012) that is the specific concern of this chapter. Before turning to explore the empirical evidence that links social mix to social cohesion, discussion first returns to the point made earlier about the importance of context in considering the relevance of US type 'neighbourhood effects' policies to Australia.

## A Note on the Importance of Context

An important aspect to take into account in any discussion of neighbourhood based problems and 'neighbourhood effects' is the differing social and political context of the US, from Australia, as it is from the former that most of the research and policy engagement with the concept of 'neighbourhood effects' originates.

### *The Urban Underclass and Questions About the Balance Between Structure and Agency*

The idea of neighbourhood effects developed from the prolonged and pejorative academic debates in the US that ensued between Wilson (1991) and Murray (1994) about the existence of an urban underclass. From their perspectives living amongst other similarly disadvantaged people detrimentally impacted on life chances and aspirations. Murray argued that concentrations of social housing caused a 'culture of poverty' to form characterised by welfare dependency, social irresponsibility and problematic tenant behaviour. The nub of the debate was the extent to which individuals who live in poverty are culpable for their own predicament (individual agency) and the degree to which societal (structural) factors affect individual capabilities, in combating or adding to disadvantage. The emphasis on an 'urban underclass' led in part to the development of policies, exemplified by the Gautreaux' and 'Moving to Opportunity' programs, that utilize housing vouchers for accessing private rental housing to 'scatter' public housing tenants across more prosperous neighbourhoods.

In Australia debates are more about 'problem neighbourhoods' linked to concentrations of disadvantaged social housing tenants situated within a policy framework of responding to social exclusion. The adoption of the term social exclusion seems to represent in part an attempt to maintain distance from US debates about neighbourhood effects and the existence of an 'urban underclass' along with the pejorative associations with social engineering of adopting these terms. Australian policies thus more closely resemble UK regeneration policies, focusing on regeneration in situ and aiming for dilution of concentrations of social housing, often through private sales of social housing to existing tenants. In both countries, social mix is often implemented as part of a broader neighbourhood regeneration framework. There are other important contextual differences which also make the use of the concept of neighbourhood effects and US policy programs less relevant in Australia.

## Where Concentrations of Disadvantage Are Located

Another key contextual difference is that in Australia, unlike the US, social housing estates are predominantly located in outer regions of cities. By comparison inner city areas are generally more advantaged with better access to employment and other opportunities. This differing situation is reflected in US mobility programs that relocate low income households from 'distressed' neighbourhoods often of inner city concentrated poverty and lacking in job opportunities to private rental housing in outer areas with wider socioeconomic mix.

The US policy direction also reflects the enforcement in law of the rights of African Americans to live in white suburbs. Friedrichs (2002) argues that in Europe minority group discrimination is lower and social integration is higher than in the US, and this renders 'neighbourhood effects' less relevant. A similar claim is applicable to Australia as compared to the US the scale of socioeconomic disadvantage, income divisions and poverty rates and levels of segregation appear lower, due to comparatively more generous provisions of the Australian welfare system.

Hence, at the heart of US policies to address neighbourhood effects and Australian attempts to ameliorate neighbourhood based problems are different explanations and social and political contexts for how disadvantage arises and is best addressed. The following section explores the contemporary empirical evidence base to investigate the perception in public policy debates that compared with neighbourhood homogeneity neighbourhood diversity is more favourable for building social cohesion, at least in relation to concentrated social housing, which is depicted as a threat to social cohesion.

## Social Cohesion and Mixed Tenure Communities – The Empirical Evidence?

Social cohesion 'is a difficult concept' to define or measure because of its numerous dimensions and utilisation by policy makers in a variety of ways that are often ill defined (van Kempen and Bolt 2009: 458). A widely used contemporary definition emphasises its critical components as the sharing of common values, moral principles and codes of behaviour through which to conduct relationships with others (Kearns and Forrest 2000: 997). It follows that at the spatial level of mixed neighbourhoods experiencing social cohesion implies a lack of tension, conflict or violence between residents across different housing tenures. In examining literature linking social cohesion to mixed communities several key themes emerged about: the differing expectations of residents and policy makers concerning the effects of recreating mixed tenure communities on levels of social cohesion; the importance of spatial scale in implementing social mix especially in avoiding conflict between different groups; the effects of rebalancing mix on existing residents' networks and social capital; the value of local schools as facilitating organisations for positive social interactions and the impact of length of time of residing in neighbourhoods.

## *Differing Expectations*

There seems little evidence to support policy makers' expectations that reordering social housing neighbourhoods to create more balanced social mix builds social cohesion. This contention appears refutable on the basis of reviews of both quantitative and qualitative research findings, at least from Australian, English and Dutch studies (van Kempen and Bolt 2009; Arthurson 2002, 2012). In quantitative studies tenure mix was not associated with reduced neighbourhood problems and income heterogeneity had a negative impact on sense of community. Likewise in studies evaluating social mix policies, residents reported that tenure differentiation leads to a decline in social cohesion (van Kempen and Bolt 2009). Baum and colleagues (2010) found that as neighbourhoods became more mixed or less homogeneous socio-economically, resident satisfaction declined. Residents appear to hold differing expectations to policy makers about neighbours and the types of relationships and benefits that might accrue in mixed tenure communities. When asked about their expectations in relation to social interaction residents generally focused on casual interactions. Owners and private renters expressed the view that there was little benefit from living in mixed tenure neighbourhoods whereas social housing tenants perceived various advantages such as security and higher self esteem (Chaskin and Joseph 2010, 2011). The findings of another study (Ziersch and Arthurson 2007) with residents of a mixed tenure neighbourhood in Adelaide suggested that housing tenure is relevant to the development of neighbourhood-based social capital and in turn social cohesion. While the majority of interviewees did not agree that differences divided the community, for some a mixed tenure community raised awareness of income and tenure differences in negative ways and private rental tenants fared worst on a number of the social capital indicators compared to other tenure groups. The findings challenged the continuing theme within the ideals set for mixed communities that they create greater feelings of acceptance and belonging between tenure groups, generating social capital and more cohesive neighbourhoods.

## *Spatial Scale*

It seems obvious that the spatial scale at which social mix is implemented within the neighbourhood is an important factor for facilitating social cohesion as it enlarges or decreases opportunities for cross-tenure contacts to occur. In general within the literature there is little evidence of social interaction occurring between different tenure groups as it is largely dependent on comparable social class proximity (Arthurson 2012). Nevertheless, a too fine-grained scale of social mix between different socioeconomic groups may lead to tension and conflict rather than the envisaged positive exchanges and social cohesion. This contention appears well founded specifically where differences in socioeconomic characteristics between residents are considerable (Chaskin and Joseph 2011). In some mixed-tenure neighbourhoods, owners and renters expressed the view that they did not mind living near each other but resistance increased exponentially as spatial proximity between residents of different housing tenures increased. Some tensions appeared due to different values and expectations about what constituted acceptable behaviours and lifestyles (Beekman et al. 2001). Other findings suggested that pepper potting different housing tenures within neighbourhoods does not lead to conflict. Nevertheless, in this particular study the extent of differences in socioeconomic characteristics between renters and homeowners was unclear and they may have been quite small (Jupp 1999). If this were the case, the findings could lend support to the inverse relationship identified in other studies of residents' socioeconomic diversity, heterogeneity of characteristics and the degree of social interaction.

## *Length of Residence*

Other literature identifies that length of residence in the neighbourhood is a critical factor in facilitating the bonding and bridging networks between residents that build trust and participation within neighbourhoods, which leads to social integration and social cohesion (Kleit and Carnegie 2011; Paranagamage et al. 2010). Findings such as these pose a challenge for urban regeneration projects that relocate long term residents to other neighbourhoods and where it is often assumed that balanced social mix is a prerequisite for social cohesion to exist. The research findings and academic debates suggest an alternative and often overlooked interpretation that strong cohesive communities already exist within what are perceived as homogeneous social housing estates (Tesoriero 2003).

## *Schools*

An interesting source of social cohesion identified in some studies of mixed-tenure neighbourhoods is the presence of children and the friendships they develop with other local children (Allen et al. 2005). Specifically in some studies social interaction was facilitated between different income groups when homeowners and renters sent their children to the same local schools (Jupp 1999; Atkinson and Kintrea 2000). This finding is, however, inconclusive, as other studies attain little evidence of social interaction taking place between different housing tenure groups, even when children attend the same schools (Beekman et al. 2001; Lees 2008). An insurmountable factor is that middle-income residents, for a variety of reasons, including judgements they make about the quality of local schools may choose to send their children to schools outside their immediate mixed-income neighbourhood. A British study of four mixed income developments found that families with previous ties to the neighbourhood were more likely to feel comfortable about sending their children to local schools than those without previous ties to the neighbourhood. In one of the developments in London where a new elementary school opened as part of the renewal project, families from across different income levels all reported that they were pleased with the school (Silverman et al. 2005). Thus, further investigation is warranted about the role good quality schools with a socioeconomic mix of students might play in the development of social cohesion.

## *Stigma*

A number of studies conclude that attempts to facilitate social cohesion through building socially diverse housing often have to deal with owner-occupier perceptions that poorer households are inherently bad neighbours (Bretherton and Pleace 2011; Arthurson 2012). Studies of estates with diversified tenure, find that owners are more likely to identify problems, such as inappropriate social behaviour, as due to social housing tenants. Economically and socially marginalised groups are viewed as an inherent 'threat' to neighbourhood social cohesion (Beekman et al. 2001). These findings raise questions about the ability of more vulnerable and chaotic households to become socially integrated within 'socially diverse' neighbourhoods where other residents both fear and seek to regulate their behaviour. This situation is not surprising given that individuals entering social housing are increasingly high need and complex tenants that without proper support and service provision may exhibit challenging behaviours for their neighbours.

## The Current Neighbourhood Study

One of the limitations of research about neighbourhood problems in the Australian context of urban studies lies in omission of widely applied integrated qualitative and quantitative approaches to advance understandings of intricate relationships and processes (Darcy 2007). With this discrepancy in mind, the current analysis initially draws on survey data but also utilises qualitative findings from in-depth interviews conducted across different housing tenure groups: social housing tenants, home owners (owned outright), homebuyers (owned with a mortgage) and private renters in three re-ordered mixed tenure estates that were previously dominated by high concentrations of social housing. Taking this approach informs the spectrum of residents' perceptions, knowledge and understandings of the dynamics of living in reordered mixed tenure communities and their views about whether the policies have effects on social cohesion.

The data collection for the research was conducted in three neighbourhoods located within the metropolitan region of Adelaide: Mitchell Park, Hillcrest and Northfield. Prior to urban renewal all three neighbourhoods were characterised by high levels of socio-economically disadvantaged households and concentrations of social housing. Post-regeneration the proportion of social housing was reduced by as much as 50 % and this also decreased the level of spatial concentration within the neighbourhoods, although the upgraded social housing was still often located in specific streets, or on one side of the street. The neighbourhoods were extensively revitalised over the previous 15–20 year period with changes made to the social mix through demolition and sales of social housing, urban infill of large, land lots and building of new housing for private sale to attract homebuyers into the neighbourhoods.

A questionnaire survey was mailed to a random sample of 800 households across the three case study neighbourhoods and 325 surveys were completed and returned. Respondents consisted of 117 males (37 %) and 199 females (63 %). After accounting for the non-deliverables (i.e. insufficient address; empty house, non residential, n = 78) the overall response rate was 45 %. Participants for the in-depth interview phase of the study were recruited through an expression of interest form that was included with the survey questionnaire. Sixty-five people returned the forms indicating their interest in participating in an interview. Forty interviews were conducted. Of these, 16 were classified as home owners living in homes that they either 'owned outright' or were 'owned with a mortgage', 14 lived in social housing and 10 were renting in the private sector. Interviews were recorded and the transcripts were collated by drawing together thematic issues in order to identify patterns, similarities and differences (Rice and Ezzy 1999).

## The Social Cohesion Scale

In the survey respondents were asked to complete a social cohesion scale[1] that required indicating their agreement with a series of nine statements about sense of community and social cohesion in the neighbourhood (on a scale of 1 strongly agree to 5, strongly disagree), as shown in Table 12.1.

As summarised in Table 12.1, private renters were least likely and social housing tenants most likely to agree with statements related to strong sense of community, close knit, and friendliness of neighbourhoods. Social housing tenants were least likely of the tenure groups to agree that people do not share the same values. None of these findings were significant. For responses to 'people can be trusted' a small significant association was found for tenure, with private renters the least likely to agree of the three groups (Chi-squared = 21.288, df = 6, n = 289, p = .002, Cramer's V = .192). A small significant association was also found between private renters (most likely to agree) and the other three tenures combined (12.9 % versus 3.1 % respectively) on 'people do not get along with each other' (Chi-squared = 7.203, df = 2, n = 292, p = .027, Cramer's V = .157). It is unclear why some private renters felt that people in the neighbourhood could not

---

[1]Questions 2,5,6,7,8, were from the social cohesion scale of Sampson and colleagues (1997) and the other questions were developed by Baum et al. (2009).

**Table 12.1** Summary findings of responses to social cohesion scale

| Level of agreement with statements | Social housing % (nos) | Private rental % (nos) | Owned outright % (nos) | Owned with mortgage % (nos) | Strength of association |
|---|---|---|---|---|---|
| Strong sense of community in neighbourhood | | | | | |
| Agree | 32.7 (16) | 12.9 (4) | 28.7 (33) | 25.5 (24) | Non Significant – social housing most likely & private rental least likely to agree |
| Neutral | 42.9 (12) | 67.7 (21) | 54.8 (63) | 57.4 (54) | |
| Disagree | 24.5 (12) | 19.4 (6) | 16.5 (19) | 17 (16) | |
| Neighbourhood is close knit | | | | | |
| Agree | 20 (10) | 9.7 (3) | 13.8 (16) | 13.7 (13) | Non Significant – social housing most likely & private rental least likely to agree |
| Neutral | 46 (23) | 51.6 (16) | 50.9 (59) | 53.7 (51) | |
| Disagree | 34 (17) | 38.7 (12) | 35.3 (41) | 32.6 (31) | |
| Have lots of friends | | | | | |
| Agree | 36 (18) | 12.9 (4) | 25.2 (29) | 22.3 (21) | Non Significant – social housing most likely & private rental least likely to agree |
| Neutral | 20 (10) | 25.8 (8) | 31.3 (36) | 22.3 (21) | |
| Disagree | 44 (22) | 61.3 (19) | 43.5 (50) | 55.3 (52) | |
| People are friendly | | | | | |
| Agree | 74.5 (38) | 58.1 (18) | 66.4 (77) | 67.7 (63) | Non Significant – social housing most likely & private rental least likely to agree |
| Neutral | 21.6 (11) | 32.3 (10) | 31.9 (37) | 30.1 (28) | |
| Disagree | 3.9 (2) | 9.7 (3) | 1.7 (2) | 2.2 (2) | |
| People do not share the same values | | | | | |
| Agree | 18 (9) | 32.3 (10) | 23.5 (27) | 26.9 (27) | Non significant – Private rental most likely & social housing least likely to agree |
| Neutral | 60 (30) | 58.1 (18) | 50.4 (58) | 50.5 (47) | |
| Disagree | 22 (11) | 9.7 (3) | 26.1 (30) | 22.6 (21) | |
| People do not get along | | | | | |
| Agree | 4 (2) | 12.9 (4) | 1.7 (2) | 4.2 (4) | Significant – Private rental most likely & owned outright least likely to agree |
| Neutral | 40 (20) | 35.5 (11) | 31.9 (37) | 31.6 (30) | |
| Disagree | 56 (28) | 51.6 (16) | 66.4 (77) | 64.2 (61) | |
| People can be trusted | | | | | |
| Agree | 28 (14) | 19.4 (6) | 49.1 (56) | 36.2 (34) | Significant – Owned outright most likely & private rental least likely to agree |
| Neutral | 60 (30) | 51.6 (16) | 44.7 (51) | 53.2 (50) | |
| Disagree | 12 (6) | 29 (9) | 6.1 (7) | 10.6 (10) | |
| People are willing to help neighbours | | | | | |
| Agree | 54 (27) | 32.3 (10) | 55.1 (65) | 51.1 (48) | Non significant – Owned outright most likely & private rental least likely to agree |
| Neutral | 34 (17) | 48.4 (15) | 35.6 (42) | 34 (32) | |
| Disagree | 12 (6) | 19.4 (6) | 14.9 (14) | 14.9 (14) | |
| People are tolerant of others | | | | | |
| Agree | 36 (18) | 35.5 (11) | 34.5 (39) | 42.6 (40) | Non significant – Owned with mortgage most likely & owned outright least likely to agree |
| Neutral | 50 (25) | 54.8 (17) | 57.5 (65) | 47.9 (45) | |
| Disagree | 14 (7) | 9.7 (3) | 8 (9) | 9.6 (9) | |

be trusted and did not get along as much as respondents in the other housing tenures. These measures were also analysed by length of time lived in the neighbourhood as this was identified as an important factor in the literature. A small significant association was found between time lived in neighbourhood and having friends. Residents of more than 10 years were more likely to agree with the statement 'I have lots of friends in the neighbourhood' than residents living there for less than 10 years (49 % versus 24 %: Chi-squared=24.622, df=2, n=301, p=<.0001, Cramer's V=.286). To explore these findings more fully and to try and gain a more nuanced understanding of whether there was a shared sense of social cohesion across residents of different housing tenure groups in the reordered mixed tenure neighbourhoods qualitative interviews were conducted.

## Qualitative Interview Findings

### *Sense of Community and Close-Knit Neighbourhood*

When respondents talked about sense of community and whether the neighbourhood was close-knit, for some an integration type of discourse was evident whereby they were positive about their mixed tenure neighbourhoods It was suggested that day-to-day life was mainly pleasant, a sense of community existed and neighbours and neighbourhoods were typically described by respondents across the four tenure groups as follows:

> Quiet, peaceful, no issue with neighbours, convenient to shops, close to city…..I'm really happy here (H108 private rental).

For others that were less positive about the neighbourhoods the decision to live there was merely based on accessing affordable rent as the view was expressed that 'it isn't a nice area' (H257 private rental).

Homeowners and social housing tenants commonly articulated a segregation type of discourse that resented the social mix in the neighbourhoods, which was depicted as working against social cohesion. Objections were raised about the increased mix of private renters and their frequent movements into and out of the neighbourhoods, which were described as making it difficult to develop a sense of community. One respondent summarised the common objections stating that the private rental house directly opposite their home was occupied by three different groups of people over the preceding two year period. "I see that the private rental people are not going to stay for very long. Like six months" (H7 owned with mortgage). These circumstances were described as detracting from opportunities to develop more stable or close knit communities comprised of longer term residents with a commitment to staying in the neighbourhood and working to enhance the sense of community. In part the problems stemmed from private landlords buying up much of the unimproved social housing and then renting it out on the private market without upgrading or improving it.

> Probably we have more trouble with the private rental ones, of the old transportable ones - one down the street here. We've had problems with various people who have been in there (H35 social housing).
>
> We have one next door [private rental] and they don't look after it, he couldn't care less (N161 owned outright).

A segregation discourse was again unequivocally expressed by some homeowners that disapproved more specifically with the presence of social housing, which like private rental tenure was depicted as working against the development of a sense of community in the mixed tenure neighbourhoods.

> I'm a little bit disappointed with the council and government who wanted to integrate [social] housing tenants amongst other normal, average down the road people. Well the idea might have sounded good but I don't think it's worked.....It makes it neighbourhood mediocrity. That's what you come out with that's the outcome because the people who are quiet and want to get on with their neighbours they become submissive to these people the way they behave. They're frightened of them. They might put a rock, like my neighbour who had a car tyre coming down the road into her bedroom (MP1 owned outright).
>
> Before there were all mainly the same types of people and now there are huge differences. Like, you're really poor and really wealthy. Not wealthy but much better off people and I think they don't mix (H7 owned with mortgage).

From the perspective of these groups of respondents social mix was perceived as an imposition that led to neighbourhood based problems and worked against social cohesion.

A different perspective from the foregoing segregation and integration discourses was expressed by other respondents, often retired or elderly residents representing the four housing tenure groups. These respondents were categorised as expressing a 'neutral' discourse; as they were basically sceptical about social mix policies, perceiving 'mix' as irrelevant to sense of community or closeness of residents. They described the realities of their circumstances as thus:

> I go to work every day early in the morning, I come home at night and I don't connect with my neighbours. So their whole theory around social mix is that it is meant to help people, but that is not necessarily happening in society today (N204, private rental tenant).

## *Friendships and Friendliness Within the Neighbourhood*

When respondents talked about friendliness within the neighbourhoods a similar neutral discourse emerged about the demands of modern lifestyles rendering social mix less relevant for day-to-day life. Whether they were in the same or a different housing tenure or socio-economic group was identified of little importance as people were too busy for mixing with neighbours. Typical responses were that:

> Over the time we've been here, it's become less friendly. But that seems to be the way of most neighbourhoods now, because neighbours just don't talk to each other. They're too busy, they have insufficient time or they're not interested ... and I think both parties are working and they don't get a lot of time (N6, owned outright).

For other respondents the age mix of residents was considered important. They pointed out that with implementation of the new mixed communities, not only were more home buyers moving into the neighbourhoods but there was a broader age mix of residents. It was argued that these diverse intergenerational characteristics impacted on opportunities to develop friendships. While many longer term residents were elderly and welcomed others into the neighbourhoods it was felt that different age groups led divergent lifestyles, which often precluded them from coming into contact with each other. As the following respondents explain:

> Some of them [neighbours] are young, some are middle-aged, some older. There's quite a mixture that are buying here. There were some young people living up the road on the right hand side, but you don't really see much of them, because they're young and they're working hard ... And then you've got elderly people living in these houses that don't get out much of the time (H2R, social housing tenant).
>
> The neighbourhood is not friendly ... because of the social mix. You've either got the really elderly people who are friendly, or you've got your cautious young couples with kids. Because of the really awful things happening with paedophiles and stuff like that, I think people are really sheltered and they hold on to kids and don't let them out of their sight. So it's not so friendly with the young (H7, owned with mortgage).

This aspect about the age mix of residents has received little attention in the literature on mixed communities or in policy discourses.

## *Shared Values, Getting Along with Each Other and Tolerating Differences*

Some of the biggest tensions that arose in interviews were about what were perceived as insurmountable differences between neighbours' values and standards of behaviour. Longer term social housing tenants (often elderly), expressed a segregation type of discourse about the present-day, complex, high need tenants entering social housing. They depicted the difficult day-to-day situations regularly encountered from these newer tenants, often their immediate neighbours:

> Oh, the language, they used to swear like anything and it was terrible, you could hear them, all the kiddies ... She couldn't care less but they were terrible children. Amazing how they get these homes, people like that (MP6, social housing tenant).
>
> The Housing Trust built them a brand new house and put them in with shutters and they have just destroyed it and they go to school and they destroy things there…Those sort of people annoy me, because they have been given a brand new beautiful house in a beautiful area and they have destroyed that whole area. Ruined it! Two people have sold their house and moved because of them…..They're just foul mouthed disgusting animals is what they're called by every neighbour that lives around them (H2R social housing).

Importantly, what this situation illustrates is that social housing tenants are not one homogenous or socially cohesive group as often depicted by the rationales underlying social mix policies. There was clearly social distancing within the social housing tenure from respondents such as these based on their perceptions of what constitutes either a 'good' or 'bad' social housing tenant.

In comparison, home owners that expressed little tolerance of social housing tenants made no such distinction portraying them as an undifferentiated and problematic tenure. This uniform image was reflected through drawing on a segregation discourse that portrayed social housing tenants as very different from themselves with undesirable and typically opposing values:

> We knew about that, we knew it had a housing trust, unfortunately a lower quality of life if you like to put it that way, and coz over the last ten years that's changed quite a bit. A lot of older people have moved into the area on the courtyard blocks. So obviously a lot of new houses are here but it will probably be another twenty years before there is a real change (MP1 owned outright).
>
> I don't think that the groups interact well. Not at all that's what I think. I don't think the low income tenants mind that there are nicer newer homes being built around the corner but I think that the high income earners wish that the housing trust units and stuff would get knocked down quicker and that more home owners would be placed here…Sometimes I think that the low income earners are kind of not happy to be poorer. But it doesn't look like they're trying to do anything to improve their situation. They don't seem to try and better themselves (H7 owned with mortgage).

Other social housing respondents and home owners utilised an integration discourse that depicted propinquity in space of different tenure groups as beneficial for all. This was expressed in terms of building awareness of similarities between groups and building up tolerance of alternative ways of life.

> It's beneficial for the kids, for everybody growing up in the area, it's more social, you meet different people in life. You get to learn respect and to value other people's opinions and property. It is a different setup and I think its working for the best. I think they should have done it a long time ago. ….No matter whether you're a Housing Trust tenant or not and 'cause I certainly don't like living next to someone that has got cars all in the front yard and crap around the backyard and it's stinking. I think it's disgusting' (H2R social housing).
>
> People need variety to start with. Where you have got areas with all public housing tenants you have got everybody's on a low income which is why they are in public housing for whatever reason and it's really easy to be demoralised by that (MP 192 social housing).
>
> For some reason or other housing trust people don't seem to have a good name and yet the people that I know here were very nice people…. Through no fault of her own, why she's in a housing trust home. She chose to leave her husband and not take anything so she had nothing. That's how she ended up being in housing trust. She's got a good education and she always keeps her house nice and tidy (H40 owned with mortgage).

## *Children and Schools*

Respondents also employed an integration discourse identifying children attending local schools as facilitating positive interactions with others from across different tenure groups, an argument that is consistent with the literature (Allen et al. 2005; Atkinson and Kintrea 2000; Holmes 2006; Jupp 1999).

> I think people tend to mix with all of the neighbours if your kids are perhaps going to kindergarten or school. That's where we got to know more people (MP45 owned outright).

> Being in the school, we have all the mixture of the Hillcrest community in there. You've got all walks of community. You've just got to go in the car park in the morning and its mind-boggling the cars you see there. There will be a sports car or there will be a beautiful four-wheel-drive and you think, my God, it's just amazing that all these different people live in this area and they all go to this one public school and they play together as one, or try to (H2R, social housing tenant).
>
> When the girls were at the primary school, and that's just down the road, there'd be lots of single mothers bringing their kids to school who lived a matter of streets away. And myself, we own our home, you've got a lot of single parents, and we all talk and get on well together (H98, owned outright).

However, a mitigating factor is that some homeowners, based on judgements made about the quality of local schools, purposely decided to send their children to schools outside of the neighbourhood:

> And unfortunately with that type of housing [social housing] I think still comes those sort of people I guess, the low income. My children don't go to the local school because of that. It was a violent school (N49 owned outright).

## *Willingness to Help Neighbours and Trust*

In relation to trust and willingness to help neighbours respondents from across the tenure groups mentioned that although often they did not have a lot to do with neighbours they knew if they needed help then "they are available" (H108 Private Rental). Other respondents provided instances of how neighbours provided unexpected but nevertheless welcomed support:

> And you know if they haven't seen us around for awhile he comes knocking on the door, you know just seeing if you're okay (MP 2 social housing)
>
> The old lady across the road doesn't speak English but brings over cakes (MP7 owned outright).
>
> Oh, I keep to myself a lot...But he calls around [neighbour] to see how I'm going (N56 social housing)

## Conclusions

It was not possible to draw comprehensive conclusions from the current study findings at least in relation to changes that may have occurred in the levels of social cohesion over time due to reordering of social mix in the neighbourhoods. The data was limited given that it was not longitudinal and did not include a before and after study of the three neighbourhoods. In spite of this, the chapter highlighted some of the processes, complexities and challenges for policy makers recreating areas of concentrated social housing into mixed tenure neighbourhoods with expectations of reducing social segregation and increasing social cohesion. In particular it showed that homogeneous social housing neighbourhoods do not have exclusive rights to 'neighbourhood based problems' with similar issues experienced in the reordered mixed tenure communities.

One group of residents expressed an integration discourse which described the heterogeneous social mix, incorporating the diversity of housing tenures in positive terms, depicting it as an important way for different tenure and income groups to integrate. They recognised that particular life experiences meant that through no fault of their own people could end up in social housing. Another group of respondents (predominately homeowners) espoused a segregation discourse that associated the presence of social housing with neighbourhood based problems, such as anti-social behaviour, and viewed it as a threat to social cohesion. The case was made at the start of this chapter that US style 'neighbourhood effects' policies are less relevant in Australia due to the different context to the US. However, it is interesting to note that this group of residents utilised a discourse that was akin to the 'culture of poverty' thesis espoused by Murray. In effect social housing tenants were depicted as a homogeneous problematic group. In comparison social housing tenants clearly argued that there were 'good' and 'bad' tenants, the latter characterizations referring mainly to new entrants. This is not surprising given that individuals entering social housing are increasingly high need and complex tenants and suggests that in the immediate term the lack of acceptance by some homeowners is likely to increase rather than dissipate.

An interesting finding was that in some instances the segregation discourse was also linked to the increased mix of private renters. Respondents objected to their high turnover, noting that houses were often not well maintained as they functioned merely to obtain rental income for absentee landlords. These aspects were associated with a lack of social cohesion in the regenerated neighbourhoods. Does this situation help to explain the survey findings whereby some private renters expressed less trust and were less positive about getting along with neighbours than residents in other housing tenures? Housing affordability issues mean entry to homeownership is increasingly delayed and in turn this situation is reflected in people renting for longer than previously. In Australia homeownership is part of national pride and a common aspiration. Perhaps private renters are disengaged from the neighbourhoods as they may perceive their situation as semi-permanent. This is supported by the finding that longer term residents were more likely to agree that they have lots of friends. Do private renters perhaps feel marginalised within these neighbourhoods and internalise some of the stigma attached to their tenure by other residents? Other research findings suggest residents do not want to be considered part of a minority group in their neighbourhood (Permentier et al. 2009). The findings raise serious questions as the balance of housing assistance in Australia is moving to favour provision of subsidies for private rental assistance, and affordable rental housing funded through private landlords as opposed to social housing predominately supplied through government.

A key finding in interviews was that residents often talked about social mix in terms of age of neighbourhood residents. Elderly and longer term residents were generally identified as more stable groups that generated a sense of community and neighbourliness leading to social cohesion. The presence of children was conducive to enhancing the sense of community and friendliness especially through local schools, which were identified as important forums for positive social interactions between different groups of people. These aspects for dealing with some neighbourhood based problems have received only limited attention in wider debates about mixed tenure communities and social cohesion.

Overall the findings raise the question of whether social mix is an outdated idea as a means for addressing neighbourhood based problems. First, widespread use of motor vehicles and new social networking technologies, such as Facebook mean local neighbourhood is less relevant for many citizens that do not rely on the geographical space of their neighbourhood or place of residence for conducting their working or social lives (Cass et al. 2005). Second, a policy conundrum exists between housing authorities' policies that attempt to reorder social mix on homogeneous estates while simultaneously constricting access to social housing to high need groups. In tandem, the policies have contradictory purposes. The gradual reversal over time of the eligibility criteria to house only those in greatest need means that low-income, working families that were housed in the past are almost certainly assured they will not currently get housed in social housing. This is because their need is not perceived as urgent relative to other groups, including those that are homeless, or with substance abuse problems, mental health issues and exiting prison. However, this situation ensures that there is not a social mix within the social housing tenure, at least in terms of socioeconomic mix and makes addressing neighbourhood based problems and enabling social cohesion objectives more difficult.

## References

Allen, M., Camina, M., Casey, R., Coward, S., & Wood, M. (2005). *Mixed tenure, twenty years on – nothing out of the ordinary*. York: Joseph Rowntree Foundation.
Arthurson, K. (2002). Creating inclusive communities through balancing social mix: A critical relationship or tenuous link? *Urban Policy and Research, 20*(3), 245–261.
Arthurson, K. (2003). Whole of government models of neighbourhood regeneration: The way forward? *Just Policy, 29*, 26–35.
Arthurson, K. (2004). From stigma to demolition: Australian debates about housing and social exclusion. *Journal of Housing and the Built Environment, 19*(3), 255–270.
Arthurson, K. (2012). *Social mix and the cities: Challenging the mixed communities Consensus in Housing and Urban Planning Policies*. Melbourne: Sustainable Cities Series, CSIRO Publishing.
Atkinson, A., & Kintrea, K. (2000). Owner occupation, social mix and neighbourhood impacts. *Policy and Politics, 28*(2004), 93–108.
Baum, F., Ziersch, A., Zhang, G., & Osborne, K. (2009). Do perceived neighbourhood cohesion and safety contribute to neighbourhood differences in health? *Health & Place, 15*(4), 925–934.
Baum, S., Arthurson, K., & Rickson, K. (2010). Happy people in mixed-up places: The association between the degree and type of local socio-economic mix and expressions of neighbourhood satisfaction. *Urban Studies, 4*(3), 467–485.
Beekman, T., Lyons, F., & Scott, J. (2001). *Improving the understanding of the influence of owner occupiers in mixed tenure neighbourhoods*. Edinburgh: ODS Ltd for Scottish Homes.
Bretherton, J., & Pleace, N. (2011). A difficult mix: Issues in achieving socioeconomic diversity in deprived UK neighbourhoods. *Urban Studies, 48*(16), 3433–3447.

Cass, N., Shove, E., & Urry, J. (2005). Social exclusion, mobility and access. *The Sociological Review, 53*(3), 539–555.
Chaskin, R. J., & Joseph, M. L. (2010). Building "community" in mixed-income developments. *Urban Affairs Review, 45*(3), 299–335.
Chaskin, R. J., & Joseph, M. L. (2011). Social Interaction in mixed-income developments: Relational expectations and emerging reality. *Journal of Urban Affairs, 33*(2), 209–237.
Darcy, M. (2007). Place and disadvantage: The need for reflexive epistemology in spatial social science. *Urban Policy and Research, 25*(3), 347–361.
Friedrichs, J. (2002). Response: Contrasting US and European findings on poverty neighbourhoods. *Housing Studies, 17*(1), 101–104.
Fulcher, H. (1999). *Determining priorities for urban/community renewal: Responding to social exclusion in public housing.* Sydney: National Housing Conference.
Holmes, C. (2006). *Mixed communities, success and sustainability.* York: Joseph Rowntree Foundation.
Jupp, B. (1999). *Living together: Community life on mixed housing estates.* London: Demos.
Kearns, A., & Forrest, R. (2000). Social cohesion and multi-level urban governance. *Urban Studies, 37*(5–6), 995–1017.
Kleit, R. G., & Carnegie, N. B. (2011). Integrated or isolated? The impact of public housing redevelopment on social network homophily. *Social Networks, 33*(2), 152–165.
Lees, L. (2008). Gentrification and social mixing: Towards an inclusive urban renaissance? *Urban Studies, 45*(12), 2449–2470.
Murray, C. (1994). *Underclass: The crisis deepens.* London: Institute of Economic Affairs.
NSW Minister for Urban Affairs and Planning and Minister for Housing (1997, July 17). *News Release, Crime hot spot to be demolished.* Sydney.
NSW Minister for Urban Affairs and Planning and Minister for Housing (1998). *News Release, New era for troubled East Fairfield housing estate*, http://www.housing.nsw.gov.au/plastvi.html. Accessed 6 Feb 2008.
Paranagamage, P., Austin, S., Price, A., & Khandokar, F. (2010). Social capital in action in urban environments: An intersection of theory, research and practice literature. *Journal of Urbanism: International Research on Placemaking and Urban Sustainability, 3*(3), 231–252.
Permentier, M., van Ham, M., & Bolt, G. (2009). Neighbourhood reputation and intention to leave the neighbourhood. *Environment and Planning A, 41*(9), 2162–2180.
Proctor, I. (1995). *Priorities for redevelopment.* Correspondence to the Department of Housing and Regional development. Adelaide: SA.
Rice, P., & Ezzy, D. (1999). *Qualitative research methods: A health focus.* Melbourne: Oxford University Press.
SA Better Cities 11 Steering Committee. (1995). *Report of the SA Better Cities 11 Steering Committee Northern Adelaide Urban Renewal Study.* Adelaide: Department of Housing and Urban Development.
Sampson, R., Raudenbush, S., & Earls, F. (1997). Neighbourhoods and violent crime: A multilevel study of collective efficacy. *Science, 277*(5328), 918–924.
Silverman, E., Lupton, R., & Fenton, A. (2005). *A good place for children? Attracting and retaining families in inner urban mixed income communities.* London/York: Chartered Institute of Housing, Joseph Rowntree Foundation.
Tesoriero, F. (2003). Housing renewal in the Parks community, South Australia. In W. Weeks, J. Dixon, & L. Hoatson (Eds.), *Community practices in Australia.* Amsterdam: Pearson Education.
van Kempen, R., & Bolt, G. (2009). Social cohesion, social mix, and urban policies in the Netherlands. *Journal of Housing and the Built Environment, 24*(4), 457–475.
Wilson, W. J. (1991). Studying inner-city social dislocations: The challenge of public agenda research. *American Sociological Review, 56*(1), 1.
Wood, M. (2003). A balancing act? Tenure diversification in Australia and the UK. *Urban Policy and Research, 21*(1), 45–56.
Ziersch, A., & Arthurson, K. (2007). Social capital and housing tenure in an Adelaide neighbourhood. *Urban Policy and Research, 25*(4), 409–432.

# Chapter 13
# Neighbourhoods: Evolving Ideas, Evidence and Changing Policies

**Duncan Maclennan**

## From Simple Certainties to New Scepticisms

This is the concluding chapter in a series of volumes (van Ham et al. 2012, 2013; and the current volume; Manley et al. 2013) that has focussed on the meaning, specification and estimation of neighbourhood effects. The volumes have brought together a considerable array of conceptual thinking and empirical evidence that demonstrates the very positive discovery processes in quantitative and qualitative neighbourhood effects research since the mid 1990s. This development of knowledge has been particularly important in Europe where neighbourhood research had lagged well behind North America. In the UK, for instance, the Rowntree Foundation Regeneration Research Programme of the late 1990s contained not one project on quantitative estimation of neighbourhood interactions and, aside from the important contribution of Buck (2001), the 1996–2001 ESRC Cities Programme ignored econometric analysis of neighbourhood related data.

Throughout these volumes some strong, and different, conclusions have been drawn about the source, scale and measurement of neighbourhood effects. There has been progress but without consensus. The salience of these findings for the relevance and structure of neighbourhood policies has also been widely argued and, again, quite different conclusions drawn about the desired relative emphases of 'people' and 'place' dimensions in policy making. Some contributions, in some places, suggest that policy design should reflect an assumption of significant neighbourhood effects, for instance Galster (2012) and Lupton and Kneale (2012). Others have argued that in some circumstances neighbourhood effects cannot be established and have minimal policy relevance, for instance Manley and van Ham (2012a) and Cheshire (2012).

---

D. Maclennan (✉)
Centre for Housing Research, School of Geography and Geosciences, University of St Andrews, St Andrews, Fife KY16 9AL, Scotland, UK
e-mail: dm103@st-andrews.ac.uk

Clearly there are no easy generalisations to be made about the likely empirical significance of neighbourhood effects and their relevance to policy making for specific places. However each of the contributors here, and in the literature more widely (for contrasting examples see Lupton (2003) and Oreopoulos (2003)) tend to assume that their own particular findings and perspective will influence policy.

This concluding chapter, based in the author's experience as a policymaker in the UK, Australia and Canada, argues rather differently.[1] The recent renaissance of neighbourhoods effects research has coincided with a period when a new emphasis in public management argued for increased bureaucratic interest in, and funding of, evidence bases for policymaking (Nutley et al. 2000). That is, many researchers have come to expect at least a closer relationship between well founded and clearly disseminated evidence and changes in policy-making. This chapter, whilst recognising the intellectual gains of the decade, takes a very different view about how neighbourhood research has actually influenced policies.

The chapter also concludes that the research findings on neighbourhood effects, even where policy responds quickly to evidence, would have been unlikely to persuade governments to abandon neighbourhood renewal programmes. There are three reasons why. First, there are, as noted above, still significant disagreements about the empirical evidence that exists, not least when different approaches to identifying neighbourhood effects are recognised, and about how to summarise the relevant policy findings and their universality.

Secondly, 'neighbourhood effects' are not the only substantive research question about neighbourhoods that lead to policy intervention. Neighbourhood choice and change processes and their outcomes are also of significance, as indicated in the second volume of this series. The problematic of neighbourhoods is not just a sum of place based externalities or spillover effects but is also obviously a direct reflection of the multi-faceted poverty of residents. It is argued below that major commitments made to neighbourhood renewal programmes in many European countries since the mid-1990s did not stem from a belief in evidence of 'neighbourhood effects' but reflected quite different aspects of the new public management, in particular the importance of integrated and preventative approaches in public services and growing roles for community voice in provision. It is argued below that these emerging management and governance approaches were more important in the 'ordinary knowledge' (Hardin 2009) of policymakers than even the best estimates of neighbourhood effects.

Thirdly, the theoretical and empirical issues involved in defining, identifying and changing neighbourhoods are extremely complex and, as noted above, heavily contested. Policymakers can often hear a cacophony rather than a concerto of ideas from the research community. It is argued below that conceptual developments in

---

[1] From 1999 until 2003 the author was Special Adviser to the First Minister of Scotland (with responsibilities for housing, neighbourhood and city policies), then 2003–2004 was Chief Economist (DSE) and Deputy Secretary for Policy and Strategy in the Government of Victoria, Australia, and from 2004 until 2008 was Chief Economist in the Canadian Federal Government Department for Cities and Communities (Infrastructure Canada).

neighbourhood policy thinking within geography, economics and planning are sufficiently inconsistent and inconclusive that they provide little clear guidance for policymakers. In consequence major changes in neighbourhood policy change can take place without what is regarded as evidence or indeed fail to change given new evidence. Put bluntly, this area of research remains sufficiently contested that new research evidence is neither necessary nor sufficient to induce place policy change. To believe otherwise is simply to ignore the observed models of innovation, change and implementation within bureaucracies.

The remainder of this chapter makes the case that neighbourhood research often does not have a coherent story to tell to policymakers. The "Neighbourhoods: The Ultimate in Fuzzy Theory?" section of the paper briefly reviews major developments in 'neighbourhoods' theory in recent decades and highlights how 'neighbourhood effects' ideas are located within what is still a relatively loose, or fuzzy, and often ungrounded set of theoretical frameworks. Section "The New Scepticism, and a Critique" highlights different kinds of evidence available to shape decisions and the real sparsity of settled evidence on most key issues that arise in making neighbourhood policies. The section "A Dissenting View About Policy Impacts!", then considers the theoretical or in-principle cases that exist for pursuing neighbourhood policies, essentially to highlight policy cases that do not require neighbourhood effects as a basis for place policy and suggests ways of classifying policy approaches. The penultimate section "Neighbourhood Research and Policy: Continuing Commitments", then considers the range of influences shaping neighbourhoods policies in a selection of countries in the last decade. A brief conclusion on research to policy connections then follows.

## Neighbourhoods: The Ultimate in Fuzzy Theory?

### *Black Boxes as Frameworks*

Efforts to measure neighbourhood effects, and a very useful review of the theoretical ideas involved and challenges in estimation can be found in Hedman and van Ham (2012), have had the benefit of producing a clear definition of what they are. In essence, neighbourhood effects are the consequences of spillovers and externalities that arise from the co-location or proximities of particular socio-economic groups or activities. However the first of the volumes in this series illustrated only too well how the focus of research on the ex post analysis of various wellbeing outcomes associated with neighbourhood concentrations of particular socio-economic groups had progressed without much effort to specify the mechanisms by which neighbourhood and selection effects actually work. Galster (2012) and Small and Feldman (2012) have stimulated a new interest in establishing the transmission mechanisms by which neighbourhood effects work. Otherwise, underpinning the increasingly clever econometric analysis of concentrations and outcomes there is a very extensive and important 'black box' of social and economic processes and cause to effect linkages. And these considerations may be important in the design of policy initiatives.

A central problem in constructing a convincing rhetoric for, or indeed against, neighbourhood policies that are based upon neighbourhood effects research is not just the absence of well-argued transmission mechanisms. It is, rather, that there is a whole series of 'black boxes' that still comprise the foundations for so much of neighbourhood research.

When problems are identified and become of interest to applied researchers in the social sciences much time is spent on developing plausible definitions, typologies of possible change mechanisms and outcomes are developed and theoretical ideas are advanced. This 'framing' process is important in and quite typical of applied social science research approaches and it leads, over time, to some new, relatively widely agreed understandings.

Arguably, progress in neighbourhood research has been different. Creative imagination within the academy in the framing of ideas is often unmatched by research effort in the field. Whilst there has been a substantial investment in better georeferenced outcome data in the UK and other countries there has been a general unwillingness by social science research funders to undertake the expensive data collection that would be required to understand the multi-sector, space-time systems that are neighbourhoods. Few governments have shown much regard for the design of data systems that are likely to capture the complex space-time pattern of impacts of the programmes they have espoused (Moving to Opportunity evaluation that has so influenced this debate is a rare exception). In consequence there is little testing of the relevance of different assumptions and models. There are few areas of empirical work on neighbourhood choices, changes and effects that are relatively uncontested so that, in contrast to many other areas of research, there are few stylised facts that will attract wide support from empirical researchers.

The proposition that neighbourhood research is a field of competing ideas weakly underpinned by a stratum of inadequate empirical evidence can be supported by examining the conclusions of decades of research on what are key aspects of neighbourhoods for socio-economic policy. Policy analysis is interested in how neighbourhoods are identified, and by implication previously defined, how they are chosen by different socio economic groups, how they change over time and the consequences of neighbourhood outcomes not just for the individuals they contain but for the functioning of wider metropolitan areas. Pattern, choice, change and consequences are critical considerations in making policy. What stylised facts about neighbourhoods can guide place policy making?

## *Research on Neighbourhoods, Uneven Patterns*

A rich tradition of research on neighbourhoods has prevailed in the USA since the 1930s. It has, until the last decade, been largely unmatched in the UK and much of

Europe. In the UK urban geography moved from a brief flirtation with factorial ecology or urban social structure studies in the 1960s (Robson 1969) to a Marxian view that argued against 'spatial fetishism' (or being too concerned with local place patterns), see Harvey (1973), Murie and Forrest (1980), (Slater 2013) through to a post-structural deconstructionist 'cultural' view of neighbourhoods and policies.

All these approaches produced interesting, theoretical insights but did little to produce coherent understandings of the functioning and dynamics of individual neighbourhoods let alone wider metropolitan systems of neighbourhoods. Above all they produced a pattern of research and knowledge in urban geography that was somewhat pathological, with the bulk of research focussed upon gentrification and, to a lesser extent, deteriorating social housing neighbourhoods, rather than focussing on overall patterns of neighbourhood change within metropolitan areas. Arguably 'upward succession' (Grigsby et al. 1987) has been the dominant neighbourhood change process in the UK for the last 30 years but is has attracted little research interest.

In other disciplines, such as social policy and planning, there has been little attention to questions of neighbourhood functioning. In housing research, where much of the European impetus for research on neighbourhoods has emanated, the dominant concern (until 2000 or so) was on defining and estimating numbers of needed homes and identifying disadvantaged areas rather than on neighbourhood choice and change (see Maclennan 2000). Until the 1990s these research emphases led to a disregard for the significance of neighbourhoods in core areas of policy.

## *Defining Neighbourhoods*

It has been American research that has led discussion of the identification of neighbourhoods and their functions. That tradition evolved from both the social ecologists of the Chicago school and the tradition of applied economics that ran through 'land economics', as reflected for instance in Hoyt's work (1939) and Grigsby's early work on 'filtering' models (Grigsby 1963). Work by Hunter (1979) stressed the different meanings of neighbourhood for different socio-economic groups, Suttles (1972) indicated that neighbourhoods operated at different spatial scales for different groups and functions and by the 1970s Downs and colleagues (1973) emphasised the perceptual influences on the identification of neighbourhoods.

More recently Galster (1986) brought together many of these, and other novel, different ideas into a coherent 'field theory' notion of neighbourhood. It allows perceptions to matter. It recognises that different 'fields' will apply to different household activities for the same household and that different households will have different field patterns around their homes.

This 'field' based notion of neighbourhood recognises that, around the home location, different households may lead more or less local lives and they have weaker and stronger interactions within the neighbourhood and to different parts of cities. This has a clear carryover into ideas of social space, where patterns of interaction between individuals can be used to identify community and weak and strong ties (Butler and Lees 2006). Galster's (1986, 2008) field theory definition can capture social interactions but also the wider activities that households undertake, engaging the economy and impacting the environment, for example (see Galster 2001). Clearly, to identify a neighbourhood, as opposed to the space-time activity pattern, or 'field' of one individual, some appropriate aggregation algorithm has to be developed.

This is a logical and inclusive framework for identifying the structure of neighbourhoods within a place. It stresses the functional nature of places. And it also allows that some households will have more place or home-based lives than others and that we do not need to be drawn into a false dichotomy that urban life is place or non-place based. Equally it should caution us that in some locations dwelling choices may be made with little reference to neighbourhood interactions (if those household groups with non-place based lifestyles are more likely, for other reasons, to be located in some neighbourhoods rather than others) but that in others they will be critical. But is this fine grain approach to identifying neighbourhoods ever used as a basis for empirical research?

The reality is that almost all government evaluation studies and the vast bulk of academic econometric analyses of neighbourhood choices and effects do not commence with a prior careful identification of neighbourhood. A clear example of this can be seen in Lebel and colleagues (2007) who, in progressing a study of health and neighbourhoods in Montreal, outline major theoretical concerns about neighbourhoods and the range of ideas for identifying them. They then use, or are simply forced to use, the small administrative areas available to them. This approach applies to almost all work on disadvantaged areas in the UK. And this approach has been apparent in many of the contributions in the volumes of this series.

How we identify areas is not incidental to how we assess neighbourhood effects. What we can identify is greatly conditioned by the scale and sensitivity of how we define and identify the phenomena we claim to be investigating. Failure to do so can see much data and fine statistics applied to approaches that have already thrown away potential babies with the bathwater. This point is well illustrated by research on the neighbourhood effects that may or may not arise from mixing incomes and tenures. Galster (2007) makes it very clear that he believes that such effects will be contingent upon the context, who is being mixed with whom, the mixing process, and the scale of any policy measures. That is, empirically testing this complex idea requires careful nuanced specification of neighbourhoods and areas of action. Few studies meet such requirements but then still draw quite major empirical and policy conclusions. There is a danger of estimation technique dominating problem specification in the way neighbourhood research is now evolving.

Household activity pattern analyses of the kind pioneered by Hägerstrand (1988) could have not only revolutionised planning for housing and transport (where the ideas have had some salience) but also shaped the ways neighbourhoods have been researched and place programmes constructed and delivered. But it has not because research funders have never had the patience nor the funds to pursue such work on a real basis. In consequence our research analyses and policy actions invariably start from an implied functional geography that is crudely identified from the formal boundaries of statistical census areas or individual data. This is akin to asking surgeons to operate without X rays, let alone MRI's. Too often we compromise our research from the outset.

## *Choosing Neighbourhoods*

Even if we were confronted with a set of adequately identified, functional neighbourhoods would we have a clear understanding of how households would choose them? This question is important in neighbourhood effects research because we need to simultaneously understand and separate neighbourhood and selection effects (see Hedman 2011; van Ham and Manley 2010, 2012; Manley and van Ham 2011). Moreover, the neighbourhoods that households choose are important policy concerns *per se*. As lifestyles and incomes change it is important, for instance, for housing planners to ensure a future supply of not just appropriate homes but preferred neighbourhoods too. And as most individuals choose neighbourhoods in market systems it is important to understand whether there are market failures in choices processes or whether outcome patterns create any wider externality effects (Maclennan 2012). Are there widely agreed, or stylised, facts about the processes and substance of neighbourhood choices?

Housing and neighbourhood choices have been subject to at least half a century of analysis with continuing interest in recent decades. In the sociological tradition, contributions such as by Mallett (2004) have explored the potential meanings of home and neighbourhood. These contributions have helped frame useful qualitative research contributions on how people choose and use homes and places (Forrest and Kearns 2001). Qualitative research contributions, often based on small localised samples of interviews, are important to research for policy in that they caution against imposing conventional wisdoms ex ante on decision taking and they are constant reminders of the complexity of the issues involved. In the competitive resource allocation decisions that confront policymakers, however, it is often difficult to use limited sample studies as a basis for decision taking. Such decisions often require simpler explanations and larger samples.

Qualitative research points up the complex influences that shape choices and the multiple aspects of household's lives impacted by housing and location decisions.

This is a position that sits easily with standard applied economics approaches to housing choices, and Galster has embedded much of this approach in indicating influences on neighbourhood choices (Galster 2001). Since the 1970s the common approach of housing economics, and neighbourhood choice economics, has been to recognise that choices of home and neighbourhood are inevitably conjoined (Segal 1979). Further, both homes and neighbourhoods are recognised as having multiple attributes (Maclennan 1982). The joint choice of complex, multiple housing and neighbourhood attributes contribute to a wide range of household consumption and, for owners, asset choices. Housing choice and demand studies have identified a range of value influencing characteristics (or attributes) that can be grouped together as:

- **Property** Features: dwelling size, type, quality and quality/amenity of structures
- **Place** Patterns: social interactions, including nature of neighbours, neighbour contact, range and intensity of social networks and environmental features, including exposure to unpriced externalities (positive and negative)
- Locational **Proximities**: ease of space/time access to local facilities, including private activity centres and public service locations as well as connections to activity sites in wider city (job location, major service centres etc.)
- Public service **Provision**: quality and range of provision, costs and governance (including neighbourhood voice capabilities) provided within/for the neighbourhood
- **Perceived** Image, status, reputation

Hedonic price approaches, as set out in Malpezzi (2002) confirm that these factors have significant impacts in many house price studies.[2] Arguably neighbourhood effects research needs to pay much more attention to which housing and neighbourhood attributes, or interacting attributes, impact the wellbeing outcomes of different consumer groups. If housing/neighbourhood choices have effects on, for example, health, education and employability it is important to establish, and separately, whether this is due to property, place, proximity, provision or perception influences, and few studies attempt to do this.

Attribute choices can be examined together with the socio-economic characteristics of households to identify demand functions for characteristics and the significance of household constraints in 'selection effects'. Estimation of hedonic price functions (that suggest, broadly, that neighbourhood attributes commonly explain between a fifth and a third of dwelling prices) is relatively commonplace in housing economics, see Malpezzi (2003). However there has been much less attention to the estimation of individual attribute demand functions. More than three decades ago Segal (1979) highlighted the possibility of estimating the demand functions for different housing and neighbourhood attributes. Few follow-up studies exist for neighbourhood attributes. In short, neighbourhood research has neglected systematically addressing the quite important question of how neighbourhood choices change as demand side features, such as household size, income and wealth change. In the UK there is not one contemporary published estimate of the income elasticity of demand for any well-defined and widely accepted neighbourhood attribute.

This omission is important. Research for housing policy and planning, for instance, invariably focusses on gross totals of households likely to locate in an area. But the real challenge as family structures, real incomes and energy prices change is to model better, then match better, the number of dwelling and neighbourhood types that

---

[2] Hedonic estimates of the (unobserved or implicit) prices of particular dwelling attributes, such as number of rooms or the presence or absence of a garage, are derived by regression analysis that estimates observed housing prices or rents as a function of the observed characteristics of a set of dwellings.

households seek. Such information, and the sense of what kind of neighbourhoods households both want and can have an effective demand for, is largely missing in policies to promote both urban renewal and new suburban development. Once again a potential Rolls-Royce of analytical techniques and possibilities is left in the garage, with increasingly expensive fuel, as we claim to understand neighbourhood choices.

Cheshire and Sheppard's impressive analysis of the effect of school quality on housing values (2005) illustrates how attributes of housing and place choices can be examined to reveal key influences on household choices. The standard housing choice framework applied to neighbourhood choice questions can provide important insights. But it also poses estimation oriented questions. First, if there is to be a systematic separation of selection and neighbourhood effects should there not need to be a better understanding of the real selection processes and influences that operate in housing markets. The standard neoclassical model assumes well informed consumers, unfettered by agent influences, in a market at or close to equilibrium. Applied research has for long suggested (Maclennan 1982) that there are significant spatial disequilibria (submarkets) within metropolitan housing markets at any point in time and that, in or out of equilibrium, housing searchers face major information limitations that require agent involvement. Agents, the gatekeepers of the housing choice and credit system, have strong neighbourhood selection influences. Further, the selection influences of agents are likely to differ within owner markets and rental systems, indeed in social rental sectors the priorities as well as the procedures of the letting landlord may dominate observed outcomes.

Secondly, if individuals, or their agents, are well informed about the set of neighbourhood choices households confront, will that knowledge extend to the 'neighbourhood' effects, say externalities, that will prevail within a place? The evidence from housing economics would suggest that some 'neighbourhood effects' or externalities can be identified prior to the start of a move process, others may be identified in residential search processes and yet more are only learned experientially (and this reinforces the importance of neighbourhood images and consumer perceptions of places in residential location decisions). There is also some evidence that housing search and choice processes are hierarchical (using something like an elimination by aspects algorithm and that for some groups tenure choices dominate other characteristics but that for some neighbourhood choice is a top level shaper of choices), see Quigley (2002).

These observations pose two questions. First, if consumers can identify and 'price' externality effects prior to a move then their willingness to pay for a particular place will reflect these valuations. This raises some issues about the possibility of separability of selection and neighbourhood effects. If all households, say, have similar preference functions for externality bundles, then the poor will inevitably have to not only live with each other but also often have to consume bundles of negative externalities. It is likely that the consumers of bundles of positive externalities will be the affluent. If externality bundles are (partly) determined by resident composition, that is neighbourhood effects are endogenously determined by those who live in a place, then there will be an inevitable collinearity between so-called selection and neighbourhood effects. And if there is such collinearity, and this is likely in the poorest and richest neighbourhoods, then separation of effects will be problematic. The second observation, is that neighbourhood choices involve risk, even in the short term.

It was noted above that individuals are influenced in their neighbourhood choices by the beliefs and behaviours of market agents (Maclennan 2012). Mortgage redlining is an obvious concern (Aalbers 2013) but other taste shaping and discriminatory processes are not uncommon in the markets for 'neighbourhood attributes'. Yet the estimation approaches in much econometric work still assumes that the market functions without distortions and is in a state of equilibrium. If it does not, then econometric estimates may either be biased or of short term relevance.

This brief review recognises the growing range of both qualitative insights regarding the importance of homes and neighbourhoods and the development of more rigorous techniques for quantitative assessment of socio-economic influences on choices. But for most nations and cities facing the question of changing neighbourhood or housing policies what it striking is not the range and quality of these methods but their near absence in informing policymaking and strategy development. Simple survey data on aspirations and satisfactions, much of it highly ambiguous, is much more widely used. For policymakers, at national and local scales, housing choice remains a black box on market processes and preferences that shape neighbourhood choices.

The multiple characteristics approach to neighbourhood choices not only conjoined issues of home and neighbourhood but it inevitably connects perspectives of space and time. Dwelling structures, not just the home but its surrounds of infrastructure and physical property are durable capital structures with potential for physical and socio-economic obsolescence. Consumer choices are not just for the short term. They reflect history and memory, attachment to places, beliefs about externalities and expectations about not just property values but neighbours and neighbourhoods too. Choice needs to be set in an understanding of change.

## *Change*

Even if there are black boxes in relation to the definition and choice of neighbourhoods are there convincing, evidenced stories on patterns and processes of neighbourhood change? Prior to the 1980s urban geography and economics were both areas that were reasonably confident that their generalised models of neighbourhood changes and outcome structures in cities, successors to the Burgess-Park model or the then 'new urban economics' access-space model, told widely relevant stories. Since the 1970s however, patterns of urban change have shaped more complex and localised, polycentric patterns of change within metropolitan areas. The recent work of Meen and Colleagues (2013) suggests that although place qualities may change over time the spatial patterns of relative qualities can be stable over very long time periods despite sustained shocks and policy changes. What changes and where are still basic, key questions in neighbourhoods research.

Urban economics has concerned itself little with overall place patterns, though there has been strengthened interest in the effects of spatial concentration of businesses (agglomeration and cluster issues for instance) or particular socio-economic groups (and hence the interest in neighbourhood effects). With some important exceptions (Bailey and Livingston 2008) UK urbanists have rarely examined overall patterns of neighbourhood change within metropolitan areas, though they have been much involved in identifying areas of disadvantage. The focus of research has either tended to be on a relatively pathological interest in either 'gentrification' or the rundown of 'social housing estates' or in assessing the impact of neighbourhood factors on health, crime and education outcomes. These are all important areas of insight. But they do not address the questions of how contemporary patterns of metropolitan systems of neighbourhoods are changing.

In Grigsby et al. (1987) set out a framework for assessing neighbourhood changes within a metropolitan housing market as a whole. It has had little application. Either those who have talked of city change as a whole have eschewed empirical research or those who have explored change dynamics have done so within the very specific confines of the two neighbourhood type change models noted above and leaned towards qualitative approaches to explain change.

A counter example that illustrates the point is the work of Hulchanski and others in Toronto (Hulchanski 2010). Studies of gentrification and other forms of neighbourhood choice and change have a long history in Toronto. What is different there is that these micro studies have been set, with great persistence, in a study of broader spatial patterns of change over time. It is possible to generalise about neighbourhood change in Toronto in a way that it is not possible for most UK cities.

The black box on change is relatively empty not just in relation to matters of pattern but in relation to processes too. The characteristics approach to neighbourhood choices, and homes, makes it clear that neighbourhoods are small, connected spaces. They have some degree of internal contact and connection but also connect to more dispersed sites within the metropolitan area and beyond. They are impacted by and also impact a variety of complex systems, the labour market, the housing system, the transport network and the ambient environment being obvious and important examples. Put briefly, neighbourhoods are spatial systems that are partly open and partly closed, they exist within a wider system of neighbourhoods, they are subject to autonomous shocks but there are also endogenous drivers of change within these recursive systems.

In that conception of neighbourhood we would expect neighbourhoods to have complex (chaotic) dynamics characteristics. These would include quite small effects, in some circumstances, tipping neighbourhood into significant non-linear change (see the Schelling (1971) model for instance). Changes are likely to have significant recursive effects that make separation of cause and effect difficult and that might manifest prolonged periods of neighbourhood disequilibrium.

Policy makers have much interest in such change. Clearly they have an interest in preventing neighbourhoods entering downward spirals, and indeed conversely ensuring that the scale of policy injections is above some threshold level sufficient to induce stable or upward succession. And with the prevalence of external shocks attention to the resilience of the individuals, institutions and investors within neighbourhoods becomes a key concern.

There also needs to be recognition that there is sufficient randomness in these recursive processes that identical triggers will, ex post, not always produce the same result. In dealing with neighbourhoods, or small, complex, human systems we can question Einstein's maxim that doing the same thing will produce the same result. To be certain of at least a probabilistic estimate of likely outcomes then it will be necessary to have analysed the dynamic trajectories of different classes of neighbourhoods across a substantial set of places. This observation is systematically disregarded in evaluations of renewal policies in most countries.

The drivers and processes of change of systems of neighbourhoods in our modern metropolises may have been more or less well theorised but there is, in most cities, an absence of an understanding of how the system of neighbourhoods is changing and we have not made it a research priority to find out. For those employed in policy making research on neighbourhood change, like definition and choice, remains conceptually interesting but largely empirically empty. The academy essentially offers policy a series of plausible stories of neighbourhood change but without garnering much empirical support or establishing ex ante which tale of dynamics will be most likely to prevail in which circumstances.

All these areas, neighbourhood definition and identification, neighbourhood choices and change patterns and processes raise important research questions for geographers, economists and others. And we understand some aspects of these processes in some places, for some groups. But for policy making we do not have, before policy is implemented, a definitive stock of real knowledge that fills the black boxes of neighbourhood identification, choice influences and the drivers and processes of change. In policymaking we intervene with worryingly low levels of system information. In research we are often driven by what is conceptually interesting rather than what is practically important to improve individual wellbeing. In consequence we face difficulty in producing a convincing, connected, theoretically based rhetoric about neighbourhood renewal that can be effectively defended. And, at the same time, the academy generates a series of competing claims as to whether and how to intervene in places.

These observations are given substance by examining how cases for neighbourhood renewal became questioned as Andersson and Musterd (2005) articulated a new scepticism about both neighbourhood effects and the salience of regeneration policies.

## The New Scepticism, and a Critique

### *After Fuzzy Theory, Fuzzy Policy?*

The new scepticism in neighbourhood research articulated by Andersson and Musterd (2005) applied not just to the likely extent of neighbourhood effects (a theme quickly reinforced by Friedrichs and colleagues (2003)) but was also directed towards the existence of area based policies. Indeed the emphasis of the language of the key Andersson and Musterd paper leans towards a view that neighbourhood renewal policies might well be misplaced. For instance, they ask 'whether it is a good idea to continue the area-based policies or whether it might be wiser to replace the area-based policies by domain or sector policies (such as city wide policies on school and adult education, job training, and citizen participation in planning practices)'. 'Policies should thus reach all people in need of support instead of just those who are living in a concentration area' (p. 379). They continued 'However it seems as if on certain occasions the balance between the attention to structural factors and the area compositional factors has been lost...many will agree there that the dominant factors impacting upon social mobility are in fact structural'.

It will be argued below that these comments on policy that can be interpreted as leaning against area renewal policies, really pose a quite false dichotomy. However at this stage it is simply noted that in their seminal paper, Andersson and Musterd shaped an association between establishing significant neighbourhood effects and the existence of area based policies. This association has already been critiqued by Lupton (2003) but it bears further critical scrutiny because this scepticism about area-based policies has permeated subsequent debate. There remain arguments voiced that in the absence of neighbourhood effects there is no real point to area based or place policies.

This line of reasoning can be resisted in at least three ways. First, as indeed Andersson and Musterd recognised, there are arguments for neighbourhood policies that are not predicated on the existence of neighbourhood effects. Secondly, that an unnecessary dichotomy has been posed between place and people policies. Thirdly, and this is a different line of reasoning, it would be potentially naïve to assume a close correspondence between hard evidence and policy shift in neighbourhood renewal (that is, the sceptics may have been right but they have been naïve about how policy changes). These three arguments are considered in turn.

## *Cases for Neighbourhood Renewal*

Neighbourhood Renewal (or regeneration) policies grew steadily in scale and diversity in Europe after the early 1970s though the rationales for policy, their extent and their scope varied from place to place and over time. Given evolution and diversity, it is pertinent to spell out some different notions of renewal or regeneration policy. Glaeser and Gottlieb (2008) are, in the main, right to observe that policies do not have policies for place per se.[3] However good policymaking involves selecting the mix of macro, sectoral and area based policies that will best achieve policy aims.

---

[3] Some cultures and nations do impute values to places per se, for instance aboriginal Australian cultures would not fit Glaeser's assumptions nor indeed would Gaelic Scots and many nations have iconic locations that they, in some sense, pay for.

That is why the people versus place debate is a false dichotomy. In most area renewal challenges the selection of often the poorest and least capable groups in obviously poor quality physical and low access locations means that the challenge is not place versus people but creating change for particular groups of people in particular kinds of places (Maclennan 2000).

It is possible to construct a typology for policies around a number of important constructs. The first is the a priori assumptions, in policymaking, about the nature of the real functional systems that are operating, that is whether the systems policy is to influence are assumed to be spatial or not. The second is the chosen level of autonomies within a polity, as clearly more localised autonomy will come closer to more localised systems. The third is that distinction has to be made between the intentions of policy and outcomes. And the fourth is whether policies are aimed at redistribution or the capacities and creativities of particular groups of places.

Some policies are intentionally *aspatial*. Much macroeconomic (and hence the label) is of this nature, though there are also local government policies that are intended to be uniform within local territories (for instance some aspects of building and planning regulations). National, or macro, policies such as interest rate measures, much fiscal policy and means-tested social security benefits are 'spatially blind' in their a priori design and essentially indifferent to their locus of incidence and to the possibility that they may trigger particular spatial effects. Clearly they may have both uneven spatial incidence and effects. Other policies, by way of contrast, are *intendedly spatial* in their structure and impacts, for example regional economic policies or neighbourhood crime prevention strategies; these policies are *area-based*.

However, *area-based* policies may have fundamentally different characters, depending on some of the constructs identified above. Some may simply be targeted redistribution, including area policies to alleviate poverty, and have no explicit aim of changing the development trajectory of a place. In this paper these policies are labelled as *palliative redistribution*. And palliative policies may arise from national, regional/provincial or local governments. Other area-based policies are intended to change the development trajectory of a place with the expectation that it will reduce its reliance on area-policy measures and revert to 'normal' policy mechanisms. These can be labelled as *area development* policies.

Area, or neighbourhood, development policies can be further subdivided. There are sectoral policies, such as better homes or more jobs for residents, with priority targeting of identified places on the assumption that this will lead to sustained regeneration (*policy bending*). These are *neighbourhood sectoral* policies. But there are also policies designed to raise the creative and competitive capacities of individuals, organisations and institutions within places. They are *creative or transformative neighbourhood* policies. They are distinguished by the belief that space and place shape development processes. In such approaches, as noted above, there is no longer a dichotomy between 'sectoral' and 'area-based' policies, or people and place policies. Rather, the approach involves integrating these alternatives where it is relevant to do so. Effective renewal will usually require the integration of not only multiple sector policies but the actions of different orders of government and, usually, communities.

Measures designed to stimulate and or capture neighbourhood effects can be then put in some perspective. They can be seen as one, but only one, of a wider set of creative/transformative policies. Andersson and Musterd assume that renewal policies in the places they examined had come to be based on neighbourhood effects. This, as is argued below, was not everywhere and always the case.

## *The Evolution of Renewal Policy Beliefs*

Neighbourhood as a spatial scale of interest in policy provision grew significantly in Europe after the 1970s. At first policies for particular sectors, especially housing, local environment and then crime were given an area basis, and that is evident in the then policy frames in, for instance, the UK, the Netherlands and France (see Maclennan 1986). In the main these areas were designed as 'inner city' measures and aimed at older, declining core neighbourhoods.

At that time area regeneration or renewal policy was largely conceived as redistributive, intended to alleviate the symptoms of decline processes in inner-city neighbourhoods (Robson 1988). These were palliative policies. They reduced the 'pain' associated with decline and were not aimed at causes. They were inherently redistributive in nature. Treasury and finance ministries saw such policies as having 'displacement effects' in the economy, that is they simply transferred tax revenues from prosperous places and people to the less affluent, and productivity gains were never the explicit rational of such policies. In the author's policy experience such a view of area renewal was dominant in finance Ministry thinking in the UK until at least 2000 and it still prevailed in, for instance, Ottawa and Canberra after 2005 (Maclennan 2006).

Andersson and Musterd are critical of area based approach to poverty alleviation as they suggest that the proportion of the poor in a metropolitan area actually concentrated into such areas may be low, even as low as 5 percent. Clearly, in such circumstances, there needs to be an emphasis on people related policies but even small area concentrations may require area policies. However their description of the overall significance of poverty concentrations within conurbations would simply be wrong for UK cities. In Glasgow for instance, by the 1980s two-thirds of the metropolitan poor lived in the poorest areas. Large scale, concentrated poverty provoked an intense, palliative response there (Maclennan 2006; Pacione 1997).

As for neighbourhood effects, in the UK at least, they did not feature explicitly in policy debate or legislation. As programmes grew practitioners, and local politicians, argued that such effects existed (and indeed faced with the bleak desolation, vandalised homes, pinched faces of poor children and all too apparent signs of drugs trading and consumption such a view was always understandable). Spending departments responsible for delivering sector policies usually showed no more than an academic interest in spillovers and the notion of 'neighbourhood effects' had, as noted above, little influence on Treasury and Cabinet offices. Andersson and Musterd, in arguing that neighbourhood effects had become the key raison d'etre for renewal policies, did not do enough to explore how views about the rationale for policies varied across different orders of government and the different kinds of government department involved. Neither around 2000 nor now, in 2012, have there generally been unitary views of place and neighbourhood effects within governments. In most instances it would appear that for area-based renewal policies the integration of cross-sectoral policies comprises the policy cake with neighbourhood effects the icing.

However it is true that the policy arguments for renewal policies have become more complex since the start of the 1980s. No more than a decade after inner city, area-based sectoral policies emerged in Europe the scope and methods of renewal policy had evolved considerably. Areas of declining social housing, often built postwar, also had become the focus of multi-sector zones of action. Academic research through the 1990s emphasised that the decline of the traditional economic base was no longer seen to be the sole process involved in expanding poverty concentrations, (Green and Owen 1998; Turok and Edge 1999; Maclennan 1998). There had also been an increasing *concentration* of the poor in disadvantaged neighbourhoods, and this appears to have been true in other countries such as Canada and Australia at the same time (Heisz and MacLeod 2004). This mirrored a range of causes involving wider spatial and structural systems. These areas are typified by poor housing, health, educational achievement and facilities and high crime, vandalism and housing turnover. Concentration of the unemployed in these areas, which were often remote from employment, located households in dwellings and social and labour market networks which reinforced *exclusion* (McGregor 1999). There was a sense, but little hard evidence, that labour market decline and housing system concentration processes could form the basis of recursive neighbourhood effects. And there was also a growing recognition that it was not enough to characterise a neighbourhood in terms of its resident characteristics and housing stock. A much more detailed understanding was required of the capacities of neighbourhood institutions to demonstrate of resilience to shocks (Maclennan 2006).

By the end of 1980s it had become recognised that single sector action was unlikely to promote sustained neighbourhood renewal and that, in particular, crime, housing, employment, education and health measures had to be locally integrated. Evolving ideas in the field of the 'new public management' emphasised the important of policies achieving outcomes rather than simply producing outputs and the importance of 'holistic' or whole-of-government policies. After the mid-1990s throughout most of Western Europe it was regarded as essential for projects to integrate a range of policy and private sector actions to achieve efficient change (Maclennan and McGregor 1992). Integrated neighbourhood renewal efforts within governments emerged from management concerns about coordination and sources of resources and not about geographies of neighbourhood effects. New management perspective argued for area renewal policies that were strategic, integrated and partnership oriented. They were often still palliative in nature, but at least addressed multiple symptoms.

When governments first stressed such multi sector approaches they did so by encouraging each policy silo to develop areas for priority action. This happened in the UK after 1997 when the Blair administration emphasised area approaches but had separate local zones for housing, economic initiatives, crime, education and health. Clearly some more locally integrated and consistent geography for policy delivery was required and such areas soon (2000) emerged with the national strategy for neighbourhood renewal (Social Exclusion Unit 2001).

The push for more locally integrated renewal policies emerged not only from new views about administrative decentralisation but in a new attention to the devolution of some powers to communities. From the 1990s onwards in much of western Europe there was both a wide commitment to subsidiarity in overall policymaking as well a recognition of the potential for enhanced community engagement in local change. And this made a clear linkage to the recognition in renewal policies that the community and institutional capacities within places had, in many instances, to be augmented for successful renewal to be likely.

These developing approaches in neighbourhood renewal, that evolved from policy experience rather than ex ante research programmes, gradually embraced more complex notions of the geographies involved, the processes of neighbourhood development and the individual and organisational capabilities that were needed for change that was not simply redistributive. By the start of this millennium renewal of places was widely understood (Social Exclusion Unit 2001) to be about multiple sectors of change, forms of governance, community dynamics, new public sector management and neighbourhood effects had also become part of the policy conversation.

Andersson and Musterd recognise much of this variety of motivation for renewal policies. However what is not clear, and why their analysis would misrepresent policy thinking in the UK, Canada and Australia, at least, is why they came to be understood as predicating neighbourhood renewal policies on primarily neighbourhood effects. There has to be a concern that as academic analysts we project on the complex evolved world of policy action our most recent academic lens and fashion.

There are related examples in this broad field of study that show a similar tendency. Atkinson and Kintrea (2001, 2002), and Graham and colleagues (2009) both analyse the development of mixed tenure housing estates in Scottish renewal projects and suggest that policy was misplaced as it did not achieve integration of the different groups housed. Lack of integration may be regrettable but it was not the core objective of the policy action involved. Quite simply, in the Scottish context, mixed estate renewal was largely funded by the national housing agency, Scottish Homes. My personal experience as a Board Member of the agency for a decade was that agency always faced a budget constraint so that renewing partly abandoned places was less expensive if home owners (with grant rates around 20 percent) could be mixed with social housing tenants (with grant rates of 80 percent). An additional policy argument was that residents succeeding on peripheral public housing schemes were leaving their estates because there were no local purchase options (Maclennan and McGregor 1992). Research confirmed (Kintrea et al. 1996) that those who purchased this housing stock often originated within the area or has strong past connections. Good mix outcomes were seen as a 'bonus'. So academia tested the icing and not the cake.

Similar comments can be levelled at studies of mixing income groups. Galster (2007) has set out a range of conditions under which income mix strategies might work, such as the dissimilarity of the households to be mixed, that scale of development, extent of mix etc. But few tests of the neighbourhood effects of 'income mix' really bother to allow for all of these crucial policy success factors. Research on the spillover effects of tenure and income mix are interesting and important per se but they cannot be marshalled to make an existential critique of place renewal policies.

In choosing, designing and delivering renewal polices it is important to know which processes are operating, which problems are being addressed and what solutions are expected. Policy in the UK, at least until 2010, tried to address decline, concentration, resilience and exclusion and this, given the historically limited conception of regeneration policy, was a demanding task. These processes have macroeconomic, urban system, specific place and person-related causes. These are not simple 'people' or 'place' problems but interacting 'people and place' difficulties (Maclennan 2000). Neighbourhood effects are only one part of the question and the solution.

## A Dissenting View About Policy Impacts!

In the introductory section it was noted that academics usually expected their well-founded results to influence policy debates. In this context there has been an assumption running forward from Andersson and Musterd that failure to identify neighbourhood effects reduces the relevance of neighbourhood renewal policies and should engender a move towards sectoral or other spatial policies. The previous section stressed that there are other clear bases for renewal policies so that rational, evidence based renewal policies might exist with zero neighbourhood effects. In this concluding section a different strand of thought is developed. Namely, in seeking better understandings of what neighbourhood renewal policy should be academia has largely ignored how policies are made and, in particular, how new ideas and evidence actually influence policy thinking.

There is now much interest in effective knowledge exchange and mobilisation. In the UK, the Research Councils have set out their interest in developing 'Pathways to Impact' and in the present REF review[4] there is an emphasis on demonstrating the impact of research. These ideas are, in some significant ways, problematic.

---

[4] The Research Excellence Framework (REF) is a review, at broadly five or six year intervals of the quality of research publications by staff and of the wider impacts of their research. The next census date is currently end 2013.

It is undoubtedly important that much research in the social sciences is applied research and that the findings of research are communicated clearly and promptly to policymakers. However it is naïve social science to assume that there is any close correlation, in any given time frame, between the delivery of an excellent idea and its use in policy. At the same time a new synthesis of old, and not all of it technically excellent, research may have great impact as the persistence of ideas is important as ideas, times and governments change.

When new ideas and evidence emerge in neighbourhoods and place policy research how can they best impact policy? It is critical to have at least one level of government with strategic power and resources interested in place issues. In the UK the national governments have had an active interest in renewal policy since the 1970s. In Canada, there have been no Federal neighbourhood renewal policies since the 1990s and only some provinces show an active interest (Manitoba, for example). Neighbourhood renewal in Canada is largely a matter of some municipalities aiming to form partnerships to change places, it is management process rather than a resource flow. In Australia, State governments kept interests in city and neighbourhood policies through the 1990s until the present. So the key first question, is which level or levels of government to aim at? Who might be interested?

Within interested governments there will be a variety of separate groups of officials and departments with different interests. For example, in most countries it has been easy to engage housing ministries with renewal issues. In contrast, economic development ministries may resist neighbourhood commitments, both because it is hard to tie economic change to very local places but also because they may be led by professionals with little interest or training in place issues. And, as noted earlier, at the Treasury level the questions will always be about how programmes might save expenditures or raise productivity. It is unusual for all of these interests within government to be simultaneously aligned to the receipt of some new idea (unless it appears to have, usually illusory, silver bullet characteristics). And it is key that messages impact policy-lead officials and not just government researchers. To make ideas count researchers need to have a better sense of how bureaucracies absorb information about the subjects they deal with, to have some persistence mechanism and to skirt, but not plunge into, the precipice of politics. Interaction directly with politics can have high returns in the short term but is problematic in competitive democracies.

That is, taking new ideas on neighbourhoods to the policy process requires very different skills and processes from these that create excellent ideas in the first instance. The higher education bureaucracies of the UK seem to have confused these two sets of issues. In the more specific context of neighbourhood effects and policy research we simply note here, that we assume too much about the making of national and local public policies if we expect policies to always shift as we discover new insights. And equally, given our limited insights about key issues regarding neighbourhoods we should not be surprised , when as this recession begins to end, that the start of the next wave of renewal policies will be unable to wait for our next new research findings.

## Neighbourhood Research and Policy: Continuing Commitments

When researchers seek to establish neighbourhood effects or the impacts of neighbourhood renewal policies they face major empirical as well as theoretical challenges. What is the scale of neighbourhood, what is the spatial range of effects, which people or groups are affected, will effects be apparent now and, or, at some time in the future, and with the dynamism and openness of neighbourhoods how will effects be exported over time (in health, human capital, attitudes, confidence etc.). Researchers are seeking to establish the effects of difficult to identify forces, but forces that shape our ideas for theory as well as policy. There is an assumption that there is some likelihood these effects exist.

Beliefs, based on reason, often serve as a basis for theory. This holds true in the sciences and social sciences too. As these volumes are written and as research on neighbourhoods slumps with public cutbacks, there is emerging news that renewed attempts at CERN, Europe's nuclear research laboratory, have finally identified the Higgs-Boson particle (at a confidence level that scientists regard as close to certainty). The existence of the particle was deduced from the 'Standard Model' of physics by Peter Higgs in 1964. If the particle had not existed then the theory will be seriously incomplete. The search for the particle commenced in 2008 and it lies in the Hadron Collider. It has cost around £700m per annum to run.

The point of this observation is not the cost of the collider, for the benefits may be inestimable. But it is simply, that if researchers are to apply rigorous standards to analysis of neighbourhood effects, if they are really to get to a consensus on methods let alone results then experiments will have to be better designed. The subtlety of the idea may now have leapt far beyond researcher capacities to actually empirically identify the weight of effects.

In such uncertain circumstances it would be risky to predicate the existence of area based policies on the existence of strong neighbourhood effects. Equally it would be unwise both to abandon area-based initiatives and not to design them to capture potential neighbourhood effects in the hope that they exist.

Looking to the future, governments intent on good evaluation and national research councils that really wish to understand how place and space shapes social and economic life will have to return to the questions of neighbourhood choices, their effects and processes of change. A lot more work needs to be done on filling some of the key black boxes or empirically empty frameworks that were discussed above. Work needs to continue on how best to define and identify neighbourhoods, the absence of contemporary understandings of the processes of housing and neighbourhood choices needs to be remedied as new demographics and new energy economics seem set to reshape the structure of neighbourhoods within metropolitan areas. Analysis has to better understand system processes rather than just rely on applying econometrics to housing and neighbourhood choice outcomes on the assumption that they reflect well ordered systems in or lose to equilibrium. Evidence on neighbourhood change suggests that non-linear and disequilibrium processes are prevalent and they need to be better understood and modelled. A new priority in urban research has to be given to how sets of neighbourhoods within metropolitan areas change in a connected fashion and this could displace a rather tired pathological interest in gentrification and the dynamics of rundown social housing estates. Within the policy arena a much better understanding of how policies evolve and how, and when, they are influenced by evidence also needs to be developed if much of the excellent research that emerges in this field is to be used to better public policy effects. And that is a worthwhile endeavour as policies for neighbourhoods usually signal a real attempt to deal with the issues confronting many of the poorest people in our still affluent economies.

**Acknowledgements** I am grateful to Alice Oldfield for comments on this chapter and to David Manley for the incisive comments he made on an earlier draft. Remaining errors are all my own.

# References

Aalbers, M. (2013). How do mortgage lenders influence neighbourhood dynamics? Redlining and predatory lending. In M. van Ham, D. Manley, N. Bailey, L. Simpson, & D. Maclennan (Eds.), *Understanding neighbourhood dynamics: New insights for neighbourhood effects research* (pp. 63–86). Dordrecht: Springer.

Andersson, R., & Musterd, S. (2005). Housing mix, social mix and social opportunities. *Urban Affairs Review, 40*(6), 761–790.

Atkinson, R., & Kintrea, K. (2001). Disentangling area effects: Evidence from deprived and non-deprived neighbourhoods. *Urban Studies, 38*(11), 2277–2298.

Atkinson, R., & Kintrea, K. (2002). Area effects: What do they mean for British housing and regeneration policy? *European Journal of Housing Policy, 2*(2), 147–166.

Bailey, N., & Livingston, M. (2008). Selective migration and area deprivation: evidence from 2001 Census migration data for England and Scotland. *Urban Studies, 45*(4), 943–961.

Buck, N. (2001). Identifying neighbourhood effects on social exclusion. *Urban Studies, 38*(12), 2251–2275.

Butler, T., & Lees, L. (2006). Super-gentrification in Barnsbury, London. *Transactions of the Institute of British Geographers, 32*, 467–487.

Cheshire, P., & Sheppard, S. (2005). The introduction of price signals into land use planning decision-making: a proposal. *Urban studies, 42*(4), 647–663.

Cheshire, P. (2012). Are mixed community policies evidence based? A review of the research on neighbourhood effects. In M. van Ham, D. Manley, N. Bailey, L. Simpson, & D. Maclennan (Eds.), *Neighbourhood effects research: New perspectives*. Dordrecht: Springer.

Downs, R. M., & Stea, D. (Eds.). (1973). *Image and environment: Cognitive mapping and spatial behaviour*. Edward Arnold: Sevenoaks, Kent.

Ellen, I. G., & Turner, M. A. (1997). Does neighborhood matter? Assessing recent evidence. *Housing Policy Debate, 8*(4), 833–866.

Forrest, R., & Kearns, A. (2001). Social cohesion, social capital and the neighbourhood. *Urban Studies, 38*(12), 2125–2143.

Friedrichs, J., Galster, G. C., & Musterd, S. (2003). Neighbourhood effects on social opportunities: The European and American research and policy context. *Housing Studies, 18*(6), 797–806.

Galster, G. C. (1986). What is neighbourhood? An externality-space approach. *International Journal of Urban and Regional Research, 10*(2), 243–261.

Galster, G. (2008). A stock/flow model of defining racially integrated neighborhoods. *Journal of Urban Affairs, 20*(1), 43–51.

Galster, G. (2001). On the nature of neighbourhood. *Urban Studies, 38*(12), 2111–2124.

Galster, G. C. (2007). Neighbourhood social mix as a goal of housing policy: A theoretical analysis. *European Journal of Housing Policy, 7*(1), 19–43.

Galster, G. C. (2012). The mechanism(s) of neighbourhood effects theory, evidence, and policy implications. In M. van Ham, D. Manley, N. Bailey, L. Simpson, & D. Maclennan (Eds.), *Neighbourhood effects research: New perspectives*. Dordrecht: Springer.

Glaeser, E. L., & Gottlieb, J. D. (2008). *The economics of place-making policies* (Working paper 14373). Cambridge: National Bureau of Economic Research.

Graham, E., Manley, D., Hiscock, R., Boyle, P., & Doherty, J. (2009). Mixing housing tenures: Is it good for social well-being? *Urban Studies, 46*(10), 139–165.

Green, A. E., & Owen, D. (1998). *The geography of poor skills and access to work*. York: Joseph Rowntree Foundation.

Grigsby, W., Baratz, M., Galster, G. C., & Maclennan, D. (1987). *The dynamics of neighbourhood change and decline*. Oxford: Pergamon.

Grigsby, W. G. (1963). *Housing markets and public policy*. Philadelphia: University of Pennsylvania Press.

Hägerstrand, T. (1978). Survival and arena: On the life-history of individuals in relation to their geographic environments. In T. Carlstein, J. Parkes, & N. J. Thrift (Eds.), *Human activity and time geography*. New York: Wiley.

Hägerstrand, T. (1988). Some unexplored problems in the modeling of culture transfer and transformation. In *The transfer and transformation of ideas and material culture* (pp. 217–232). College Station, TX: A & M University Press.

Hardin, R. (2009). *How do you know. The economics of ordinary knowledge*. Princeton: Princeton University Press.

Harvey, D. (1973). *Social justice and the city*. Melbourne: Edward Arnold Press.

Hedman, L. (2011). The impact of residential mobility on measurement of neighbourhood effects. *Housing Studies, 26*(4), 501–519.

Hedman, L., & van Ham, M. (2012). Understanding neighbourhood effects: Selection bias and residential mobility. In M. van Ham, D. Manley, N. Bailey, L. Simpson, & D. Maclennan (Eds.), *Neighbourhood effects research: New perspectives* (pp. 79–100). Dordrecht: Springer.

Heisz, A., & MacLeod, L. (2004). Low income in census metropolitan areas. *Perspectives on Labour and Income, 5*(5). Ottawa: Statistics Canada.

Hoyt, H. (1939). *Structure and growth of residential neighbourhoods in American cities*. Washington, DC: FHA.

Hulchanski, D. (2010). *The three cities within Toronto. Income polarization among Toronto's neighbourhoods, 1970–2005*. Toronto: University of Toronto.

Hunter, A. (1979). The urban neighbourhood: Its analytical and social contexts. *Urban Affairs Quarterly, 14*(3), 267–288.

Kintrea, K., Gibb, K., & Conchua, C. H. (1996). *An evaluation of GRO grants for owner occupiers*. Edinburgh: CRU, Scottish Office.

Lebel, A., Pampalon, R., & Villeneuve, P. Y. (2007). A multi-perspective approach for defining neighbourhood units in the context of a study on health inequalities in the Quebec City region. *International Journal of Health Geographics, 6*, 27.

Lupton, R. (2003). *'Neighbourhood Effects': Can we measure them and does it matter?* (CASE paper 73). London: Centre for Analysis of Social Exclusion London School of Economics.

Lupton, R., Kneale, D. (2012). Theorising and measuring place in neighbourhood effects research: the example of teenage parenthood in England. In M. van Ham, D. Manley, N. Bailey, L. Simpson, & D. Maclennan (Eds.), *Neighbourhood effects research: New perspectives* (pp. 121–146). Dordrecht: Springer.

Maclennan, D. (1982). *Housing economics: An applied approach*. London: Longman.

Maclennan, D. (1986). *Maintenance and modernization of urban housing*. OECD Urban Affairs Programme (mimeo.).

Maclennan, D. (1998). Urban regeneration in Britain: new times, new challenges. In: B. Badcock and K. Harris, *Proceedings of the National Urban Renewal Seminar*, Department of Human Services, Adelaide.

Maclennan, D. (2000). *Changing places, engaging people*. York: JRF.

Maclennan, D. (2006). *Remaking neighbourhood renewal: Towards creative neighbourhood renewal policies for Britain*. Ontario: Caledon Institute.

Maclennan, D. (2012). Understanding housing markets: Real progress or stalled agendas. In D. Clapham, W. Clark, & K. Gibb (Eds.), *The SAGE handbook of housing studies*. London/Thousand Oaks: SAGE.

Maclennan, D., & McGregor, A. (1992). *Strategic approaches to urban regeneration in Scotland*. Edinburgh: Scottish Homes.

Mallett, S. (2004). Understanding home: A critical review of the literature. *The Sociological Review, 52*(1), 62–89.

Malpezzi, S. (2002). Hedonic pricing models: A selective and applied review. In T. O'Sullivan & K. Gibb (Eds.), *Housing economics and public policy*. New York: Wiley.

Malpezzi, S. (2003). Urban regulation, the "new economy", and housing prices. *Housing Policy Debate, 13*(2), 323–349.

Manley, D., & van Ham, M. (2011). Choice-based letting, ethnicity and segregation in England. *Urban Studies, 48*(14), 3125–3143.

Manley, D., & van Ham, M. (2012a). Occupational mobility and neighbourhood effects. In M. van Ham, D. Manley, N. Bailey, L. Simpson, & D. Maclennan (Eds.), *Neighbourhood effects research: New perspectives* (chap. 7, pp. 147–174). Dordrecht: Springer.

Manley, D., & van Ham, M. (2012b). Neighbourhood effects, housing tenure and individual employment outcomes. In M. van Ham, D. Manley, N. Bailey, L. Simpson, & D. Maclennan (Eds.), *Neighbourhood effects research: New perspectives*. Dordrecht: Springer.

Manley, D., van Ham, M., Bailey, N., Simpson, L., & Maclennan, D. (Eds.). (2013). *Neighbourhood effects or neighbourhood based problems? A policy context*. Dordrecht: Springer.

McGregor, S. L. T. (1999). Socializing consumers in a global marketplace. *Journal of Consumer Studies & Home Economics, 23*(1), 37–45.

Meen, G., Nygaard, C., & Meen, J. (2013). The causes of long-term neighbourhood change. In M. van Ham, D. Manley, N. Bailey, L. Simpson, & D. Maclennan (Eds.), *Understanding neighbourhood dynamics: New insights for neighbourhood effects research* (pp. 43–62). Dordrecht: Springer.

Murie, A., & Forrest, R. (1980). *Housing market processes and the inner city*. New York: Social Science Research Council.

Nutley, S. M., Davies, H. T., & Smith, P. C. (2000). *What works: Evidence-based policy and practice in public services*. Bristol: Policy Press.

Oreopoulos, P. (2003). The long-run consequences of living in a poor neighbourhood. *Quarterly Journal of Economics, 118*, 1533–1575.

Pacione, M. (1997). Local Exchange Trading Systems as a response to the globalisation of capitalism. *Urban Studies, 34*(8), 1179–1199.

Quigley, J. M. (2002). Transaction costs and housing markets. In A. O'Sullivan & K. Gibb (Eds.), *Housing economics and public policy*. Oxford: Blackwell.

Robson, B. T. (1969). *Urban analysis: A study of city structure with special reference to Sunderland CUP archive*. Cambridge: Cambridge University Press.

Robson, B. (1988). *Those Inner Cities*. Oxford: Oxford University Press.

Slater, T. (2013). Capitalist urbanization affects your life chances: Exorcising the ghosts of 'Neighbourhood Effects'. In D. Manley, M. van Ham, N. Bailey, L. Simpson, & D. Maclennan (Eds.), *Neighbourhood effects or neighbourhood based problems? A policy context* (chap. 6). Dordrecht: Springer.

Suttles, G. D. (1972). *The social construction of communities*. Chicago: The University of Chicago Press.

Segal, D. (Ed.). (1979). *The economics of neighbourhood*. New York: Academic press.

Schelling, T. C. (1971). Dynamic models of segregation. *Journal of Mathematical Sociology, 1*(2), 143–186.

Small, M. L., & Feldman, J. (2012). Ethnographic evidence, heterogeneity, and neigbourhood effects after moving to opportunity. In M. van Ham, D. Manley, N. Bailey, L. Simpson, & D. Maclennan (Eds.), *Neighbourhood effects research: New perspectives* (pp. 57–77). Dordrecht: Springer.

Social Exclusion Unit. (2001). *A new commitment to neighbourhood renewal*. London: Cabinet Office.
Turok, I., & Edge, N. (1999). *The jobs gap in Britain's cities*. York: Joseph Rowntree Foundation.
van Ham, M., & Manley, D. (2010). The effect of neighbourhood housing tenure mix on labour market outcomes: A longitudinal investigation of neighbourhood effects. *Journal of Economic Geography, 10*(2), 257–282.
van Ham, M., & Manley, D. (2012). Neighbourhood effects research at a crossroads. Ten challenges for future research. *Environment and Planning A, 44*, 2787–2793.
van Ham, M., Manley, D., Bailey, N., Simpson, L., & Maclennan, D. (Eds.). (2012). *Understanding neighbourhood dynamics: New insights for neighbourhood effects research*. Dordrecht: Springer.
van Ham, M., Manley, D., Bailey, N., Simpson, L., & Maclennan, D. (Eds.). (2013). *Neighbourhood effects research: New perspectives*. Dordrecht: Springer.

# Index

**A**

Area based initiatives (ABIs), 5, 10, 11, 21, 25–39, 52, 53, 56, 60, 161, 192, 288
Attainment, 9, 10, 25, 26, 28–32, 54, 125, 149
Australia, 2, 5, 10, 19, 99, 124, 125, 133, 141, 142, 251–267, 270, 281, 284, 285, 287

**B**

Black boxes, 2, 20, 21, 271–272, 278–280, 288
British cohort study, 16, 181

**C**

Canada, 10, 15, 16, 102, 157–174, 270, 284, 285, 287
Capabilities theory, 96
Chicago school, 20, 69, 118, 273
Choice, 18, 26, 34, 59, 118, 119, 140, 143, 144, 150, 205, 209, 216, 220–223, 235, 243, 272, 275–278
City strategy pathfinders (CSP), 57
Community, 3, 27, 53, 67, 89, 122, 138, 157, 189, 201, 216, 253, 270
Community policing, 11, 67, 70, 77, 163, 189
Concentrated poverty, 4, 11, 14, 74, 114, 122, 124, 134, 137, 143, 150, 159, 228, 243, 244, 255, 283
Crime, 3, 27, 46, 67–83, 93, 115, 134, 168, 185, 221, 252, 279
Crime reduction policies, 68, 77–81
CSP. *See* City strategy pathfinders (CSP)
Cultural relativism, 199
Curriculum, 10, 26, 33–35, 37, 39, 124

**D**

Deconcentrate poverty, 217, 239–244
Deconcentration, 6, 18, 128, 149, 196, 215–244
Defensible space, 12, 73–74
Deindustrialisation, 45
Demolition, 7, 8, 14, 19, 114, 123, 128–130, 142, 143, 145, 149, 196, 197, 208, 225, 226, 228, 252, 253, 259
Deprivation/deprived, 5, 8, 11, 13, 28, 34, 37, 43–46, 50, 54, 69, 79, 80, 92, 94, 96–98, 100–104, 106, 113, 135, 136, 138, 141, 145, 181–183, 186, 198, 203, 242, 267
Deprived neighbourhoods, 15, 18, 43–59, 62, 80, 99, 100, 138, 150, 177, 182, 192
Desegregation, 17, 114, 133, 139, 141, 142, 145, 151, 196, 200–204, 209, 217–219, 223–226, 235–237
Displacement, 7, 8, 129, 130, 205–206, 283
Distressed neighbourhoods, 157, 168, 170, 221, 228, 238, 255
Dutch, 17, 126, 141, 145, 149, 195, 197–201, 204, 206, 209, 210, 256

**E**

EAZ. *See* Education action zones (EAZ)
Economics, 96, 118, 189, 271, 273, 275–279, 288
Education, 2, 25–39, 46, 77, 102, 115, 135, 161, 183, 202, 222, 253, 276
Education action zones (EAZ), 27, 29
Educational priority areas (EPA), 27
Engels, 14, 91, 116–118, 122, 123

England, 25–29, 31, 43–62, 68, 75, 79, 80, 91, 113, 114, 116, 119, 134, 136, 141, 143, 145, 147, 149
Environmental justice, 13, 89–106
EPA. *See* Educational priority areas (EPA)
Ethnic integration, 201
Ethnicity, 34, 37, 50, 71, 77, 98, 141, 186, 203, 242
Ethnic minorities, 17, 45, 54, 58, 98, 104, 159, 195–200, 204–206, 209
Evidence, 2, 4, 6, 7, 12, 14, 16, 17, 21, 25, 27, 29, 31, 32, 44, 47–49, 52, 54, 56, 58–60, 73, 74, 78–81, 83, 91, 92, 95, 97–102, 104, 115, 125, 129, 133–135, 138–140, 147–149, 159–161, 167, 172, 177–192, 196, 219, 230, 234–239, 243, 251–267, 269–289
Excellence in cities (EiC), 27, 29

## G

Gautreaux, 18, 142, 215, 223–224, 226, 236, 254
Geography, 45, 46, 75, 89, 91, 98, 106, 119, 202, 215, 223, 226, 237, 241–244, 271, 273, 275, 278, 285
Global, 1, 2, 35, 36, 44, 61, 62, 92, 94, 95, 104, 150, 157, 158, 235
Green book, 16, 179, 185, 191

## H

Health, 3, 26, 46, 80, 89–106, 113, 140, 158, 177, 225, 251, 274
Health geographies, 93
Health inequalities, 13, 90–103, 105, 106
HMR. *See* Housing market renewal (HMR)
Homelessness, 168, 235
HOPE VI. *See* Housing opportunities for people everywhere (HOPE VI)
Housing, 3, 36, 43, 69, 106, 113, 136, 159, 177, 196, 215–244, 251, 270
Housing choice voucher (HCV), 18, 19, 143, 216, 218–238, 240, 243, 244
Housing market renewal (HMR), 114, 129
Housing opportunities for people everywhere (HOPE VI), 18, 114, 143, 146, 216–219, 225, 228–230, 234, 240
Housing tenure, 141, 147, 181–182, 242, 252, 253, 255–258, 261–263, 266, 267
HUD. *See* U.S department of housing and urban development (HUD)

## I

Impact, 2, 5, 9, 15, 16, 20, 31, 34–36, 44–46, 54–56, 58–62, 69, 71–73, 75–77, 80–83, 96, 97, 101, 104, 124, 135–138, 140, 145, 148–150, 160, 161, 163, 172, 177, 204, 205, 218, 227, 228, 234, 240, 241, 243, 244, 254–256, 263, 271, 272, 274–276, 279, 281, 282, 286–288
Income, 7, 34, 46, 91, 113, 134, 160, 186, 195, 215, 251, 274
Income redistribution, 134
Incrementalism, 15, 161–163, 169–170, 173
Inequalities, 2–5, 7, 8, 10, 13–15, 25, 26, 31, 32, 35, 37, 38, 48, 61, 80, 89–106, 115, 119, 120, 122, 123, 125, 127–129, 134, 136, 150, 151, 186, 189, 197
Intergenerational worklessness, 47
Interscalar links, 15, 161, 162, 165–166, 170–171, 173

## J

Job seekers, 48, 56, 58, 59, 183–184
Justice, 8, 10, 13, 25, 31, 37, 89–106, 113, 115, 116, 120, 134, 151, 165, 166, 196, 198

## L

Labour markets, 11, 17, 32, 43–53, 55–62, 92, 124–126, 129, 135, 138, 159, 177, 183, 184, 198, 200, 201, 279, 284
Learning from the local, 15, 161, 162, 166–167, 170–173
LIHTC. *See* Low income housing tax credit (LIHTC)
Local, 1, 27, 43, 67, 89, 135, 157, 183, 199, 216, 253, 273
Local strategic partnerships (LSP), 57
Low income housing tax credit (LIHTC), 18, 19, 146, 217, 218, 222, 226–228, 231–233, 238–240, 243, 244
Low income neighbourhoods, 98, 125, 126, 149, 196
Low level disorder, 12, 70, 72–73, 78
LSP. *See* Local strategic partnerships (LSP)

## M

Marx/Marxist, 98, 118, 120, 122, 123, 127, 129
Mixed communities/mixed community, 14, 15, 133, 137, 138, 142, 143, 147, 188–191, 252, 255, 256, 263

Index                                                                                                                                     295

Mixed income, 18, 128, 140, 147–148, 216, 228–230, 239, 242, 257
Moving to opportunity (MTO), 74, 143, 149, 150, 218, 223–226, 228, 235–238, 254, 272
MTO. *See* Moving to opportunity (MTO)
Multiculturalism, 199
Multilevel model, 11, 67, 68, 74, 76, 81, 93

**N**

NDC. *See* New deal for communities (NDC)
Neighbourhood
 change, 2, 16, 59, 104, 119, 158, 162, 168–173, 273, 278–280, 288
 choice, 75, 201, 205, 228, 243, 270, 272–280, 288
 renewal, 11, 29, 36, 53–55, 57, 79, 80, 114, 208, 270, 281–288
 revitalization, 157–174
Neoclassical, 118–120, 122, 125, 277
TheNetherlands, 2, 8, 10, 17, 133, 141, 143, 146, 149, 150, 195–210, 283
New deal for communities (NDC), 29, 53, 55–56, 58, 61, 68, 77, 80–82, 143
New Zealand, 91, 92, 98–101, 104

**O**

Omitted-variable problem, 201

**P**

Pathogens, 90, 96–102
Perception of crime, 11, 12, 67–83, 93
Physical isolation, 11, 44, 46, 49–50
Planning, 18, 55, 95, 104, 118, 139–142, 146–147, 162, 168, 197–199, 203, 210, 240–242, 252, 271, 273, 274, 276, 281, 282
Policy makers, 5, 7, 9, 16, 17, 19–21, 39, 115, 134, 139–142, 146, 162, 168, 169, 172, 177, 180, 186, 251, 253, 255, 256, 265, 279
Poverty, 4, 26, 46, 74, 91, 113, 134, 159, 177, 195, 215–244, 270

**R**

Recognition, 10, 25–39, 52, 53, 57, 68, 77, 80, 81, 83, 97, 162, 163, 167, 168, 252, 280, 284, 285
Redistribution, 10, 25–39, 134, 282

Renewal, 11, 29, 36, 53–55, 57, 79, 80, 114, 116, 129, 139, 161, 195–197, 199, 251–267, 270, 277, 280–288
Right to buy, 6, 199, 209, 210
Rotterdam, 17, 126, 145, 202–204, 207–210
Rotterdam law, 17, 203, 204, 209, 210

**S**

Salutogens, 97–101
Scattered site (housing), 18, 145, 216, 218–220, 240
Segregation, 17, 18, 37, 69, 95, 114, 134, 136, 141, 144–147, 149, 160, 195–200, 202–210, 215, 233, 237, 255, 261–266
Selection bias, 2, 68, 74, 221, 223, 244
Selective migration, 3, 7, 103
Social cohesion, 19, 136, 137, 139, 159, 161, 185, 195, 202, 251–267
Social control, 69–73, 78, 83, 135, 137
Social difference, 90
Social disorganisation, 11, 69–71
Social groups, 6, 15, 20, 139
Social housing, 3, 46, 113, 141, 181, 197, 241, 251, 273
Social injustice, 105
Social isolation, 17, 72, 127, 200, 201
Social justice, 10, 13, 25, 31, 113, 115, 120, 134, 151
Socially disadvantaged groups, 96, 101
Social mix, 3, 5, 6, 10, 15, 19, 20, 133–151, 187, 195, 197, 206, 243, 252–257, 259, 261–263, 265–267
Social mobility, 2, 32, 69, 134, 150, 195, 200, 203, 281
Social networks, 14, 15, 17, 47–48, 50, 52, 69, 70, 114, 122, 125–127, 135–137, 149, 151, 164, 201, 202, 234, 238, 243, 267, 276
Social ties, 137, 200–201, 242, 244
Spatial dispersion, 193
Spatial exclusion, 3
Spatial fetishism, 273
Subcultural diversity, 12, 71–72

**T**

Toronto, 16, 168, 170–172, 279

**U**

Underclass, 124, 125, 128, 140, 254
Uneven development, 120–122
United Kingdom, 2, 10, 11, 91, 158–160

United States of America (USA), 2, 6, 10, 18, 20, 91, 96–99, 102, 114, 133, 141–143, 149, 150, 158–160, 169, 244, 272
Urban renewal, 116, 195–197, 199, 251–267, 277
Urban restructuring, 197, 204–206, 208, 209
USA. *See* United States of America (USA)
U.S department of housing and urban development (HUD), 18, 216–219, 223, 226–228, 231, 232, 243

**V**
Vancouver, 15, 16, 158, 162–168

**W**
Western Europe, 19, 142, 241–242, 244, 284, 285
Worklessness, 9, 11, 43–62, 80, 135
World health organisation (WHO), 91

Printed in Poland
by Amazon Fulfillment
Poland Sp. z o.o., Wrocław